Recent Advancements in Multidimensional Applications of Nanotechnology

(Volume 1)

Edited by

Virat Khanna
University Centre for Research & Development
Chandigarh University
Punjab, India

Suneev Anil Bansal
ELFROU Inc
Gurgaon, India

Vishal Chaudhary
Bhagini Nivedita College
University of Delhi, Delhi
India

&

Reddicherla Umapathi
NanoBio High-Tech Materials Research Center
Department of Biological Engineering
Inha University, Incheon
South Korea

Recent Advancements in Multidimensional Applications of Nanotechnology

(Volime 1)

Editors: Virat Khanna, Suneev Anil Bansal, Vishal Chaudhary and Reddicherla Umapathi

ISBN (Online): 978-981-5238-84-6

ISBN (Print): 978-981-5238-85-3

ISBN (Paperback): 978-981-5238-86-0

First published in 2024.

General:

1. Any dispute or claim arising out of or in connection with this License Agreement or the Work (including non-contractual disputes or claims) will be governed by and construed in accordance with the laws of the U.A.E. as applied in the Emirate of Dubai. Each party agrees that the courts of the Emirate of Dubai shall have exclusive jurisdiction to settle any dispute or claim arising out of or in connection with this License Agreement or the Work (including non-contractual disputes or claims).
2. Your rights under this License Agreement will automatically terminate without notice and without the need for a court order if at any point you breach any terms of this License Agreement. In no event will any delay or failure by Bentham Science Publishers in enforcing your compliance with this License Agreement constitute a waiver of any of its rights.
3. You acknowledge that you have read this License Agreement, and agree to be bound by its terms and conditions. To the extent that any other terms and conditions presented on any website of Bentham Science Publishers conflict with, or are inconsistent with, the terms and conditions set out in this License Agreement, you acknowledge that the terms and conditions set out in this License Agreement shall prevail.

Bentham Science Publishers Ltd.
Executive Suite Y - 2
PO Box 7917, Saif Zone
Sharjah, U.A.E.
Email: subscriptions@benthamscience.org

BENTHAM SCIENCE

CONTENTS

PREFACE

In this rapidly evolving field, nanotechnology has emerged as a powerful tool with endless possibilities. This book aims to provide a comprehensive overview of the latest advancements and applications of nanotechnology across various dimensions. It covers a wide range of topics, from electron microscopy to biogenic synthesis methods, from energy applications to agro-nanotechnology, and from nanotherapeutic strategies to nanosensors for virus detection.

Chapter 1 explores the remarkable capabilities of the electron microscope for qualitative and quantitative analysis of nano-materials. It sets the stage for subsequent chapters that delve into specific applications of nanotechnology. Advancements in perovskite nanomaterials for advanced energy applications are discussed in Chapter 2, emphasizing their significance in the quest for sustainable energy solutions. Chapter 3 explores the current developments in the use of copper oxide nanoparticles in the oil and gas industries, highlighting their potential for enhancing efficiency and performance. In Chapter 4, the application of nano-coatings to combat hot corrosion of metallic substrates is examined, presenting an innovative approach to protect materials subjected to extreme conditions. Agro-nanotechnology, discussed in Chapter 5, presents a promising pathway towards sustainable agriculture, where nanotechnology is harnessed to enhance crop production and mitigate environmental challenges. The impact of economic natural dyes on the performance and efficiency of TiO_2 nano-structure solar cells is explored in Chapter 6, uncovering new possibilities for greener and more efficient solar energy generation. Chapter 7 investigates the effect of annealing conditions on chemical bath-deposited CdTe thin films, offering insights into optimizing thin-film deposition processes. Chapter 8 highlights the biomedical necessities and green future of metallic nanoparticles, revealing their potential in various healthcare applications. Chapter 9 focuses on the production of silver nanoparticles with enhanced cytotoxicity and biological activity from Kalanchoe Gastonis-Bonnieri leaf extract, opening avenues for novel therapeutic approaches. Chapter 10 provides an overview of recent biogenic synthesis methods of metal nanoparticles and their applications, showcasing the potential of nature-inspired approaches in nanotechnology. Chapter 11 takes a step further by benchmarking different CNN architectures on a COVID-19 dataset, highlighting the role of nanotechnology in addressing public health challenges. Finally, Chapter 12 explores the application of novel nanotherapeutic strategies in treatment through herbal medicines, presenting an exciting fusion of traditional and modern medicine.

Readers of this book will gain a comprehensive understanding of the recent advancements in nanotechnology and its multidimensional applications. From the fundamentals of electron microscopy to cutting-edge developments in nanotherapeutics and biogenic synthesis methods, this book offers a broad perspective on the field. It equips readers with the knowledge to explore new possibilities, drive innovation, and contribute to the advancement of nanotechnology across various domains.

We hope that this book serves as a valuable resource for researchers, scientists, academicians, and students interested in nanotechnology and its applications. It is our sincere belief that the knowledge shared within these pages will inspire further research, foster interdisciplinary collaborations, and contribute to the realization of a more sustainable and technologically advanced future.

Virat Khanna
University Centre for Research & Development
Chandigarh University
Punjab, India

Suneev Anil Bansal
ELFROU Inc
Gurgaon, India

Vishal Chaudhary
Bhagini Nivedita College
University of Delhi, Delhi
India

&

Reddicherla Umapathi
NanoBio High-Tech Materials Research Center
Department of Biological Engineering
Inha University, Incheon
South Korea

List of Contributors

Aquib Khan	Department of Polytechnic, Integral University, Kursi road Lucknow, Uttar Pradesh, India
Anand Singh	DST – Centre of Interdisciplinary Mathematical Sciences, Institute of Science, Banaras Hindu University (BHU)Varanasi, Varanasi, Uttar Pradesh, India
Ayushi Rastogi	Scitechesy Research and Technology Private Limited, Central Discovery Centre, BioNEST BHU, Banaras Hindu University, Varanasi – 2210035, India
	Department of Humanities and Applied Sciences, School of Management Sciences, College of Engineering, Lucknow – 226001, Uttar Pradesh, India
Che Azurahanim Che Abdullah	Centre for Diagnostic Nuclear Imaging, Universiti Putra Malaysia, 43400, Serdang, Selangor, Malaysia
Celin. S. R.	Department of Chemistry, WCC (Affiliated to MS University, Abishekapatti, Tirunelveli-627012), Nagercoil, Tamilnadu, India
Chesta Mehta	Department of Chemistry, M.L.Sukhadia University, Udaipur, Rajasthan, 313001, India
Deepshikha Verma	Department of Chemistry, M.B.S. College of Engineering and Technology, 181101, Jammu and Kashmir, India
Faria Fatima	Department of Agriculture, IIAST, Integral University, Kursi road Lucknow, Uttar Pradesh, India
Giriraj Tailor	Department of Chemistry, Mewar University, Chittorgarh, Rajasthan, 31290, India
Harsh Kumar Mishra	DST – Centre of Interdisciplinary Mathematical Sciences, Institute of Science, Banaras Hindu University (BHU)Varanasi, Varanasi, Uttar Pradesh, India
Jyoti Bhattacharjee	Department of Chemical Engineering, University of Calcutta, Kolkata 700009, India
Jyoti Chaudhary	Department of Chemistry, M.L.Sukhadia University, Udaipur, Rajasthan, 313001, India
Kumar Anurag	School of Energy Materials, Mahatma Gandhi University, Kottayam, Kerala-686560, India
Lankipalli Krishna Sai	School of Electronics Engineering, Vellore Institute of Technology, Chennai-600127, India
Muhammad Salman Habib	Department of Metallurgical & Materials Engineering, University of Engineering Technology G.T. Road Lahore, Pakistan
Muhammad Asif Rafiq	Department of Metallurgical & Materials Engineering, University of Engineering Technology G.T. Road Lahore, Pakistan
Mhd Hazli Rosli	Nanomaterial Synthesis and Characterization Lab, Institute of Nanoscience and Nanotechnology, Universiti Putra Malaysia, 43400, UPM Serdang, Selangor, Malaysia
Mohamed Abdelmonem	Department of Physics, Faculty of Science, Universiti Putra Malaysia, 43400, UPM Serdang, Selangor, Malaysia

Manahil E. Mofdal	Qassim University, Faculty of Science, Department of Physics, Buraydah, KSA
Nur Farahah Mohd Khairuddin	Centre of Foundation Studies for Agricultural Science, Universiti Putra Malaysia, 43400, UPM Serdang, Selangor, Malaysia
Nada M. O. Sid Ahmed	Computer Engineering Department Computer Science and Engineering College, University of Hail, Hail, KSA
Nodar. O. Khalifa	Department of Physics, Sudan University of Science and Technology, Khartoum, Sudan
Nada H. Talib	Solar Energy Department, National Energy Research Center, Khartoum, Sudan
R. Ajitha	Department of Chemistry, WCC (Affiliated to MS University, Abishekapatti, Tirunelveli-627012), Nagercoil, Tamilnadu, India
Sunil Kumar Pradhan	School of Electronics Engineering, Vellore Institute of Technology, Chennai-600127, India
Santosh Kumar	Department of Mechanical Engineering, Chandigarh Group of College, Landran, Mohali, Punjab, India
Sudeshna Surabhi	Brindavan College of Engineering, Yelahanka, Bengaluru, Karnataka-560063, India
S.R. Kumar	Thin Film Laboratory, National Institute of Advanced Manufacturing Technology(Formerly NIFFT), Ranchi-834003, India
Subhasis Roy	Department of Chemical Engineering, University of Calcutta, Kolkata 700009, India
Saurabh Singh	M.L.V. Government College, Bhilwara, Rajasthan, 311001, India
Sumanta Bhattacharya	Maulana Abul Kalam Azad University of Technology, West Bengal, India
Tadisetti Taneesha	School of Electronics Engineering, Vellore Institute of Technology, Chennai-600127, India

Electron Microscope: The Tool for Qualitative and Quantitative Analysis of Nano-Materials

Lankipalli Krishna Sai[1], **Tadisetti Taneesha**[1] and **Sunil Kumar Pradhan**[1,*]

[1] *School of Electronics Engineering, Vellore Institute of Technology, Chennai-600127, India*

Abstract: An electron microscope is a highly advanced sophisticated tool where high energy electron beam is used as the source. Since an electron beam has a shorter wavelength than visible light photons, it may expose the structure of tiny objects and has a higher resolving power than a light microscope. While most light microscopes are constrained by diffraction to around 500 nm resolution and usable magnifications below 2000, a scanning electron microscope (SEM) may attain 5 nm resolution and magnifications up to roughly 10,000,000. Electromagnetic lenses, which are similar to the glass lenses of an optical light microscope, are used in electron microscopes to create electron optical lens systems. Large molecules, biopsy samples, metals, crystals, and other biological and inorganic specimens, among others, can all have their ultra-fine structure studied using electron microscopes. Electron microscopes are frequently used in industry for failure analysis and quality control. The images are captured using specialised digital cameras and frame grabbers by modern electron microscopes to create electron micrographs. To create an appropriate sample from materials for an electron microscope, processing may be necessary. Depending on the material and the desired analysis, a different procedure is needed. Transmission electron microscopes (TEM), scanning electron microscopes (SEM), reflection electron microscopes (REM), scanning tunnelling microscopes (STM), and other types of electron microscopes are commonly employed in academic and research institutions. The initial and operating costs of electron microscopes are higher and they are also more expensive to construct and maintain. High-resolution electron microscopes need to be kept in sturdy structures (often underground) with specialised amenities like magnetic field cancelling devices.

Keywords: Cryogenic transmission electron microscopy, Electron mapping, Energy-filtered transmission electron microscopy, Electron energy loss spectroscopy, Electron microscope, Environmental electron microscope, Low-voltage electron microscope, Magnification, Nano-materials, Scanning transmission electron microscope.

* **Corresponding author Sunil Kumar Pradhan:** School of Electronics Engineering, Vellore Institute of Technology, Chennai-600127, India; E-mails: sunilpradha@gmail.com, sunilkumar.pradhan@vit.ac.in

Virat Khanna, Suneev Anil Bansal, Vishal Chaudhary and Reddicherla Umapathi (Eds.)

INTRODUCTION

Electron microscopy for nanotechnology is the use of electron microscopes to observe, analyse and manipulate materials at the nanoscale [1 - 5]. This field plays a crucial role in the development of modern nanotechnology and materials science, as it enables scientists and engineers to observe and study the structure, composition, and properties of nanoscale materials and devices [6, 7]. There are several types of electron microscopes used for nanotechnology, including Transmission Electron Microscopy (TEM), Scanning Electron Microscopy (SEM), Scanning Transmission Electron Microscopy (STEM), and Cryogenic Transmission Electron Microscopy (Cryo-TEM) [8, 9]. TEM works by passing a beam of electrons through a thin sample, producing an image of the internal structure of the material. SEM, on the other hand, uses electrons to scan the surface of a sample and produce a high-resolution image. STEM uses a beam of electrons to probe the sample and obtain chemical and structural information. Cryo-TEM operates at cryogenic temperatures, allowing for the study of delicate biological samples. Electron microscopy for nanotechnology has numerous applications, including the study of materials for electronic and energy applications, the development of new drug delivery systems, the investigation of cellular and molecular structures, and the characterization of nanoscale devices and materials [7 - 10]. The electron microscope, which uses electrons instead of light to magnify images, was first developed in the 1930s and revolutionized the field of microscopy [11, 12]. In the early days of electron microscopy, the technology was primarily used for imaging biological samples, but as the field developed, researchers began to apply the technology to the study of materials at the nanoscale. The development of electron microscopy was closely tied to advancements in the field of physics [13 - 16]. In the late 19th and early 20th centuries, scientists were exploring the properties of electrons and the way they interacted with matter. This research laid the foundation for the development of the electron microscope, which would use electrons to image samples and reveal their structure at an incredibly high level of detail [17]. In the 1930s, two German scientists, Max Knoll and Ernst Ruska, independently developed early versions of the electron microscope. These first-generation electron microscopes were large, complex devices that required a high level of expertise to operate, but they were capable of producing images of biological samples with a much higher level of detail than was possible with light microscopes [18 - 20]. In the decades that followed, advances in technology allowed for the development of smaller, more accessible electron microscopes. These microscopes made it possible for researchers to study a wider range of samples, including inorganic materials and materials at the nanoscale. In the 1960s and 1970s, the field of electron microscopy underwent a major expansion as researchers began to develop new techniques for imaging materials at the nanoscale [21]. This was a critical development, as it allowed scientists to study materials in much greater detail than was previously possible. With the ability to see materials at the nanoscale, scientists were able to discover new properties and behaviours that could not be observed at the macroscale. One of the key applications of electron microscopy in nanotechnology is imaging materials at the atomic scale. This allows researchers to study the structure and composition of materials at the smallest possible level, which can provide important insights into their properties and behaviour [22, 23]. In recent years, electron microscopy has also been used to study materials in various states, including liquids, gases, and even plasmas. Electron microscopy has also been critical in the development of other important technologies, such as nanolithography and nanofabrication. Nanolithography involves the patterning of materials at the nanoscale, and electron microscopy is used to verify the accuracy and precision of these patterns. Nanofabrication involves the creation of nanoscale structures and devices, and electron microscopy is used to study and refine these structures during the fabrication process [24, 25]. Today, electron microscopy is an essential tool in the field of nanotechnology and materials science, allowing researchers to study materials at the smallest possible scale and uncover new properties and behaviours. The field of electron microscopy continues to evolve and advance, with new techniques and innovations being developed all the time.

TYPES OF ELECTRON MICROSCOPY

Electron microscopes are a key tool in nanotechnology, allowing scientists and engineers to visualize, analyse and manipulate materials at the nanoscale. There are several types of electron microscopes, each with its advantages and limitations. Here are the most common types:

1. Transmission Electron Microscope (TEM): This type of microscope uses a beam of electrons to form an image of a thin sample. The electrons pass through the sample and are scattered, forming an image of the internal structure. TEMs are used for high-resolution imaging and analysis of a variety of materials, including metals, ceramics, and biological samples [26].

2. Scanning Electron Microscope (SEM): This type of microscope uses a beam of electrons to scan the surface of a sample, producing a topographical image. SEMs are used for surface imaging and analysis, and can also be used to obtain information about the chemical composition of a sample [27]. The schematic representation of the Field Emission Scanning Electron Microscope (FE-SEM) is illustrated in Fig. (**1**).

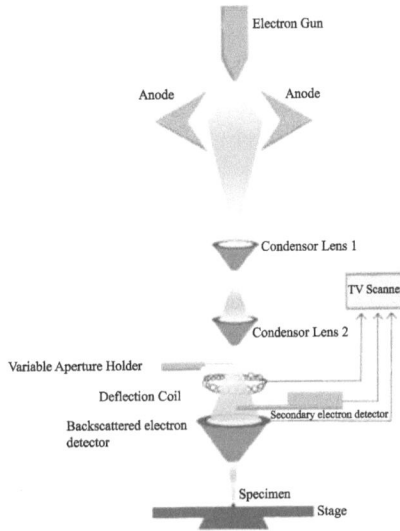

Fig. (1). Schematic diagram of Field Emission Scanning Electron Microscope (FE-SEM).

3. Scanning Transmission Electron Microscope (STEM): This type of microscope combines the features of a TEM and an SEM, allowing for high-resolution imaging of both the surface and internal structure of a sample. STEMs are often used for imaging and analysis of materials in the fields of electronics, materials science, and biology [28]. The schematic diagram of the Transmission Electron Microscope (TEM) is shown in Fig. (**2**) and the Schematic diagram of Energy Dispersive X-ray analysis in the TEM is illustrated in Fig. (**4**).

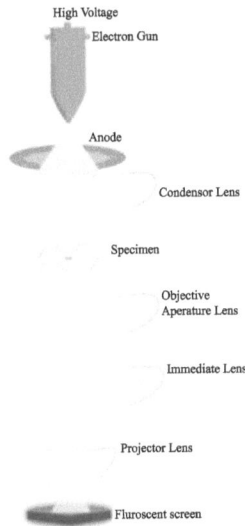

Fig. (2). Schematic diagram of Transmission Electron Microscope (TEM).

4. High-Resolution Transmission Electron Microscope (HRTEM): This type of TEM is specifically designed for high-resolution imaging, allowing scientists to visualize the internal structure of materials at the atomic scale. HRTEMs are used for a variety of applications, including the study of materials for electronic and optical applications, as well as the study of biological materials [29].

5. Environmental Electron Microscope (EEM): This type of electron microscope allows scientists to study materials under controlled environmental conditions, such as temperature, pressure, and gas composition. EEMs are used for a variety of applications, including the study of materials for energy and environmental applications, as well as the study of materials for biological and medical applications [30]. The schematic diagram of the Environmental Scanning Electron Microscope (ESEM) is shown in Fig. (**6**).

6. Electron Spectroscopy for Chemical Analysis (ESCA): This type of electron microscope uses electrons to analyse the chemical composition of a sample. ESCA is used for a variety of applications, including the study of materials for electronic and optical applications, as well as the study of biological materials [31]. The schematic diagram of photoemission electron microscopy is illustrated in Fig. (**3**).

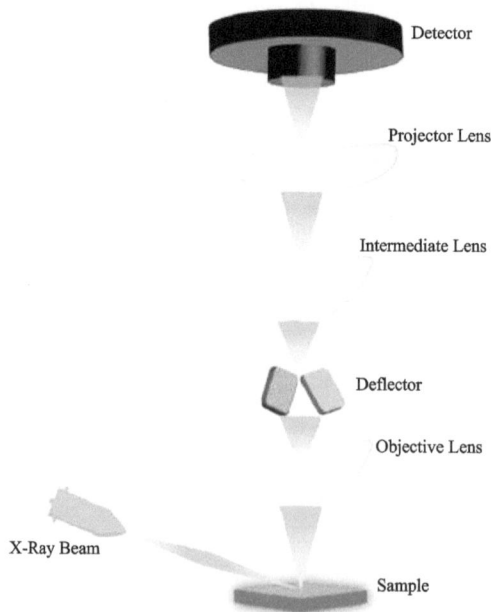

Fig. (3). Schematic diagram of photoemission electron microscopy.

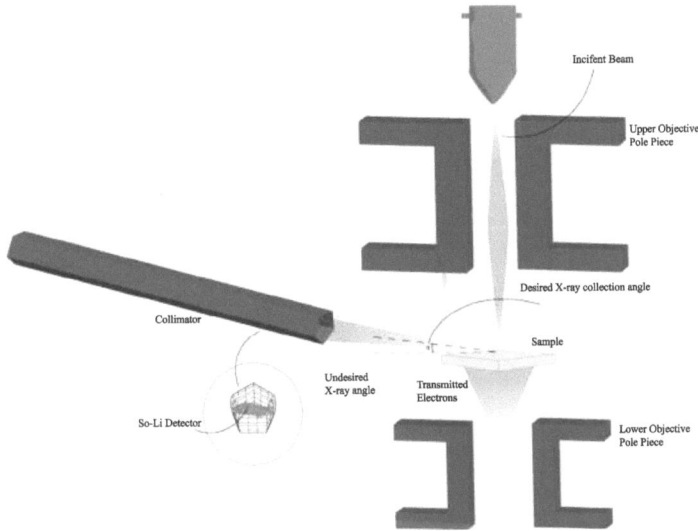

Fig. (4). Schematic diagram of energy dispersive X-ray analysis in the TEM.

7. Low-Voltage Electron Microscope (LVEM): This type of electron microscope uses a lower voltage electron beam, allowing for the imaging and analysis of biological materials without damage to delicate structures. LVEMs are used for a variety of applications in the fields of biology and medicine [32].

SAMPLE PREPARATION

Preparation of the sample is critical to obtaining accurate and high-resolution images with an electron microscope. This chapter will discuss various methods of preparing samples for electron microscopy, including thin-film preparation, embedding, and staining. It will also discuss the importance of controlling the sample's environment, including temperature and pressure, during imaging. Sample preparation is a critical step in the use of electron microscopes for nanotechnology [33]. The quality of the sample preparation will have a direct impact on the quality of the images and data obtained from the electron microscope. One of its kind is EPMA (Electron Probe Micro Analyser). The schematic diagram of the electron-electron probe micro analyser is illustrated in Fig. (**5**). Here are some key steps in the sample preparation process:

1. Sample Selection: The first step in sample preparation is selecting a suitable sample. The sample should be representative of the material being studied and should be of sufficient size and quality to allow for imaging and analysis in the electron microscope.

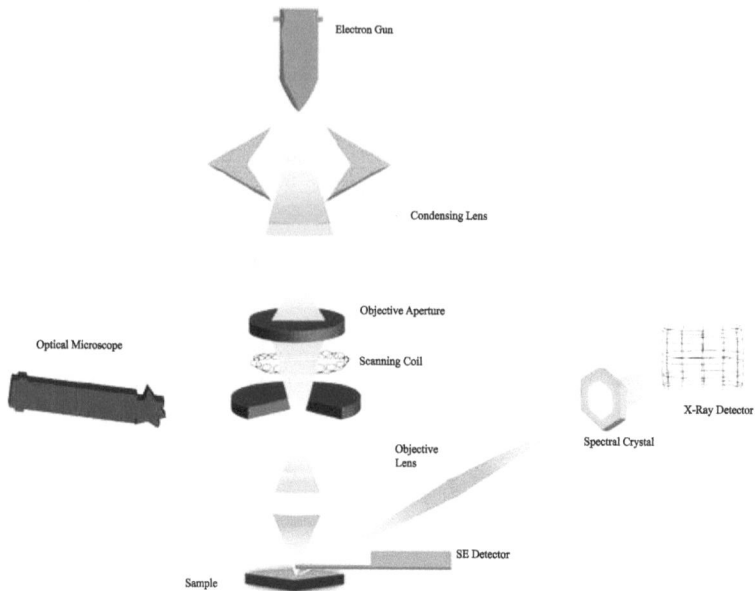

Fig. (5). Schematic diagram of electron probe micro analyser (EPMA).

2. Sample Preparation: Once a suitable sample has been selected, it must be prepared for analysis in the electron microscope. This often involves cutting, polishing, and thinning the sample to create a thin, electron-transparent section that can be imaged in the electron microscope.

3. Specimen Preparation for SEM: For samples to be imaged in a Scanning Electron Microscope (SEM), they must be coated with a conductive material, such as gold or platinum, to prevent charging and to improve the quality of the images.

4. Specimen Preparation for TEM: For samples to be imaged in a Transmission Electron Microscope (TEM), they must be prepared as ultra-thin sections that are electron transparent. This often involves preparing a "shadow" replica of the sample using a variety of techniques, such as the replication of the surface of the sample with a thin layer of material or the use of a focused ion beam to mill a thin section from the sample.

5. Specimen Preparation for STEM: For samples to be imaged in a Scanning Transmission Electron Microscope (STEM), they must be prepared as ultra-thin sections that are electron transparent. This often involves preparing a "shadow" replica of the sample using a variety of techniques, such as the replication of the surface of the sample with a thin layer of material or the use of a focused ion beam to mill a thin section from the sample [34].

6. Specimen Preparation for EEM: For samples to be imaged in an Environmental Electron Microscope (EEM), they must be prepared in a way that allows for the controlled exposure of the sample to specific environmental conditions, such as temperature, pressure, and gas composition. This often involves the use of specialized sample holders and the preparation of the sample in a controlled environment.

7. Specimen Preparation for LVEM: For samples to be imaged in a Low-Voltage Electron Microscope (LVEM), they must be prepared as thin sections that are electron transparent, but are also able to withstand the low-energy electron beam used in LVEMs. This often involves the use of special sample preparation techniques, such as the use of low-energy beam damage mitigation techniques, to minimize damage to delicate biological structures [35]. Fig. (**6**) represents the schematic of Environmental Scanning Electron Microscope (ESEM)

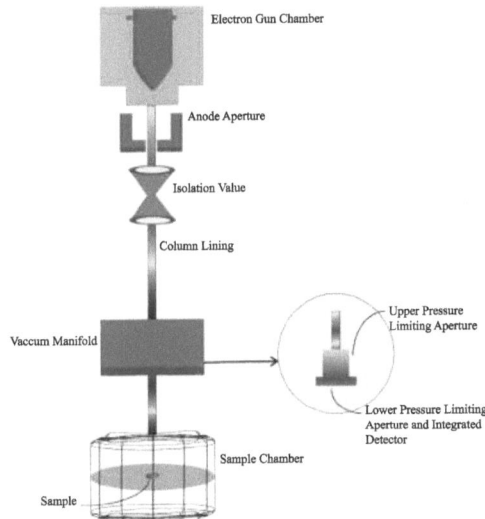

Electron Gun Chamber

Anode Aperture

Isolation Value

Column Lining

Vaccum Manifold

Upper Pressure Limiting Aperture

Lower Pressure Limiting Aperture and Integrated Detector

Sample Chamber

Sample

Fig. (6). Schematic of Environmental Scanning Electron Microscope (ESEM).

IMAGING TECHNIQUES

We will discuss various imaging techniques that can be used with electron microscopes, including bright-field imaging, dark-field imaging, and annular dark-field imaging. It will also cover advanced imaging techniques, such as electron tomography and holography. Image techniques play a crucial role in the use of electron microscopes for nanotechnology. The schematics of low-energy electron microscopy are shown in Fig. (**7**). Electron microscopes provide high-resolution images that can reveal the detailed structure and composition of materials at the nanoscale [36]. Here are some of the key image techniques used in electron microscopy for nanotechnology:

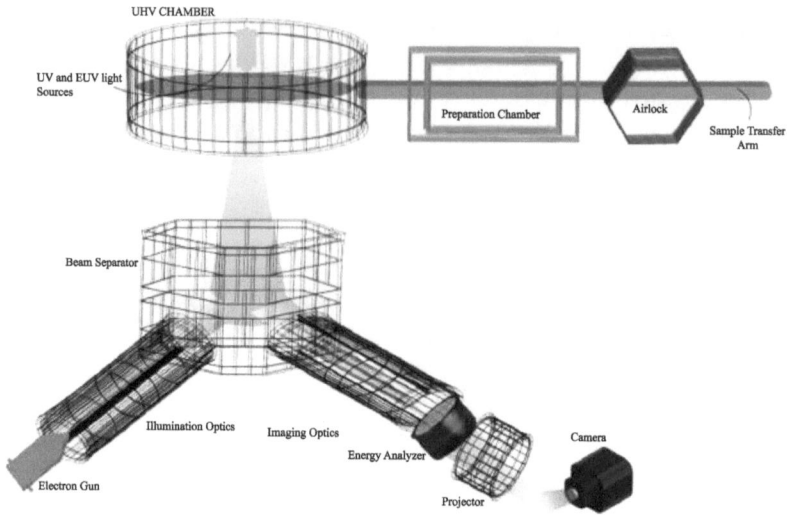

Fig. (7). Schematics of low-energy electron microscopy.

1. Bright Field Transmission Electron Microscopy (TEM): Bright field transmission electron microscopy (BF-TEM) is a widely used imaging technique in the field of nanotechnology. BF-TEM works by transmitting electrons through a thin sample and imaging the electrons that are transmitted through the sample. The resulting image provides information about the sample's composition and structure. In BF-TEM, a beam of electrons is directed through a thin sample, and the electrons that are transmitted through the sample are collected by a detector. The electrons that are transmitted through the sample are the ones that are not absorbed or scattered by the sample's atomic structure. This results in a bright background and dark contrast for the atomic columns in the sample, hence the name "bright field." BF-TEM provides high-resolution images of materials and is commonly used for imaging materials in the nanoscale, including metals, ceramics, polymers, and biological materials. BF-TEM is particularly useful for imaging materials with a high atomic number, as the electrons are more likely to be absorbed by the material, providing better contrast in the image. One of the main advantages of BF-TEM is its high resolution, which allows researchers to observe the details of the atomic structure of a sample [37]. This high resolution is possible because of the small wavelength of electrons, which is much smaller than the wavelength of light used in other imaging techniques, such as optical microscopy. BF-TEM also provides information about the chemical composition of the sample, which is particularly useful for the study of materials at the nanoscale. Another advantage of BF-TEM is that it is a non-destructive technique, meaning that the sample is not altered or damaged during the imaging process. This is important for the study of delicate and sensitive materials, such as biological samples or materials with complex structures. BF-TEM also provides information about the three-dimensional structure of the sample, as the electrons are transmitted through the entire thickness of the sample, rather than just the surface. However, BF-TEM also has some limitations. One of the main limitations is that it requires a thin sample to be prepared, as electrons can only be transmitted through thin samples. This can be a time-consuming and complex process, particularly for samples that are not naturally thin or that have complex structures. Additionally, BF-TEM requires the sample to be in a vacuum, as the electrons cannot be transmitted through air. This can also present challenges for the study of biological samples, as they are sensitive to the vacuum conditions required for BF-TEM.

2. Dark Field Transmission Electron Microscopy (TEM): Darkfield transmission electron microscopy (DF-TEM) is an imaging technique used in the field of nanotechnology to study the structure and composition of materials. It is similar to bright field transmission electron microscopy (BF-TEM), but instead of imaging the electrons transmitted through the sample, DF-TEM images the electrons that are scattered by the sample's atomic structure [38]. In DF-TEM, the electrons are directed towards the sample at an angle and those that are scattered by the sample's atomic structure are collected by a detector, while the electrons transmitted through the sample are blocked by an aperture. This results in a dark background and bright contrast for the atomic columns in the sample. DF-TEM provides high-resolution images of materials and is commonly used for imaging materials in the nanoscale, including metals, ceramics, polymers, and biological materials. DF-TEM is particularly useful for imaging materials with low atomic numbers, as the electrons are less likely to be absorbed by the material, providing better contrast in the image. One of the main advantages of DF-TEM is its ability to image materials that are not visible in BF-TEM, such as interfaces between two materials, grain boundaries, or defects in the sample. DF-TEM provides a clearer representation of the sample's structure and can provide information about the sample's composition and crystal structure. Another advantage of DF-TEM is its non-destructive nature, which is particularly important for the study of delicate and sensitive materials, such as biological samples. DF-TEM also provides information about the three-dimensional structure of the sample, as the electrons are scattered through the entire thickness of the sample [39].

However, like BF-TEM, DF-TEM also has some limitations. One of the main limitations is that it requires a thin sample to be prepared, as electrons can only be scattered through thin samples. This can be a time-consuming and complex process, particularly for samples that are not naturally thin or that have complex structures. Additionally, DF-TEM also requires the sample to be in a vacuum, as electrons cannot be scattered through air [40].

3. High-Angle Annular Dark Field Scanning Transmission Electron Microscopy (HAADF-STEM): High-Angle Annular Dark Field Scanning Transmission Electron Microscopy (HAADF-STEM) is a type of imaging technique used in nanotechnology to study the structures of materials at the nanoscale [41]. The technique involves the use of a transmission electron microscope (TEM) to probe the sample with a beam of electrons and image the resulting scattered electrons. In HAADF-STEM, the scattered electrons are collected using an annular detector, which captures electrons that are scattered at high angles with respect to the incident electron beam. HAADF-STEM is particularly useful for imaging heavy atoms in a sample, as these atoms scatter electrons more strongly than lighter atoms. This makes it possible to produce high-resolution images of material structures that contain heavy atoms, such as interfaces and grain boundaries in metals and alloys. The technique is also useful for imaging materials with low atomic numbers, such as carbon-based materials, where other imaging techniques, such as bright field TEM, may not provide enough contrast for clear imaging. HAADF-STEM imaging can be performed in either high-resolution or low-resolution mode, depending on the desired imaging goal. High-resolution mode provides images with high spatial resolution, making it possible to resolve fine details in the material structure. Low-resolution mode provides images with lower spatial resolution but a higher signal-to-noise ratio, making it possible to image larger structures or material distributions. The HAADF-STEM imaging process starts with preparing a thin sample of the material to be imaged. This can be done using a variety of techniques, including mechanical thinning, chemical etching, or ion beam milling. Once the sample has been prepared, it is loaded into the TEM and positioned in the electron beam. The incident electron beam is then focused onto the sample, and the resulting scattered electrons are collected using the annular detector. The detector collects electrons scattered at high angles with respect to the incident electron beam, providing a high-resolution image of the material structure [42]. HAADF-STEM images can be analysed to determine the chemical composition and crystal structure of the material, as well as its electrical and mechanical properties. This information is critical for understanding the properties of materials at the nanoscale and for developing new materials with improved properties. In addition to imaging, HAADF-STEM can also be used for spectroscopic analysis, such as energy-dispersive x-ray spectroscopy (EDS) and x-ray fluorescence (XRF), which provide information on the chemical composition of the sample. EDS and XRF use X-rays emitted from the sample to determine the elements present and their distribution. This information can be combined with the HAADF-STEM imaging data to produce a more complete picture of the material structure and properties. One of the major advantages of HAADF-STEM is its ability to provide high-resolution imaging of materials with low atomic numbers, such as carbon-based materials. This is particularly important in the field of nanotechnology, where the development of new materials with improved properties is a key goal. By providing detailed information on the structure and properties of these materials, HAADF-STEM can help researchers develop new materials with improved properties and performance.

4. Scanning Electron Microscopy (SEM): Scanning electron microscopy (SEM) is a type of electron microscopy that generates images of a sample surface by scanning the sample with a focused beam of electrons and detecting the electrons that are emitted from the sample. It is widely used in nanotechnology for imaging and analysis of materials at a nanoscale resolution. In SEM, the electrons interact with the sample to produce a signal that contains information about the sample's composition and topography. This signal is then processed to produce an image. SEM images typically have a high resolution, with the ability to resolve features as small as a few nanometres. There are different imaging modes in SEM that are used to generate images with different information content, such as secondary electron imaging, backscattered electron imaging, and X-ray energy-dispersive spectroscopy (EDS). Secondary electron imaging provides information about the sample's surface topography, while backscattered electron imaging provides information about the sample's composition and bulk structure. EDS provides a chemical analysis of the sample, with the ability to identify the elemental composition and chemical state of the sample. SEM is a powerful tool for imaging and analysing materials at the nanoscale and has wide applications in materials science, biology, electronics, and many other fields. The schematics of wavelength dispersive spectroscopy (WDS) is shown in Fig. (**8**).

5. Low-Voltage Electron Microscopy (LVEM): Low-voltage electron microscopy (LVEM) is a type of electron microscopy used for imaging samples at low electron beam energies, typically less than 10 kV. This allows for the imaging of delicate or biological samples without damaging them. The imaging techniques used in LVEM include:

a) Bright Field Imaging: This is the most basic form of LVEM imaging, where the electrons are transmitted through the sample to form an image based on the transmission and absorption of electrons by the sample.

b) Dark Field Imaging: This is an alternative to bright field imaging, where electrons that are scattered by the sample are collected to form an image. This can be useful for detecting structures that are not visible in bright-field imaging.

c) Z-Contrast Imaging: This imaging technique is used to determine the composition of the sample based on the contrast between different elements. This can be useful for imaging samples with different atomic numbers or for imaging samples with light elements.

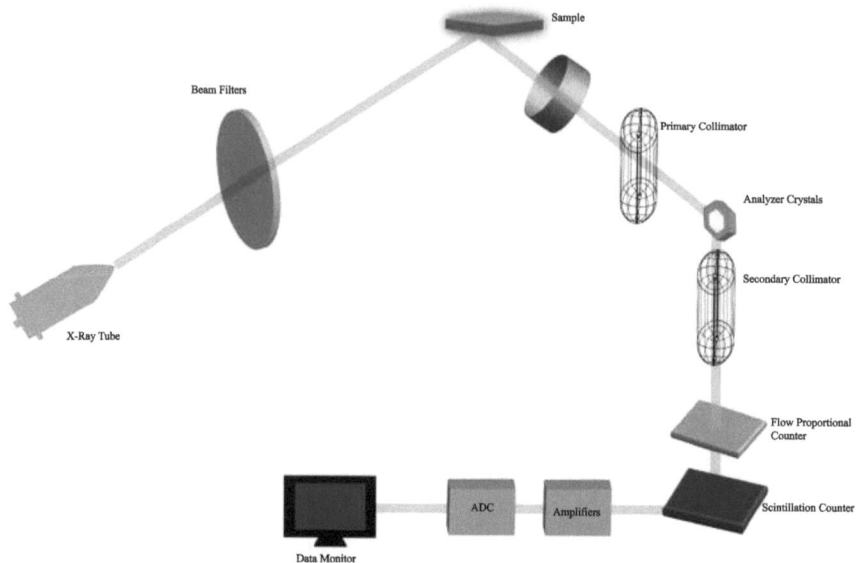

Fig. (8). Schematics of wavelength dispersive spectroscopy (WDS).

6. Scanning Transmission Electron Microscopy (STEM): This is a type of LVEM that uses a focused electron beam to scan across a sample and collect the electrons that are transmitted through the sample. STEM can provide high-resolution images of the sample structure, including atomic arrangements.

7. Energy-Dispersive X-Ray Spectroscopy (EDS): This is a technique used in conjunction with LVEM to determine the elemental composition of the sample. EDS measures the energy of the x-rays emitted by the sample as the electron beam is scanned across it. These techniques allow LVEM to provide high-resolution images of delicate or biological samples with minimal damage, making it a valuable tool for nanotechnology research and development.

8. Energy-Filtered Transmission Electron Microscopy (EFTEM): Energy-filtered transmission electron microscopy (EFTEM) is an imaging technique used in nanotechnology to study the chemical composition of materials at the nanoscale. It works by filtering the transmitted electrons through a monochromator and detecting the electrons that have lost a specific amount of energy due to interactions with the sample [43]. This energy loss information is then used to produce an elemental map of the sample, providing information about the distribution of different elements in the material. EFTEM can be combined with other imaging modes, such as dark field and bright field imaging, to obtain additional information about the sample. EFTEM is a powerful tool for the study of nanoscale materials, as it allows for chemical imaging with sub-nanometre resolution.

9. Electron Energy Loss Spectroscopy (EELS): Electron Energy Loss Spectroscopy (EELS) is a widely used imaging technique in nanotechnology for characterizing the composition, electronic structure, and chemical bonding of materials at the nanoscale. It works by analysing the energy loss of electrons as they pass through a material and interact with its atoms. In EELS, a focused electron beam is directed at the sample, and the energy loss of the electrons is measured as they pass through the material. This information can be used to produce images of the material's composition and atomic structure, as well as to obtain spectra that reveal its electronic structure and chemical bonding. The schematic of Electron Energy Loss Spectroscopy (EELS) is depicted in Fig. (**9**). There are several different types of EELS, including low-loss EELS that focuses on the low-energy loss events and can reveal information about the unoccupied electronic states, and high-loss EELS that focuses on the high-energy loss events and can provide information about the inner-shell excitations and chemical bonding. EELS is commonly used in combination with transmission electron microscopy (TEM), which provides high-resolution imaging of the material's structure and morphology. The combination of TEM and EELS provides a powerful tool for investigating the nanoscale structure and composition of materials [44]. The schematic diagram of Reflection Electron Microscope is shown in Fig. (**10**).

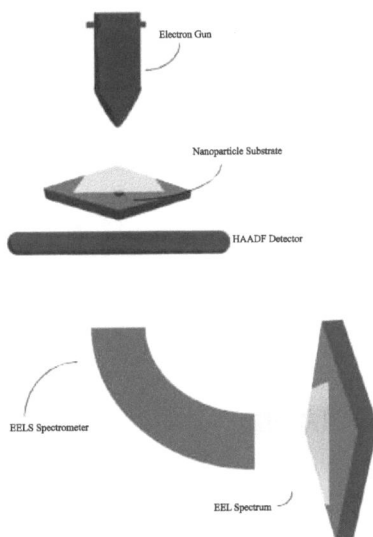

Fig. (9). Schematic diagram of electron energy loss spectroscopy (EELS).

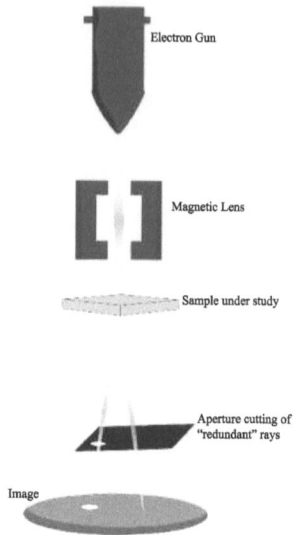

Fig. (10). Schematic diagram of reflection electron microscope.

QUANTITATIVE ANALYSIS BY ELECTRON MICROSCOPY

Quantitative analysis of nano-materials can be possible with the aid of electron microscopy. In order to achieve this, analytical techniques for elemental analysis or chemical characterization of a sample are performed using energy-dispersive X-ray spectroscopy (EDS, EDX, or XEDS), also known as energy dispersive X-ray analysis (EDXA) or energy dispersive X-ray microanalysis (EDXMA). The elemental composition of individual spots or the lateral distribution of elements from the imaged area can be mapped out using the EDS analysis. Additionally, it can be utilised to determine the composition of quasi-bulk specimens (high accelerating voltage, low SEM magnification), as well as particular particles, morphologies, or isolated regions on filters or inside deposits. The EDS possesses an analytical capacity that may be integrated with multiple applications, such as transmission electron microscopy (TEM), scanning electron microscopy (SEM), and more. In numerous earth and material science applications, it is essential to comprehend the spatial distribution of components in solid materials. SEM-EDS mapping combines the capability of X-ray spectroscopy with the spatial resolution of a contemporary electron microscope. High-resolution elemental maps can be gathered over an area of interest by actively collecting X-rays and scanning an electron beam across the sample surface. So, by combining compositional precision from methods like EDS microanalysis with high-resolution imaging, elemental mapping presents useful data in an aesthetically appealing and easily comprehensible format that helps the scientific community effectively and persuasively convey complex information. In this consideration, the foundation of elemental mapping is the collection of incredibly detailed elemental composition data over a sample's surface. Usually, EDS analysis is used for this in a TEM or SEM. Together with the EDS data, a high-resolution picture of the region of interest is gathered, and the two are correlated. A complete elemental spectrum is also acquired for each pixel that is obtained from the digital image. These spectra can be processed to produce computed colorization layers, which colour code the electron photomicrograph to represent layers and sites of elemental compositional information in the sample based on the relative intensity of spectral features associated with several elements of interest.

APPLICATIONS OF ELECTRON MICROSCOPY

Electron microscopy has a wide range of applications, including materials science, biology, and medicine. This chapter will discuss how electron microscopes are used to study the structure of materials, including metals, ceramics, and polymers. It will also discuss how electron microscopes are used in biological research to study cells, tissues, and proteins. Electron microscopes are essential tools for nanotechnology research, providing high-resolution images that reveal the detailed structure and composition of materials at the nanoscale. Here are some of the key applications of electron microscopes in nanotechnology:

1. Material Science: Electron microscopes are used extensively in material science to study the structure and properties of materials. For example, electron microscopes are used to study the microstructure of metals and alloys, to determine the crystal structure of materials, and to study the arrangements of atoms and molecules in materials. Electron microscopes are also used to study the properties of materials, such as their mechanical, electrical, and optical properties.

2. Nanotechnology: Electron microscopes are used to study the properties and behaviour of materials at the nanoscale, including the size and shape of nanoparticles, the arrangement of atoms and molecules in nanomaterials, and the interactions between nanoparticles and other materials. Electron microscopes are also used to study the behaviour of nanoscale materials under various conditions, including high-temperature and high-pressure conditions, and to determine the performance and stability of nanoscale materials.

3. Biotechnology: Electron microscopes are used extensively in biotechnology to study the structure and function of biological systems. For example, electron microscopes are used to study the structure of proteins, viruses, and cells, and to determine the arrangement of atoms and molecules in biological samples. These are also used to study the interactions between biological samples and other materials, including drugs and other therapeutic agents.

4. Energy and Environmental Science: Electron microscopes are used to study the structure and behaviour of materials and systems that are relevant to energy and environmental science. For example, these are used to study the structure of catalysts and other materials used in energy conversion and storage and to determine the interactions between these materials and other substances. Electron microscopes are also used to study the behaviour of materials in environmental conditions, such as exposure to high temperatures and pressures, and to determine the stability and performance of materials in these conditions.

5. Semiconductors: Electron microscopes are used to study the structure and properties of semiconductors, including the arrangement of atoms and molecules in these materials, the size and shape of semiconductor particles, and the behaviour of semiconductors under various conditions. These are also used to study the performance and stability of semiconductors and to determine the interactions between semiconductors and other materials [45].

6. Nano-electronics: Electron microscopes are used to study the structure and behaviour of nanoscale electronic devices, including transistors, diodes, and other nanoelectronics components. Electron microscopes are also used to study the behaviour of nanoscale materials used in electronic devices, such as graphene and other two-dimensional materials.

7. Materials Characterization: Electron microscopes are used to determine the composition, structure, and properties of materials, including the size and shape of particles, the arrangement of atoms and molecules in materials, and the distribution of different phases within materials. These are also used to study the behaviour of materials under various conditions, including high-temperature and high-pressure conditions, and to determine the stability and performance of materials in these conditions [47, 48].

In conclusion, electron microscopes are critical tools for nanotechnology research, providing high-resolution images that reveal the detailed structure and composition of materials at the nanoscale. Electron microscopes are used in a wide range of applications, including material science, nanotechnology, biotechnology, energy and environmental science, semiconductors, nano-electronics, and materials characterization.

LIMITATIONS

Electron microscopes are powerful tools for studying materials at the nanoscale, providing high-resolution images that reveal the detailed structure and composition of materials. However, electron microscopes also have a number of limitations that impact their utility in nanotechnology research. Here are some of the key limitations of electron microscopes:

1. Sample preparation: Electron microscopes require samples to be prepared in a specific way in order to obtain high-quality images. This can be a time-consuming and complex process, especially for biological samples, which require special preparation techniques to preserve their structure. In addition, the preparation process can introduce artifacts into the sample, affecting the accuracy of the images obtained [46].

2. Radiation damage: Electron microscopes use high-energy electrons to generate images, which can cause radiation damage to delicate samples. This can result in changes to the sample's structure and composition, making it difficult to obtain accurate images. In addition, the radiation damage can cause the sample to degrade over time, making it difficult to study the sample over a long period of time.

3. Low signal-to-noise ratio: Electron microscopes typically have a low signal-to-noise ratio, which can make it difficult to distinguish between noise and signal in images. This can make it challenging to obtain accurate images, especially for samples that are difficult to visualize, such as those that are low contrast or contain weak signals.

4. Sample size: Electron microscopes are typically limited in terms of the size of the sample that can be imaged, with some electron microscopes only capable of imaging samples on the order of tens of nanometres in size. This can limit their utility in studying larger samples or materials that are not well-suited to electron microscopy.

5. Image resolution: Although electron microscopes are capable of providing high-resolution images, the image resolution is dependent on a number of factors, including the electron beam energy, the lens system, and the sample preparation. In addition, the image resolution can be limited by the size of the electron probe, which is typically on the order of a few angstroms in size [49].

6. Cost: Electron microscopes are typically very expensive, and the cost of operating and maintaining these instruments can be significant. In addition, the

specialized training and expertise required to operate an electron microscope can also be a significant barrier to entry for many researchers.

7. Specialized equipment: Electron microscopes typically require specialized equipment, such as electron detectors and sample preparation systems, which can be expensive and difficult to obtain. In addition, the specialized nature of electron microscopes can make it difficult to obtain consistent and accurate results across different systems and instruments [50].

FUTURE OF ELECTRON MICROSCOPY

The future of electron microscopes for nanotechnology is very promising, with numerous advances and improvements being made in the field. Here are some of the key trends and developments that are shaping the future of electron microscopes:

1. Increased resolution: One of the key trends in the field of electron microscopy is the development of electron microscopes with increased resolution. This is being achieved through the development of new lens systems, electron detectors, and image-processing algorithms. For example, researchers are exploring the use of holographic imaging techniques to increase the resolution of electron microscopes, which could result in images with unprecedented detail and accuracy.

2. Better imaging of biological samples: Electron microscopes are commonly used to study biological samples, but the preparation of these samples can be challenging, as they are often delicate and prone to radiation damage. In the future, advances in sample preparation techniques and the development of new imaging methods, such as cryo-electron microscopy, will likely result in better images of biological samples and an improved understanding of their structure and function [48, 56, 57].

3. Integration with other imaging techniques: Another trend in electron microscopy is the integration of electron microscopes with other imaging techniques, such as scanning probe microscopes, X-ray crystallography, and mass spectrometry. This integration will likely result in the ability to obtain a more complete picture of the materials being studied, providing insights into their structure, composition, and properties that would not be possible with a single imaging technique.

4. Automation and user-friendliness: Electron microscopes are highly specialized instruments that require a significant amount of training and expertise to operate. In the future, advances in automation and user-friendly interfaces will likely make electron microscopes more accessible and user-friendly, making it possible for a wider range of researchers to use these instruments.

5. Cost reduction: Despite their importance in the field of nanotechnology, electron microscopes are often cost-prohibitive for many researchers. In the future, advances in manufacturing and technological improvements may result in the production of more affordable electron microscopes, making these instruments more accessible to a wider range of researchers.

6. Virtual electron microscopy: Another trend in the field of electron microscopy is the development of virtual electron microscopy, which allows researchers to visualize and manipulate images of materials at the nanoscale using computer-generated models. This could have significant implications for nanotechnology research, as it would allow researchers to study materials in a controlled environment, free from the limitations and limitations of physical experiments [47].

7. Advancements in 3D imaging: Electron microscopes have traditionally been used for 2D imaging, but advances in 3D imaging are expected to revolutionize the field of electron microscopy in the coming years [50 - 55]. 3D imaging will provide researchers with a complete and more accurate picture of materials, enabling the study of their structure and properties in a way that was not possible before.

REFERENCES

[1]   R.F. Egerton, *Electron Energy-Loss Spectroscopy in the Electron Microscope.* Plenum Press: New York, 1996.
[http://dx.doi.org/10.1007/978-1-4757-5099-7]

[2]   M.J. Kelly, "Scanning transmission electron microscopy", In: *Nanoscale Imaging.,* M.J. Le Roux, S.W. Wilkins, P.J. Worsley, Eds., vol. 3. Springer: Cham, 2017.

[3]   Z. Zhang, and C.M. Lieber, "Nanoscale imaging, spectroscopy, and manipulation with scanning probe microscopy", *Chem. Rev.,* vol. 111, pp. 3889-3930, 2011.

[4]   R.J. Clift, and A.J. Cullis, "Transmission electron microscopy and electron energy-loss spectroscopy in materials science", Royal Society of Chemistry: Cambridge, 2010.

[5]   D.B. Williams, and C.B. Carter, *Transmission Electron Microscopy: A Textbook for Materials Science.* Springer: Berlin, 1996.
[http://dx.doi.org/10.1007/978-1-4757-2519-3]

[6]   G. Sánchez-Santolino, and J.M. González-Calbet, "Electron microscopy in nanotechnology: principles, applications and trends", *Mater. Today,* vol. 18, no. 7, pp. 392-405, 2015.

[7]   X. Wang, and W. Li, "Electron microscopy in nanotechnology", Royal Society of Chemistry, 2017.

[8]   D.H. Kim, and H. Lee, *Electron microscopy for nanotechnology.* Springer, 2017.

[9]   X.D. Li, and J. Mao, "Nanotechnology and the electron microscopy revolution", *Microsc. Microanal.,* vol. 16, no. 5, pp. 794-808, 2010.

[10]   Y. Niimi, and T. Ishizuka, "Electron microscopy of nanostructures: recent advances and future prospects", *Nanoscale Res. Lett.,* vol. 12, no. 1, p. 330, 2017.
[PMID: 28476085]

[11]   J. Xu, and Q. Shen, *Scanning transmission electron microscopy in nanotechnology.* Royal Society of Chemistry, 2017.

[12]   R. Kumar, and N. Singh, "Application of electron microscopy in nanotechnology", *Journal of Nanotechnology and Nanomedicine,* vol. 7, no. 1, pp. 1-9, 2017.

[13]   X.Q. Pan, Y. Wang, and J. Zhang, "Electron microscopy in nanotechnology", *Nanoscale Res. Lett.,* vol. 5, no. 10, pp. 1457-1464, 2010.

[14]   B. Schaffer, and D. Knopp, "Electron microscopy in nanotechnology: innovations, challenges and trends", *J. Microsc.,* vol. 265, no. 1, pp. 1-12, 2017.

[15]   X. Shi, and J. Zhang, "Electron microscopy in nanotechnology", *J. Nanopart. Res.,* vol. 12, no. 6, pp. 2145-2154, 2010.

[16]   C. Belton, "Electron microscopy in nanotechnology", *Wiley Interdiscip. Rev. Nanomed. Nanobiotechnol.,* vol. 2, no. 1, pp. 67-78, 2010.

[17]   L.H. Chen, and J.J. Tsai, *Electron microscopy and nanotechnology.* Springer, 2017.

[18]   J. Zhang, and X. Shi, "Electron microscopy and its applications in nanotechnology", *J. Nanopart. Res.,* vol. 12, no. 5, pp. 1605-1613, 2010.

[19]   K. Blum, and V. Skakalova, *Scanning electron microscopy in nanotechnology.* Springer, 2015.

[20]   Y.K. Mishra, and P.K. Mishra, "Electron microscopy in nanotechnology", *J. Nanopart. Res.,* vol. 12, no. 1, pp. 1-10, 2010.
       [PMID: 21170117]

[21]   S. Ahmed, M. Ahmad, B.L. Swami, and S. Ikram, "A review on applications of electron microscopy in nanotechnology", *J. Nanosci. Nanotechnol.,* vol. 16, no. 2, pp. 1441-1462, 2016.

[22]   K. Singh, V. Khanna, A. Rosenkranz, V. Chaudhary, Sonu, G. Singh, and S. Rustagi, "Panorama of physico-mechanical engineering of graphene-reinforced copper composites for sustainable applications," Materials Today Sustainability, vol. 24, p. 100560, Dec. 2023.,
       [http://dx.doi.org/10.1016/j.mtsust.2023.100560]

[23]   Z. Chen, M. Chen, W. Wang, L. Wang, and X. Zhang, "Advances in electron microscopy techniques for nanomaterials research", *Nanomaterials (Basel),* vol. 9, no. 5, p. 771, 2019.
       [PMID: 31137475]

[24]   R. Dahal, and S. Chaudhary, "Electron microscopy: A powerful tool for nanotechnology", *J. Microsc. Ultrastruct.,* vol. 6, no. 3, pp. 115-124, 2018.

[25]   X. Duan, C. Wang, and A. Pan, "Electron microscopy in the study of nanomaterials", *J. Mater. Sci. Technol.,* vol. 35, no. 5, pp. 692-702, 2019.

[26]   P. Ercius, and A.P. Alivisatos, "Electron microscopy of nanocrystals and their assemblies", *Annu. Rev. Mater. Res.,* vol. 42, pp. 411-431, 2012.

[27]   D. Fujita, K. Ueda, and Y. Takahashi, "Scanning transmission electron microscopy for nanotechnology: from imaging to probing", *Nanotechnology,* vol. 30, no. 46, p. 462001, 2019.

[28]   H. Guo, S. Yuan, and L. Liu, "Advanced electron microscopy techniques for nanomaterials research", *Microsc. Res. Tech.,* vol. 81, no. 6, pp. 624-632, 2018.
       [PMID: 29528159]

[29]   L. Houben, P. Hovington, and C. Lavoie, "Characterization of nanomaterials by transmission electron microscopy", *Mater. Sci. Semicond. Process.,* vol. 76, pp. 43-49, 2018.

[30]   Y. Jiang, M. Wang, Y. Chen, X. Li, Y. Wang, and S. Guo, "Applications of scanning electron microscopy in nanotechnology", *Microsc. Res. Tech.,* vol. 81, no. 10, pp. 1059-1067, 2018.

[31]   A.K. Katiyar, and R.S. Tiwari, "Advances in transmission electron microscopy for nanotechnology", *Mater. Today Proc.,* vol. 18, pp. 43-50, 2019.

[32]   G. Kothleitner, and F. Hofer, "Advanced transmission electron microscopy techniques for nanotechnology", *Micron,* vol. 116, pp. 20-38, 2019.

[33]   X. Li, Y. Li, and H. Jiang, "Recent advances in electron microscopy for nanotechnology", *Microsc. Res. Tech.,* vol. 81, no. 6, pp. 598-606, 2018.

[34]   Y. Li, S. Zhang, S. Hu, and H. Zeng, "Scanning electron microscopy techniques for nanotechnology research", *Micron,* vol. 126, pp. 30-39, 2019.

[35]   H. Liao, J. Zhang, and H. Yu, "Electron microscopy studies of nanomaterials for energy and environmental applications", *Nanoscale Adv.,* vol. 1, no. 7, pp. 2609-2627, 2019.

[36]   J. Lin, X. Chen, Z. Chen, H. Lin, and W. Zhou, *Electron microscopy studies of nanomaterials,* 2019.

[37]   J.J. Boonstra, *The Electron Microscope and its Applications in Nanotechnology.* CRC Press, 2018.

[38]   P.M. Ajayan, and T.W. Ebbesen, "Nanometre-scale science with carbon nanotubes", *Nature,* vol. 441, pp. 265-268, 2006.

[39]   M.F. Chisholm, and S.J. Pennycook, "High-Resolution Electron Microscopy of Nanomaterials: An Overview", *Microsc. Microanal.,* vol. 16, pp. 687-698, 2010.

[40]   J.C.H. Spence, *High-Resolution Electron Microscopy.* 3rd ed. Oxford University Press, 2023.

[41]   P.R. Buseck, and E.C.T. Chao, "Electron Microscopy in the Study of Nanoscale Materials", *Elements,* vol. 3, pp. 401-406, 2007.

[42]   L.M. Peng, and X. Chen, "The application of electron microscopy in nanotechnology", *J. Nanosci. Nanotechnol.,* vol. 10, pp. 4771-4782, 2010.

[43]   M.R. McCartney, P.D. Nellist, and J.M. Zuo, "Applications of aberration-corrected STEM to materials science", *Ultramicroscopy,* vol. 108, pp. 163-169, 2008.

[44]   J.H. Li, Y. Li, and J. Li, "Advances and challenges of electron microscopy for characterizing nanomaterials", *J. Mater. Sci. Technol.,* vol. 34, pp. 1-14, 2018.

[45]   A.H. Zewail, "Femtochemistry: Atomic-Scale Dynamics of the Chemical Bond Using Ultrafast Lasers (Nobel Lecture)", *Angew. Chem. Int. Ed.,* vol. 39, no. 15, pp. 2586-2631, 2000.
        [http://dx.doi.org/10.1002/1521-3773(20000804)39:15<2586::AID-ANIE2586>3.0.CO;2-O] [PMID: 10934390]

[46]   C.H. Ahn, A. Bhattacharya, M. Di Ventra, J.N. Eckstein, C.D. Frisbie, M.E. Gershenson, A.M. Goldman, I.H. Inoue, J. Mannhart, A.J. Millis, A.F. Morpurgo, D. Natelson, and J.E. Spencer, "Electronics and optoelectronics of two-dimensional transition metal dichalcogenides", *Nat. Nanotechnol.,* vol. 10, pp. 156-169, 2015.

[47]   S.J. Pennycook, and P.D. Nellist, *Scanning Transmission Electron Microscopy: Imaging and Analysis.* Springer, 2011.
        [http://dx.doi.org/10.1007/978-1-4419-7200-2]

[48]   G. Chen, Z. Fu, H. Guo, S. Kumar Pradhan, and P. Hao, "Study of accumulation behaviour of tungsten based composite using electron probe micro analyser for the application in bone tissue engineering", *Saudi J. Biol. Sci.,* vol. 27, no. 11, pp. 2936-2941, 2020.
        [http://dx.doi.org/10.1016/j.sjbs.2020.07.022] [PMID: 33100849]

[49]   S.K. Pradhan, "Design and development of thermionic emission microscope for the characterization of multi-beam cathode", *Ultramicroscopy,* vol. 202, pp. 140-147, 2019.
        [http://dx.doi.org/10.1016/j.ultramic.2019.04.012] [PMID: 31030108]

[50]   P.K. Sharma, S.K. Pradhan, M. Pramanik, M.V. Limaye, and S.B. Singh, "MXene Based Electrospun Polymer Electrolyte fibers: Fabrication and Enhanced Ionic Conductivity", *ChemistrySelect,* vol. 7, no. 40, p. e202201986, 2022.
        [http://dx.doi.org/10.1002/slct.202201986]

[51]   K. Singh, V. Khanna, S. Sonu, S. Singh, S.A. Bansal, V. Chaudhary, and A. Khosla, "Paradigm of state-of-the-art CNT reinforced copper metal matrix composites: processing, characterizations, and applications", *J. Mater. Res. Technol.,* vol. 24, pp. 8572-8605, 2023.

[http://dx.doi.org/10.1016/j.jmrt.2023.05.083]

[52]   V. Khanna, V. Kumar, S.A. Bansal, C. Prakash, M. Ubaidullah, S.F. Shaikh, A. Pramanik, A. Basak, and S. Shankar, "Fabrication of efficient aluminium/graphene nanosheets (Al-GNP) composite by powder metallurgy for strength applications", *J. Mater. Res. Technol.,* vol. 22, pp. 3402-3412, 2023. [http://dx.doi.org/10.1016/j.jmrt.2022.12.161]

[53]   M. Dahiya, V. Khanna, and S. Anil Bansal, "Effect of graphene size variation on mechanical properties of aluminium graphene nanocomposites: A modeling analysis", *Mater. Today Proc.,* vol. 73, no. Part 2, pp. 249-254, 2023. [http://dx.doi.org/10.1016/j.matpr.2022.07.259]

[54]   P. Gupta, N. Ahamad, D. Kumar, N. Gupta, V. Chaudhary, S. Gupta, V. Khanna, and V. Chaudhary, "Synergetic Effect of CeO 2 Doping on Structural and Tribological Behavior of Fe-Al 2 O 3 Metal Matrix Nanocomposites", *ECS J. Solid State Sci. Technol.,* vol. 11, no. 11, p. 117001, 2022. [http://dx.doi.org/10.1149/2162-8777/ac9c92]

[55]   K. Singh, S.A. Bansal, V. Khanna, and S. Singh, "Effects of Performance Measures of Non-conventional Joining Processes on Mechanical Properties of Metal Matrix Composites", *Metal Matrix Composites,* no. Aug, pp. 135-165, 2022. [http://dx.doi.org/10.1201/9781003194897-7]

[56]   A.T. Mohamed, and K.E.A. Ahmed, "Controlling on attraction forces of water droplets on surfaces of polypropylene nanocomposites coatings", *Transactions on Electrical and Electronic Materials,* vol. 19, no. 5, pp. 387-395, 2018. [http://dx.doi.org/10.1007/s42341-018-0054-4]

[57]   A. Thabet, "Synthesis and Measurement of Optical Light Characterization for Modern Cost-fewer Polyvinyl chloride Nanocomposites Thin Films", *Transactions on Electrical and Electronic Materials Journal,* vol. 24, Nature Springer, no. 6, pp. 98-109, 2023.

CHAPTER 2

Amelioration of Perovskite Nanomaterials for Advance Energy Applications

Muhammad Salman Habib[1,*] and **Muhammad Asif Rafiq**[1]

[1] *Department of Metallurgical & Materials Engineering, University of Engineering Technology G.T. Road Lahore, Pakistan*

Abstract: The demand of energy highlight the need to explore new energy resources with less emissions without depleting the environment. With this perspective, novel perovskite lead-free materials are taking over the conventional energy systems of fossil fuels that produce carbon in the environment. It has been years of struggle that scientists are working on materials for more energy with less waste materials. The challenge was readily accepted by perovskite nanomaterials that can generate energy, store it, and use it when required. The development of these nanomaterials with their promising properties such as dielectric coefficient, superconductivity, and sustainability at high temperatures, withstand high mechanical properties and can be coated, pasted, or in the form of thin and thick films. This can be done by the solid-state reaction (SSR) mixing the metallic oxides in a fixed ratio in ball milling by wet or dry method. The composites prepared were calcined, pressed, and sintered at high temperatures. Following the characterization to check the properties make them superior for high-energy advanced applications. The perovskite nanomaterials' composites can be utilized perfectly for hydrogen generation and production, photocatalysis reactions, photovoltaic solar cells, solid oxide fuel cells, electrolysis, supercapacitors, sensors, actuators, structural health monitoring applications and metal-air batteries. This chapter covers the application-based synthesis, characterizations, and properties of the perovskite nanomaterials for high-energy applications.

Keywords: Actuators, Batteries, Dielectric, Superconductivity, Sensors.

INTRODUCTION

Energy demands are increasing day by day as the world population increases. For this, the non-renewable energy is not sufficient. In this regard, scientists and researchers are working on renewable energy sources. Improvement of new materials and their co-relation for the betterment of human beings is always an

---

* **Corresponding author Muhammad Salman Habib:** Department of Metallurgical & Materials Engineering, University of Engineering Technology G.T. Road Lahore, Pakistan; E-mail: salmanhabib2000@gmail.com

Virat Khanna, Suneev Anil Bansal, Vishal Chaudhary and Reddicherla Umapathi (Eds.)

important area of research in the modern era of science and technology. It is always a desire to develop materials using advanced and latest technology at lower expense. Everyday life demands materials with producing energy. Therefore, study and development of energy materials is important. The materials performing multiple tasks are under highest consideration at present all over the world. Smart energy materials are those materials, which change their properties (*i.e.*, physical, and electrical) on the application of external source *i.e.*, electrical or mechanical. Piezoelectric, magneto strictive, electro strictive and shape memory materials are some types of smart materials [1].Among all the smart materials, piezoelectric materials are most common and widely used all over the world because of their good sensing and response by applying external source [2 - 4]. Many applications of these materials are found written in Table **1**.

Table 1. Piezo-electric devices & their uses [5]

Devices	Uses
Piezoelectric Sensors	Micro Phones, Force Sensors, Strain Sensors, Pressure Sensors, Micro Balances, Acceleration and Acoustic emission sensors
Piezoelectric Actuators	Loudspeakers, Piezoelectric motors, Acousto-optic modulator, Atomic force microscopes, Scanning, tunnelling microscopes, Inkjet Printers, Diesel engines, CT, MRI Scanners
Piezoelectric Transducers	Non Destructive Testing, Mega Sonic Cleaning, Vibration Monitoring, Doppler Probes, Industrial and Process control, Automotive engine management Systems, Medical imaging.

Among all the perovskites lead based ceramics, the solid solution $PbZrO_3$-$PbTiO_3$,*i.e.*, Lead Zirconium Titanate (PZT), is the most considerable material because of its piezoelectric and dielectric properties. Having matchless electromechanical coupling coefficients (Kp), PZT based piezo ceramics are the main materials for actuators, motors, and sensors of the present era. A few examples of piezo ceramics are included.

• $Pb\,(Nb_{2/3}\,Mg_{1/3})\,O_3$-$PbTiO_3$ (PT- PMN)

• $Pb(Nb_{2/3}\,Zn_{1/3})O_3$-$PbTiO_3$ (PT- PZN) [6].

However, the major concern is that the PZT-based piezoelectric ceramics contains toxic lead oxide (PbO), which has very adverse effect on environment and human health because of its adverse effect on intellectual and neurological development. Therefore, extensive study is being done to find out the materials free of lead but having properties comparable to PZT [7, 8]. Some of the lead-free classes of materials includes.

- Perovskites (ABO_3)

- Aurivillus Oxide type layered structure,

- Tungsten Bronze families

The first class, being extensively used due to good energy generation by piezoelectric effect and dielectric properties as compared to the last two [9 - 11].

Piezoelectric ceramics are one of the excellent sources to convert energy from mechanical to many other forms of energy. The piezoelectric effect can be utilized by number of compositions from bismuth-based ceramics, barium-based ceramics, and by mixing the compositions of these two of the solid solutions [12 - 17]. Among the perovskite class, $BaTiO_3$ (BT), KNN and NBT are considered the best replacement of PZT. Moreover, BT family is considered as a best replacement of PZT because of its high piezoelectric properties [11, 18, 19]. However, the lower Curie temperature (Tc ~ 120°C) limits its applications, because polarization disappears as it reaches and crosses its curie temperature T_c [18, 20, 21]. Asymmetric crystal structure creates polarization in crystalline materials. As T_c reaches, crystal becomes symmetric and polarization disappears [22, 23].

Another important class of materials used in multi-effect nanogenerator is inorganic ferroelectrics. The main prominent features which make them highly competitive and attractive materials for multi-effect nanogenerators are improved opto-electricity, piezoelectricity, die-electricity, pyroelectricity and multifunctional nature. Generally, they possessed better and higher dielectric constant, piezoelectric constant, and anomalous photovoltaic features. Multiple materials have been developed to use as hybrid nanogenerators like Barium Titanate (BTO), Lead Zirconate Titanate (PZT) and Bismuth Ferrites [24, 25].

Multiferroic materials are those materials that have both ferroelectric and ferromagnetic behavior at the same time, also have magnetoelectric coupling (ME) between ferroelectric and ferromagnetic materials. These materials are being extensively used in electronic devices such as tunable filters, phase shifters and generators. As single-phase multiferroics does not have both properties (ferromagnetic & ferroelectric) at the same time so this behavior limits their applications. So, to overcome this problem, multiferroic composites are suggested because they overcome those problems that are faced using single phase multiferroics [26 - 28]. Few composite multiferroics have been successfully reported in the recent past. Perovskite oxides are usually suggested to prepare ferroelectric type material *e.g.*, barium titanate ($BaTiO_3$/BT). BT based multiferroic composite systems are not amply reported in the literature. Owing to its modified piezoelectric properties, barium calcium zirconium titanate

$(Ba_{0.85}Ca_{0.15})$. $(Zr_{0.1}Ti_{0.9})$ O_3 (BCZT) is selected as a ferroelectric phase. On the other hand, $BiFeO_3$ (BFO) possessing excellent magnetic properties is chosen as a ferromagnetic phase to develop multiferroic composite [29]. Sintering aid (CuO) is used to achieve high density [30, 31].

Electrical properties differentiate depending on the difference in their structure *i.e.* grain boundary, grain itself, oxygen and also defects like imperfection in material under consideration strongly affect electrical properties. Because polycrystalline ceramics are mostly being used at higher temperature, therefore, these types of defects are unescapable [32]. Electrical properties of material under consideration are very well affected by above mentioned effects. We can engineer the properties of the said material by having the knowledge about the behavior of these defects.

Complex Impedance spectroscopy (CIS) is a new method developed to dissect the electrical properties of single crystal ceramics as well as polycrystalline ceramics using conducting electrodes in its principle [33, 34]. The graphs obtained from experimental work give much information, and also tells how grain and grain boundaries contribute to enhance the said properties [35]. It is very much important to know the contribution, which involves the presence of defect, concentration of defects, chemical modification, preparation conditions and methods and temperature range of analysis to study the microstructure as well as to develop a material keeping in view the newly market demand [33, 36].

The relation of dielectric properties at higher temperature and Complex Impedance Spectroscopy (CIS) of BCZT-BFO system is not completely studied yet, so to study this phenomenon, and investigate the electrical properties of undoped and dopped CuO at higher temperature and at higher frequencies is the main subject of this work [37]. Also, the effect of frequency and higher temperature on the dielectric properties has been studied. The relation of activation energy between grain and grain boundaries has also been studied. XRD and SEM analyses have been conducted to authenticate our argument.

POLARIZABILITY OF PEROVSKITE NANOMATERIALS

Polarizability is the ability to form dipoles without external source Fig. (**1**) gives its classification.. Every crystalline material has dipoles, which upon application of external field give conduction. The properties of these materials are changed at a certain critical temperature and materials show hysteresis loop after the application of an external source. When external source is removed, the materials still show a positive value. This external source may be electric or magnetic. At present, a lot of materials exist that show electric or magnetic polarization. The fig shows the summary of different possible materials. Other than electrically and magnetically polarizable materials, some other materials exist that display ferroelectric and ferromagnetic ordering. Magneto electric materials show electric and magnetic polarization at the same time. Ferroics material is the class of material in which at least one phase-transition occurs while multiferroics is the class of material that shows more than one ferroic property [38, 39].

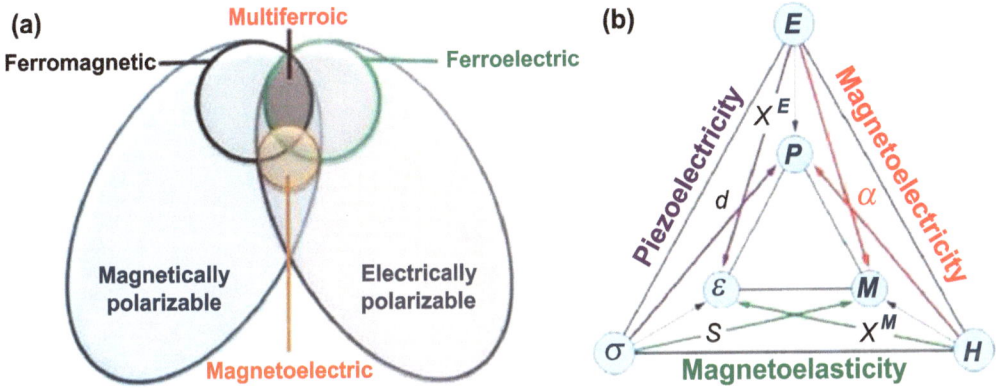

Fig. (1). Classification of polarizable materials [40].

Ferroelectric Behavior of Perovskite Nanomaterials

Perovskite materials have symmetry and sometimes these materials change their symmetry naturally when external sources of environment change. In crystallography, the study of symmetry changes occurring because of pressure, temperature etc. is a very well-developed subject. The change of symmetry is the process of phase transition. Ferroics crystal is that class of crystal in which at least one change in phase occurs, which results in the change of direction of symmetry. The term "Ferroics materials" covers ferromagnetic materials, ferroelectric materials, ferroelastic materials, ferrogyrotopic materials, *etc* [41].

• Ferroics materials *i.e.* ferroelectric, ferro-magnetic or ferro-elastic are the materials that show a change in their properties at a certain temperature range, and also has a unique hysteresis loop having two equal values even at zero electric field [38] (Fig. **2**).

• As shifting of ferroic phase from higher temperature to slightly lower temperature occurs, symmetry is disturbed as well as ferroic phase splits itself into domains [42].

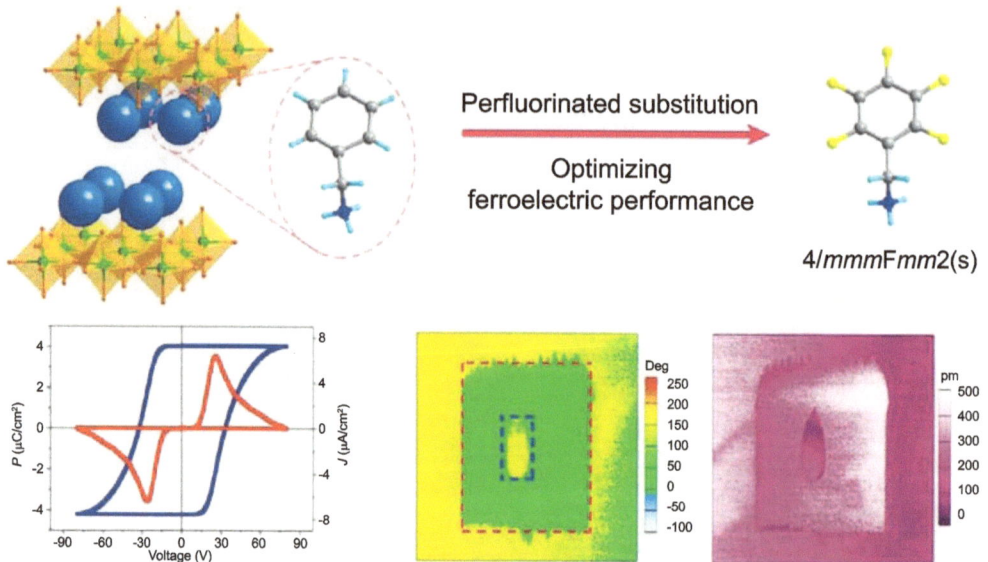

Fig. (2). Ferroelectric behavior of materials [43].

• A characteristic hysteresis loop for any ferroic order is formed, which is a result of the nature of order parameter as well as dynamics of domains of that ferroic material. This characteristic loop for any ferrioc material, depends strongly upon the external causes *i.e.* electric, magnetic or mechanical [38].

• Mostly known orderings of ferroic materials are ferroelectric (results as combination of charge polarization and electric field) ferromagnetic (results as combination of magnetic moment and magnetic field) and ferroelastic (results by combining stress and strain). Apart from these three orderings, another one, naming ferrotordoicity (results by arranging magnetic vertices) exists [42].

Multiferroic Behavior of Perovskite Nanomaterials

Multiferroics materials are those materials that have minimum two ferroic properties (ferroelectric, ferromagnetic or ferroelastic). As these have a great share in applications in multifunctional devices, multiferroic materials have found much interest at present. In these materials, coupling interactions between different order parameters (electrical, magnetic, and elastic or mechanical) produce a new effect called magneto electric (ME) effect. In primary ferroics, at least one of these effects is responsible for a long range ordering at and below ferroic phase transition. But multiferroics, having two or all three interactions compete in a directly balanced manner are of special interest for functional materials at present [44 - 47].

Magnetoelectric effect (ME) is anewly generated effect, which results from combining at least two different ferroic parameters. In direct ME effect, electric polarization is generated when magnetic field is applied while in reverse ME effect, magnetization is generated when electric field E is applied. A new effect, naming magnetoelectric effect, appears inherently in a few materials, especially at lower temperature, is a topic of current research due to its excellent properties for devices such as in memory storage devices, spintronic, and multiple-state memories [48, 49].

As an alternative, and having excellent synthesis techniques, multiferroic ME composites produced by mingling piezoelectric and magnetic materials unruffled have found noteworthy interest in current time because of having multi functionality. Coupling collaboration among piezoelectric and magnetic substances can produce a large ME as compared with a single phase material at room temperature. These ME composites have found their usage in modern devices such as magnetoelectric transducers, sensors and actuators [50] (Fig. **3**).

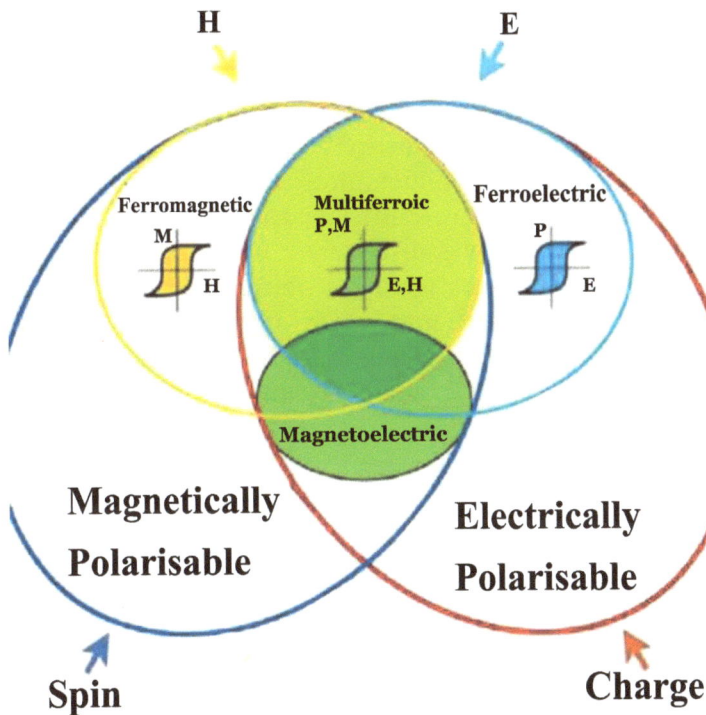

Fig. (3). Multiferroic effect of materials by applying heat, electricity, and charge [51].

The ME effect in composite materials is the result of cross collaboration of two-phase ordering, which are present in composite. Piezoelectric and magnetic materials separately do not have the ME effect, but when they are combined in the form of a composite, they have a noteworthy ME effect. So ME effect is obtained as by-product of the magneto strictive effect present in the magnetic-phase and the piezoelectric effect in the piezoelectric phase [47, 52].

ME_H effect = mechanical / electric x magnetic/mechanical

ME_E effect = mechanical / magnetic x electric / mechanical

Combined Activity of Ferromagnetism and Piezoelectric Effect

All magnets have the same basic principle of magnetism because of the presence of localized electrons in d or f shell (partially filled) of rare earth metals as well as transition metals, which on the other hand, have magnetic moment. Interaction between these magnetic moments results in magnetic order. Ferroelectrics show different behavior and ferroelectricity is generated through a lot of microscopic sources, and so we can achieve several diverse types of multiferroics. Broadly, multiferroics are classified in two groups. The 1st group, called type-I multiferroics, comprises those materials in which ferroelectricity and magnetism are produced from several resources and are not affected by one another, seem largely individually of one another, as some coupling exists between them. In such types of materials, ferroelectricity occurs at high temperature, while magnetism appears at lower temperature, also the spontaneous polarization is comparatively large (10-100 $\mu C/cm^2$) *i.e.*, $BiFeO_3$ and $YMnO_3$. The 2nd group, called type-II multiferroics, was discovered in recent times and for this class of materials, magnetism produces ferroelectricity. One drawback is that for these materials, the value of polarization is smaller (10-2$\mu C/cm^2$) as compared with type I materials. A lot of research is being done to find out composite multiferroics consisting of known values of magnetism and ferroelectric like multilayers and self-organized nanostructures [53].

Multiferroic Composites

According to the original definition, multiferroic are those materials that possess at minimum two properties at the same time, *i.e.*, ferroelectricity, ferromagnetism, or ferroelectricity. As multiferroic composites are materials that consist of different phases, but neither phase supports the ME effect [54, 55]. Principally, multiferroic composite contains ferroelectric and ferromagnetic phase. The ferroelectric part gives piezoelectricity while ferromagnetic part gives magnetic property to the composite. It is the coupling, which tells how much the magnetic field has effect on the generation of polarization. This coupling may result as interaction in-between the two order parameters or it may occur because of stress/strain in ME composites [56].

Multiferroics having $Pb(Ti_xZr_{1-x})O_3$ or $(1-x)Pb(Nb_{2/3}Mg_{1/3})O_3$-xPbTiO as piezoelectric portion has excellent larger values of ME-effect, but the presence of lead (toxic element for health) restricts its use and also motivates to generate lead-free materials that can replace the old ones [57]. BT-based ceramics have become a strong candidate to replace lead-based ceramics because of their much better properties *i.e.* d_{33}= 191 pC / N and direct magnetoelectric coefficients α_E, below 80. Intensive study is being done to find out the replacement of lead-based piezoelectric materials. A number of ways are being adopted to optimize the properties of BT- based piezoelectric ceramics *i.e.* ME effect can be optimized by changing connectivity type, microstructure and by using materials with greater piezoelectric coefficients. As illustrated schematically in Figure, ME effects of composites results by joining magneto strictive effect (part of magnetic phase) and piezoelectric effect(part of piezoelectric phase). Many bulk ME composites exist which produce ME effect above the room temperature because of the strain present there. A step further, multiferroic films have an advantage as compared with multiferroic composites. In bulk composites, phases such as ferroelectric and magneto strictive can be combined at the nano scale, also these two phases are usually mixed by solid oxide method, mix-sintering or by adhesive bonding, which results in the loss of properties at interface of both phases. On the other hand, in composite films, phases can be combined at atomic level, so interface losses are reduced. Also by joining different phases with the same crystal lattice, superlattice composite films are produced, which make it easy to understand ME compound at atomic level, so in integrated devices *i.e.* sensors, high density memories and spintronics, mostly ME films are being used [29, 58] (Fig. **4**).

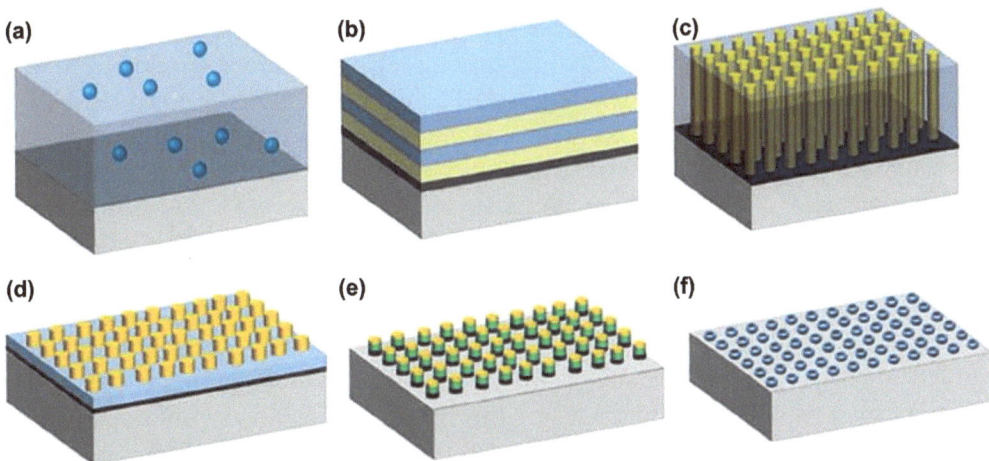

Fig. (4). Promising multiferroic-composite structures. (**A**) Homogeneous mixture of electric & magnetic phase; (**B**) laminated bi-layer structure; (**C**) laminated multi-layer structure; (**D**) composite made of particles mixed in a matrix; and (**E**) Multiferroic [59].

SYNTHESIS OF PEROVSKITE NANOMATERIALS

The basic method of making perovskite materials for energy applications is being discussed in detail. The series of steps that are followed by different equipment's are discussed below.

Ball Mill

A ball mill is used to grind powders for the preparation of perovskite materials. The stoichiometric ratio of precursors is weighted. The precursors are put in zirconia jars with zirconia balls of different sizes in a liquid medium for wet milling The milling is done for at least 4-6 hours for the powder to be completely homogenized and mixed with the nanosized. There are different types of ball mills like planetary ball mill, traditional ball mill, and batch ball mill. The forces of shear at the same time impact forces that will create between the balls and the powder particles that will help decrease the powder size to nanometers. The difference in the speeds and the ball to mass ratios of the powders will make the powders to get mixed homogenously. The wet milling process must be at least 200-400 rpms with the solvents usually using ethanol, methanol, and acetone. After milling the powders for the preparation of perovskite materials, they will be dried to evaporate the solvent and then nano sized powders will be stored.

Calcination

The process of removal of small molecules at different temperatures ranging from 400-600 deg when heated in the absence of air is called calcination. This temperature is ordinarily characterized as the temperature at which the standard Gibbs free vitality for a specific calcination response is reduced to even zero. The decomposition reaction in which the molecules like water, ammonia, carbon dioxide and other nonmetallic oxides if present in the perovskite will be removed. The secondary phases that are low temperature melting will also be removed from the nanomaterial's powders. The calcination temperature will be around 0.5Tm ranging from 400-600 deg for most of the perovskite materials but it is higher for some composition of barium bases, potassium based and niobium-based materials. The thermal degradation of the materials will be an important phenomenon as the temperature is selected using different techniques like Thermogravimetric Analysis (TGA), Differential Thermal Analysis (DTA) and Dilatometry. The gases removed from the solid phases of the solid oxides and the change in color, shrinkages and other parameters will be the findings of proper calcination of the powders for perovskite nanomaterials for energy applications.

Pressing

The powders obtained after calcination is put in the metallic stainless steel die for the pressing of the samples into disc shaped pellets. The hydraulic press machine with pressure around 6000-12000 psi should be used for this action. The green pellets will be handled with care so that it can be fed into the furnace at high temperature for sintering. Sometimes lubricating materials or additives will be used for the proper strength of the pellets to form (Fig. **5**).

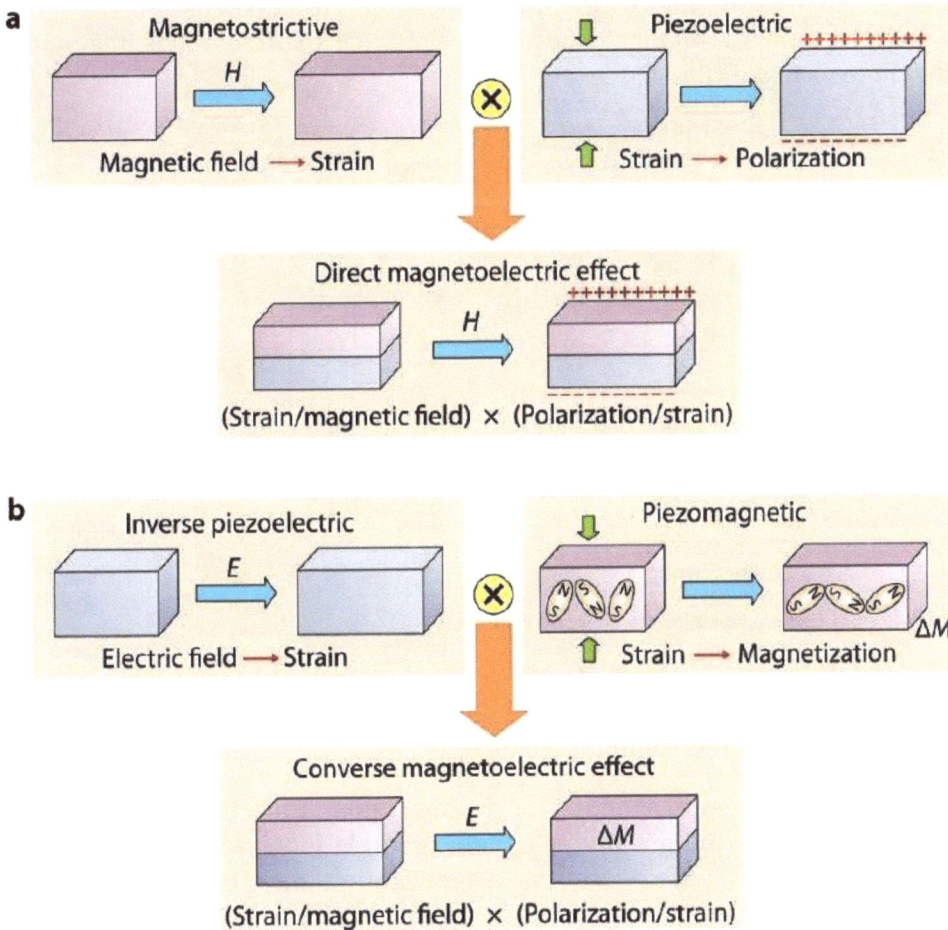

Fig. (5). Graphical illustration of strain-arbitrated ME effect in a composite system comprising a magnetic-layer represented as purple color and ferroelectric represented as pink color (**a**) Direct ME effect (**b**) Converse ME effect [59].

Sintering

When the thermal energy is applied to the disc shaped pellets inside the furnace, the densified pellets are obtained which increases the grain size for nanoparticles. It is a process used to produce density control materials or compounds from metal or ceramic powder by applying thermal energy. Sintering focuses on producing a sintered part with a reproducible and preferably designed microstructure by controlling the sintering variables. Microstructural control means controlling the amount of gain, the sintered density, and the distribution of size of phases that include pores. Mostly, microstructural inspection will make a high-density body with a fine-grained structure. Due to the presence of the dopant, the sintering temperature changed (Fig. **6**).

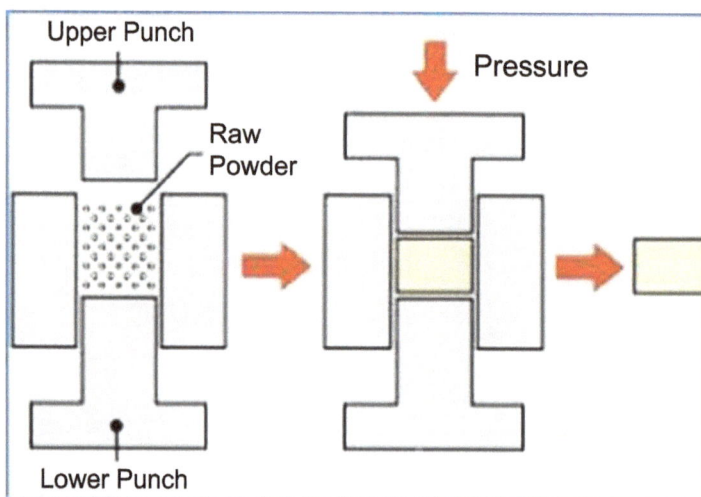

Fig. (6). The basic phenomenon of pressing powder in metal die.

EFFECT OF DOPING ON PEROVSKITE MATERIALS

Effect of Neodymium (Nd) and Cobalt (Co)

After preparation by solid state mixed oxide method, it has been found that grain size decreases while the density of ceramic sample increases as the quantity of Co-doping increases. Also remnant magnetization increases while coercive-magnetic field reduces while increasing Co-doping [60].

Effect of Manganese

Mn-doping spontaneously increases magnetization, and also magnetic hysteresis loop gives more clarity. This increase of magnetization is because of the change of long-range anti-ferromagnetism to collinear anti-ferromagnetism [61].

Effect of Strontium

Magnetic properties of $BiFeO_3$ have been enhanced by chemical manipulation. To produce a high magnetic coupling order in multiferroic is the main objective, and this objective can be obtained by substituting a chemical agent at site-A(Bi) and site-B(Fe). When Sr is doped at Bi-site, the sample becomes oxygen sub-stoichiometric and also magnetization reduces with Sr-doping [62].

Effect of Barium (Ba) and Calcium (Ca)

$BiFeO_3$ doped with barium shows ferroelectricity and ferromagnetism instantaneously and$BiFeO_3$ doped with Ca shows coupling of ferromagnetism and ferroelectricity.

Effect of Copper (Cu)

As a result of Cu addition to Bismuth ferrite, the density and the electrical properties improved. Due to Cu doping, dielectric constant rises and the dielectric loss, which, also known as loss tangent, decreases. Cu acts as a sintering aid for $BiFeO_3$ [60].

Effect of Titanium (Ti)

The magnetic and electric properties are improved by co-doping of BFO with Ti. The remanent magnetization increases by the substitution of Ti to BFO.

ENERGY HARVESTING THROUGH PEROVSKITE NANOMATERIALS

Perovskite materials have drawn a lot of interest for energy harvesting applications due to their high catalytic activity, good electrical conductance, and durability. Energy storage systems must be compacted and made smaller to accommodate new autonomous electronic gadgets. Perovskite lattices allow for ion movement, making it possible to use such materials as battery electrodes. When it is in comparison with the bulk materials with the same crystal structure, nanoparticle perovskite systems have shown improved reduction of catalytic activity, oxygen as well as a greater discharge threshold and specific capacity. For energy harvesting purposes, perovskite oxide nanocrystals have been used in batteries since 2014. Different properties influence the electrochemical performance of perovskite oxides: size, shape, porosity in nanocrystals, lattice defects or quality of nanocrystals. It has been observed that nanocrystals obtained by processing in ball mills show improved catalytic working than without this process. It is because of changes in the crystal structure and defects in the production of crystal system. The effect of doping in B-site manganite perovskite oxides (La0.8Sr0.2Mn1-xNixO3) has been studied and showed the enhanced efficiency and performance due to an increase in the vacancies of oxygen at the surface. In metal- air batteries, catalytic activity is improved by using a layer of $NiCo_2O_4$ on nanorods of $La0.8Sr0.2MnOM_3$. Other than this, perovskite nanorods with Ag nanocrystals, perovskite nanofibers with RuO_2 nanoparticles, perovskite porous nanofibers with RuO_2 nanosheets are nowadays beingused to increase the efficiency of energy harvesting devices and also to improve the catalytic activity [63].

Mechanical energy can be harvested in the systems created in a variety of ways during the last 20 years to support or power small size electronics. The problem of developing a basic and economical method to improve the current and charge the performance of piezoelectric energy harvesters remains a difficult subject. Here, a perovskite $BaTiO_3$ (BT) nanowire (NW)-based poly (vinylidene fluoride-c--trifluoroethylene) (P(VDF-TrFE)) is used to create a 1D-3D completely piezoelectric nanocomposite for a high-performance hybrid nanocomposite generator (hNCG) device. The flexible hNCG's harvested output is higher (up to 14 V and 4 A) than that of even previous flexible energy harvesters. According to simulations using the finite element analysis approach, the excellent performance of hNCG devices is partly due to the piezoelectric combination of tightly controlled BT NWs, the P(VDF-TrFE) matrix, as well as the efficient stress transferability of the piezopolymer. The flexible hNCG is directly attached to a hand to scavenge energy utilizing human motion in various biomechanical frequencies for applications such as self-powered wearable patch devices. This is a unique method to develop biocompatible, wearable electronics with good performance [64].

Energy-harvesting nanogenerators can be generated by adding formamidinium lead halide nanoparticles with perovskite to the PVDF polymer. The result obtained by this energy harvesting nanogenerator is the output voltage of 30V and a 6.2 mA/cm2 current density. Light-emitting diode systems can be made using the ac this system generates. Also, this polymeric system can be generated with a lead-free methodology, and by adding barium titanate to the PVDF-TrFE membrane, it can produce extremely effective piezoelectric materials that might be used in intelligent wearable electronics and self-sufficient nano-systems. They are also capable of efficiently capturing biomedical energy. Their energy output is quite similar to that of PZT materials. Barium titanate nanowires can be used to create flexible energy harvesters with one-dimensional nanostructures that are nicely aligned. As a result of repeated bending deformations, these arrays were able to successfully convert a maximum open-circuit voltage of around 15 V, a maximum short-circuit current of over 400 nA, and an effective power of 0.27 mW [65].

COATED PEROVSKITE MATERIALS ON METALLIC SUBSTRATES

Ceramic coatings give high performance oxides to be coated on suitable metals and alloys for the solutions of corrosion, wear, heat, and friction. This increases the lifetime of the material. Besides this, it has certain disadvantages as coating is brittle, it is difficult to coat because the methods are expensive, and it forms cracks just after its formation. But currently, the piezoelectric properties of ceramics are discussed [66].

While coating a ceramic layer on metal, it must be suited to all applications and it also favors all processes, and it must pass all the phases to achieve the final product. There are various methods which can be used depending on the metal compatibility and the availability of the metal as it must be cheap and has high melting point due to sinter ability of ceramics. The metal should intact the ceramic layer and the process should be cheap. There are some metals that are used commonly but having limited applications like platinum and gold. Copper can be used but due to low melting point, it cannot sinter BCZT so it was ruled out while stainless steel may be used but due to the carbon content, it cannot be utilized [67].

Thick layered lead free coatings are used for several applications in instruments, detectors, UV devices and dielectrics. Besides this, the methods of coatings vary from time to time and are dependent upon application to application to achieve the required properties. Most of the common methods of coatings are used by different scientists, which include dip coating, pulsed laser deposition, spray coating, solution casting technique and PVD. CVDs are also used for some of the compositions. These techniques require possible substrates as discussed above for the most suitable applications and the vast coating thickness to be attained for energy harvesting low density production.

SOLAR CELLS PEROVSKITE NANOMATERIALS

Solar energy is the most abundant, inexhaustible, and clean renewable energy source. Large amounts of solar energy can be converted into electrical energy through solar cells. Perovskite solar cells have grabbed attention due to the considerable increase in efficiency from 3.8% in 2009 to 25.7% in 2021. In 2012, researchers came to know that solar cells having perovskite structure as active component have 10% greater power conversion efficiency than silicon solar cells. Two years later, this number increased to 17%. The combination of perovskite solar cells and conventional solar cells to convert sunlight into electrical energy is also being researched. Power conversion efficiency above 29% can be achieved by nanotextured designs. Perovskite solar cells are low-cost and have high power conversion efficiency *i.e.*, more than 20%. Due to these attributes, PSCs can be commercialized in the future. Graphene-based solar cells have outstanding physiochemical properties due to their unique 2D structure, high conductivity and fast electron mobility. The optoelectronic properties of graphene are further enhanced by the addition of various nanomaterials. The addition of graphene quantum dots (GQD) in TiO_2 nanocomposite PSCs enhances it performance. The obtained power conversion efficiency is 10.15%, which is more than the PSC without the GQD layer [68].

Due to the complicated excitonic nature of hybrid organic-inorganic semiconductors, simultaneously achieving high open-circuit voltage and high short-circuit current is a major hurdle in the development of high-efficiency perovskite solar cells. This problem is solved by adding Au nanoparticles to a mesoporous TiO_2 film, on which he deposits a MgO passivation film. This method yielded a power conversion efficiency of 16.1%. An open circuit voltage of 1.09 V and a short circuit current density of 21.76 mA cm-2 were obtained. The power conversion efficiency of the designed device improved by 34.2 times over the pure TiO_2 device [69].

In perovskite solar cells, mesoscopic TiO_2 was modified with $LaFeO_3$ to form a nanocomposite of $LaFeO_3$:TiO_2 as electron transport layer. An optimal amount addition of $LaFeO_3$ in TiO_2 electron transport layer reduces charge transfer resistance. Charge carrier recombination also decreased. The electron transport layer modification of solar cell with 6mg inorganic perovskite nanomaterial $LaFeO_3$ has power conversion efficiency of 17.05%. The obtained results justify that the power conversion efficiency of $LaFeO_3$-modified electron transport layer is higher than the solar cell containing pure TiO_2 electron transport layer. It has been observed that $LaFeO_3$-modified electron transport layer solar cell maintains >70.1% of its initial power conversion efficiency after 800 h [70]. The problem of presence of defects from ionic charges on the grain boundaries from the surface to perovskite film has been treated by the introduction of inorganic perovskite nanomaterials shaped into nanowires in the film. The presence of nanowires passivated the defects at grain boundaries and facilitated charge transport. The nanowire modified perovskite film reduced the defect sites and increased solar cell life. It has a power conversion efficiency of 21.56% and device stability of 3500-hour without encapsulation [71].

Addition of inorganic perovskite material AlMo0.5O3 into mesoporous TiO_2 layer photoanode of die-sensitized solar cell enhances the power conversion efficiency of solar cell. The addition of different amounts of AlMo0.5O3 in TiO_2 layer yields that the cell containing 0.5% AlMo0.5O3 has the highest efficiency. The results show that the TiO_2 + 0.5%AlMo0.5O3 composite for die sensitized photo anode increases device efficiency by 2.9% than pure TiO_2 photoanode. The presence of

AlMo0.5O3 increases charge transfer and declines the recombination of charge carriers simultaneously [72].

Methyl ammonium tin bromide (CH3NH3SnBr3), a lead-free perovskite, acts as absorber layer. CH3NH3SnBr3-based perovskite solar cells have the power conversion efficiency of 21.66% with open-circuit voltage of 0.80V. Methyl ammonium tin bromide perovskite solar cells are environmentally friendly in nature [73].

PHOTOVOLTAIC CELLS APPLICATIONS OF PEROVSKITE NANOMATERIALS

Perovskite nanomaterials have a wide range of applications in photovoltaic cells. Photovoltaic technology can help reduce climate change by reducing the use of fossil fuels. Highly efficient monocrystalline silicon cells are dominating the market. On the laboratory scale, crystalline silicon photovoltaic cells reached 25% efficiency two decades ago. The first generation of photovoltaic cells are silicon-based wafers. Using processing knowledge and raw ingredients given by the microelectronics sector, silicon-based photovoltaic cells were the first segment of photovoltaics to hit the market. More than 80% of silicon-based solar cells have the world's installation capacity and 90% market share. A thick layer of silicon is present in the first-generation photovoltaic cells. First-generation photovoltaic cells can be mono-, poly-, and multi-crystalline silicon and in single III-V junctions (GaAs). Monocrystalline silicon cells have an efficiency of 12-24%, a band gap of 1.1 eV, and a life cycle of 25 years while polycrystalline cells have an efficiency of 10% to 18% and a life span of 14 years. GaAs-based solar cells have 28-30%, a band gap is 1.43eV, and a life cycle is 18 years. Single III-V junctions have a great conversion efficiency of 29.1% when having GaAs single junction while under concentrated sunlight, 47.1% efficiency increased. A lot of labour intensive, expensive and often ineffective process stages are involved in finishing including photolithography, manual spin coating application, contact alignment, metal evaporation and lifting [74].

Second-generation photovoltaic cells are thin films based on CdTe, gallium selenide, copper, and amorphous silicon. They have advantages such as low cost, non-toxic, and high absorption coefficient, but also have disadvantages such as low efficiency. Amorphous silicon is used in the replacement of crystalline silicon to reduce the cost, but it runs the risk of losing efficiency. CdTe-based solar cells have an efficiency of 15-16%, a band gap is 1.45eV and a life span is 20 years while copper indium gallium-based solar cells have an efficiency of 20% and a life cycle is 12 years.

The third-generation has dye-sensitized photovoltaic cells with an efficiency of 5-20%. Quantum dots solar cells have an efficiency of 11-17%, organic and polymeric photovoltaic cells have an efficiency of 9-11%, and solar cells based on perovskite have an efficiency of 21%. Multi-junction solar cells have 36% efficiency, high performance, and are expensive.

The fourth generation of photovoltaic cells is called a hybrid inorganic cell because it has the flexibility of polymers, low cost, and stability of metal nanoparticles and metal oxides, graphene, and their derivatives. In case of applications, polymers are more advantageous due to negligible vapor pressure. No vaporization takes place before decomposition. Nanometer-thick semiconductor films have robust mechanical properties and are compatible with roll-to-roll fabrication on flexible substrates. Photovoltaic cells are used in thin film transistors, which increase the range of applications. Also, they are used in low-end microelectronics such as radio frequency identification lags. Renewable clean energy is the conversion of conventional solar energy to electrical energy [76] (Fig. 7).

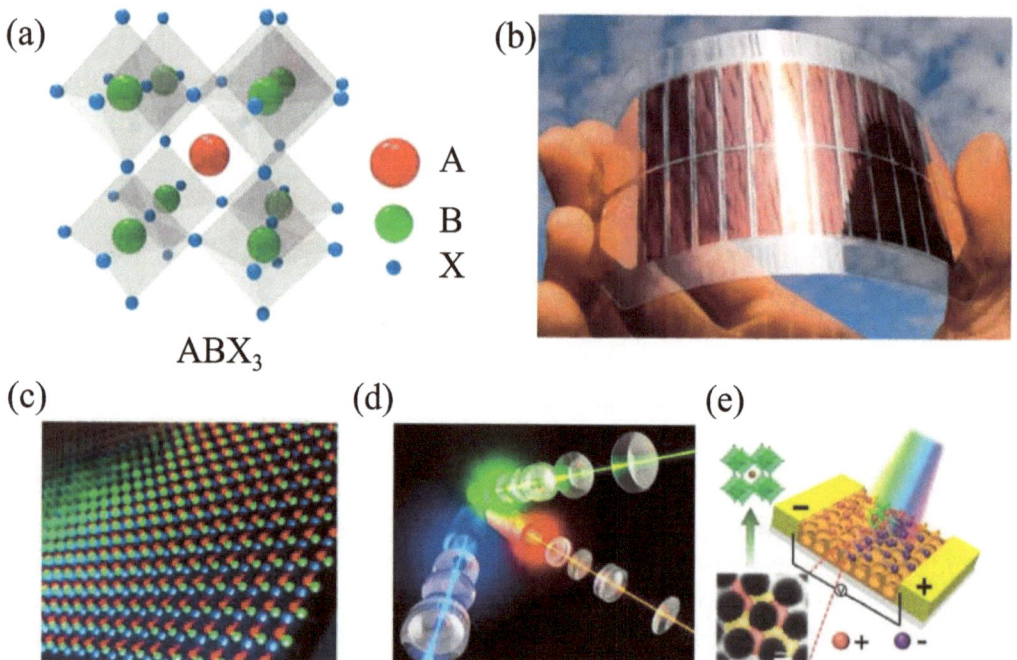

Fig. (7). (a) Structure of perovskite materials for solar cells, (b) Flexible coated PV solar cell, (c) Energy generation from solar cell, (d) Effect of light on the solar cell, (e) Schematic of generation of hole-electron when light falls on perovskite [75].

During the last three years, bulk heterojunction photovoltaic cells have experienced tremendous progress in performance, and it is a potential competitor to the amorphous silicon-based photovoltaic cell. New market applications of polymeric semiconductors play a vital role in their success. These devices are used in concentrator systems. From a future perspective, they aim to produce highly efficient and inexpensive photovoltaic devices. Being low cost is an important characteristic of these devices when thin film technologies are used directly on building materials. Photovoltaic approaches can be improved with the use of multifunctional nanomaterials and nanotechnology, which can enhance solar energy generation and conversion [77].

CONCLUSION

The perovskite materials utilized in the applications of photovoltaics, solar cells, energy harvesting, smart materials, sensors, and actuators will take over the world with boom. These materials will in the coming days replace the fossil fuel sector which have been the only source of energy since the beginning of the earth. The efficiency of the materials is increasing day by day which is the advancement of the energy sector. The materials generating energy either in bulk form or in coating both have their own characteristics features. The development of structure property relationship in the synthesis route helped scientists and engineers to design the electrical microstructure of perovskite materials. The synthesis route can be modeled for the preparation of materials with increasing efficiencies like the addition of dopants, composite preparation and these can be further characterized, for the applications highlighted in the chapter. This chapter gives an insight into the selection of proper lead-free perovskite material, its preparation technique and the possible characterizations to design the electrical microstructure of the perovskite nanomaterials for specific energy generation and harvesting applications.

LIST OF NOMENCLATURE

PZT Lead zirconium Titanate

Kp   Coupling Coefficient

BZT Barium Zirconium Titanate

BCZT Barium Calcium Zirconium Titanate

CuO Copper Oxide

Tc   Curie Temperature

LCR Inductance Capacitance, Resistance meter

CIS Complex Impedance Spectroscopy

XRD X-Ray Diffraction Analysis

SEM	Scanning Electron Microscopy
$\mathbf{d_{33}}$	Dielectric Constant
DF	Dielectric Loss
Z	Impedance
Z'	Real component of Impedance
Z"	Imaginary component of Impedance
kB	Boltzmann constant
A.E	Activation Energy
eV	Electron volt
BFO	Bismuth Ferrite
I	Current
V	Voltage

REFERENCES

[1]   A. McWilliams, *Smart materials and their applications: technologies and global markets* vol. 161. BCC Research Advanced Materials Report, 2011.

[2]   Z.L. Wang, and Z.C. Kang, *Functional and smart materials-Structural evolution and structure analysis* Academic press: Peking, China, 2002.

[3]   P.M. Vilarinho, Y. Rosenwaks, and A. Kingon, "Scanning Probe Microscopy: Characterization, Nanofabrication and Device Application of Functional Materials", *Proceedings of the NATO Advanced Study Institute on Scanning Probe Microscopy: Characterization, Nanofabrication and Device Application of Functional Materials,* vol. 186, 2006 Algarve, Portugal Springer Science & Business Media

[4]   J. Buchaca, *Smart Materials: A technical and market assessment.* BCC Research, 2005.

[5]   L. Ajdelsztajn, J.A. Picas, G.E. Kim,, "Oxidation behavior of HVOF sprayed nanocrystalline NiCrAlY powder", *Mater. Sci. Eng. A.,* vol. 338, pp. 33-43, 2002.

[6]   S.E. Park, and T.R. Shrout, "Ultrahigh strain and piezoelectric behavior in relaxor based ferroelectric single crystals", *J. Appl. Phys.,* vol. 82, no. 4, pp. 1804-1811, 1997. [http://dx.doi.org/10.1063/1.365983]

[7]   H. Needleman, "Low level lead exposure: history and discovery", *Ann. Epidemiol.,* vol. 19, no. 4, pp. 235-238, 2009. [http://dx.doi.org/10.1016/j.annepidem.2009.01.022] [PMID: 19344860]

[8]   H. Needleman, "Lead Poisoning", *Annu. Rev. Med.,* vol. 55, no. 1, pp. 209-222, 2004. [http://dx.doi.org/10.1146/annurev.med.55.091902.103653] [PMID: 14746518]

[9]   T.R. Shrout, and S.J. Zhang, "Lead-free piezoelectric ceramics: Alternatives for PZT?", *J. Electroceram.,* vol. 19, no. 1, pp. 113-126, 2007. [http://dx.doi.org/10.1007/s10832-007-9047-0]

[10]   B. Malič, A. Benčan, T. Rojac, and M. Kosec, "Lead-free Piezoelectrics Based on Alkaline Niobates: Synthesis, Sintering and Microstructure", *Acta chimica slovenica,* vol. 55, no. 4, pp. 719-726, 2008.

[11]   J. Rödel, W. Jo, K.T.P. Seifert, E.M. Anton, T. Granzow, and D. Damjanovic, "Perspective on the development of lead-free piezoceramics", *J. Am. Ceram. Soc.,* vol. 92, no. 6, pp. 1153-1177, 2009. [http://dx.doi.org/10.1111/j.1551-2916.2009.03061.x]

[12]   A. Maqbool, A. Hussain, R.A. Malik, J.U. Rahman, A. Zaman, T.K. Song, W-J. Kim, and M-H. Kim, "Evolution of phase structure and giant strain at low driving fields in Bi-based lead-free incipient piezoelectrics", *Mater. Sci. Eng. B,* vol. 199, pp. 105-112, 2015. [http://dx.doi.org/10.1016/j.mseb.2015.05.009]

[13]   X. Liu, Z. Chen, D. Wu, B. Fang, J. Ding, X. Zhao, H. Xu, and H. Luo, "Enhancing pyroelectric properties of Li-doped (Ba 0.85 Ca 0.15)(Zr 0.1 Ti 0.9)O 3 lead-free ceramics by optimizing calcination temperature", *Jpn. J. Appl. Phys.,* vol. 54, no. 7, p. 071501, 2015. [http://dx.doi.org/10.7567/JJAP.54.071501]

[14]   V. G. Kostishyn, L. V. Panina, L. V. Kozhitov, A. V. Timofeev, and A. N. Kovalev, "Synthesis and multiferroic properties of M-type $SrFe_{12}O_{19}$ hexaferrite ceramics", *Journal of Alloys and Compounds,* vol. 645, pp. 297-300, 2015. [http://dx.doi.org/10.1016/j.jallcom.2015.05.024]

[15]   D. Chen, I. Harward, J. Baptist, S. Goldman, and Z. Celinski, "Curie temperature and magnetic properties of aluminum doped barium ferrite particles prepared by ball mill method", *J. Magn. Magn. Mater.,* vol. 395, pp. 350-353, 2015. [http://dx.doi.org/10.1016/j.jmmm.2015.07.076]

[16]   Y. Bai, A. Matousek, P. Tofel, V. Bijalwan, B. Nan, H. Hughes, and T.W. Button, "$(Ba,Ca)(Zr,Ti)O_3$ lead-free piezoelectric ceramics—The critical role of processing on properties", *J. Eur. Ceram. Soc.,* vol. 35, no. 13, pp. 3445-3456, 2015. [http://dx.doi.org/10.1016/j.jeurceramsoc.2015.05.010]

[17]   A. Awadallah, S.H. Mahmood, Y. Maswadeh, I. Bsoul, and A. Aloqaily, "Structural and magnetic properties of vanadium doped M-type barium hexaferrite ($BaFe_{12-x}V_xO_{19}$)", *IOP Conference Series: Materials Science and Engineering,* 2015p. 012006 Irbid, Jordan [http://dx.doi.org/10.1088/1757-899X/92/1/012006]

[18]   A. Thabet, and N. Salem, "Experimental investigation on dielectric losses and electric field distribution inside nanocomposites insulation of three-core belted power cables", *Advanced Industrial and Engineering Polymer Research,* vol. 4, pp. 19-28, 2021. [http://dx.doi.org/10.1016/j.aiepr.2020.11.002]

[19]   A. Thabet Mohamed, and K. Ebnalwaled, "Controlling on Attraction Forces of Water Droplets on Surfaces of Polypropylene Nanocomposites", *Transactions on Electrical and Electronic Materials,* vol. 1, pp. 1-9, 2018.

[20]   L.M. Levinson, *Advances in Ceramics.* vol. 1. Grain Boundary Phenomena in Electronic Ceramics, 1979.

[21]   D.C. Hill, and H.L. Tuller, "Ceramic sensors: theory and practice", *Ceramic Materials for Electronics,* vol. 272, p. 335, 1991.

[22]   Q.K. Muhammad, M. Waqar, M.A. Rafiq, M.N. Rafiq, M. Usman, and M.S. Anwar, "Structural, dielectric, and impedance study of ZnO-doped barium zirconium titanate (BZT) ceramics", *J. Mater. Sci.,* vol. 51, no. 22, pp. 10048-10058, 2016. [http://dx.doi.org/10.1007/s10853-016-0231-y]

[23]   A. Thabet, and N. Salem, "Experimental Progress in Electrical Properties and Dielectric Strength of Polyvinyl Chloride Thin Films Under Thermal Conditions", *Transactions on Electrical and Electronic Materials,* vol. 21, no. 2, pp. 165-174, 2020. [http://dx.doi.org/10.1007/s42341-019-00163-1]

[24]   J. Qi, N. Ma, X. Ma, R. Adelung, and Y. Yang, "Enhanced photocurrent in $BiFeO_3$ materials by coupling temperature and thermo-phototronic effects for self-powered ultraviolet photodetector system", *ACS Appl. Mater. Interfaces,* vol. 10, no. 16, pp. 13712-13719, 2018. [http://dx.doi.org/10.1021/acsami.8b02543] [PMID: 29619823]

[25]   N. Ma, and Y. Yang, "Boosted photocurrent in ferroelectric $BaTiO_3$ materials via two dimensional

planar-structured contact configurations", *Nano Energy,* vol. 50, pp. 417-424, 2018.
[http://dx.doi.org/10.1016/j.nanoen.2018.05.069]

[26]   J. Valasek, "Piezo-electric and allied phenomena in Rochelle salt", *Phys. Rev.,* vol. 17, no. 4, pp. 475-481, 1921.
[http://dx.doi.org/10.1103/PhysRev.17.475]

[27]   H. Schmid, "Multi-ferroic magnetoelectrics", *Ferroelectrics,* vol. 162, no. 1, pp. 317-338, 1994.
[http://dx.doi.org/10.1080/00150199408245120]

[28]   X. Qi, J. Zhou, Z. Yue, Z. Gui, L. Li, and S. Buddhudu, "A ferroelectric ferromagnetic composite material with significant permeability and permittivity", *Adv. Funct. Mater.,* vol. 14, no. 9, pp. 920-926, 2004.
[http://dx.doi.org/10.1002/adfm.200305086]

[29]   C.A.F. Vaz, J. Hoffman, C.H. Ahn, and R. Ramesh, "Magnetoelectric coupling effects in multiferroic complex oxide composite structures", *Adv. Mater.,* vol. 22, no. 26-27, pp. 2900-2918, 2010.
[http://dx.doi.org/10.1002/adma.200904326] [PMID: 20414887]

[30]   A. Kamal, M.A. Rafiq, M.N. Rafiq, M. Usman, M. Waqar, and M.S. Anwar, "Structural and impedance spectroscopic studies of CuO-doped (K0.5Na0.5Nb0.995Mn0.005O3) lead-free piezoelectric ceramics", *Appl. Phys., A Mater. Sci. Process.,* vol. 122, no. 12, p. 1037, 2016.
[http://dx.doi.org/10.1007/s00339-016-0564-z]

[31]   H. Sun, Y. Zhang, X. Liu, Y. Liu, and W. Chen, "Effects of CuO additive on structure and electrical properties of low-temperature sintered Ba0.98Ca0.02Zr0.02Ti0.98O3 lead-free ceramics", *Ceram. Int.,* vol. 41, no. 1, pp. 555-565, 2015.
[http://dx.doi.org/10.1016/j.ceramint.2014.08.104]

[32]   S.M. Neirman, "The Curie point temperature of Ba(Ti1?x Zr x)O3 solid solutions", *J. Mater. Sci.,* vol. 23, no. 11, pp. 3973-3980, 1988.
[http://dx.doi.org/10.1007/BF01106823]

[33]   D.C. Sinclair, and A.R. West, "Impedance and modulus spectroscopy of semiconducting BaTiO3 showing positive temperature coefficient of resistance", *J. Appl. Phys.,* vol. 66, no. 8, pp. 3850-3856, 1989.
[http://dx.doi.org/10.1063/1.344049]

[34]   J.R. Macdonald, and E. Barsoukov, "Impedance spectroscopy: theory, experiment, and applications", *History (Lond.),* vol. 1, pp. 1-13, 2005.

[35]   J. Suchanicz, "The low-frequency dielectric relaxation Na0.5Bi0.5TiO3 ceramics", *Mater. Sci. Eng. B,* vol. 55, no. 1-2, pp. 114-118, 1998.
[http://dx.doi.org/10.1016/S0921-5107(98)00188-3]

[36]   D. Kobor, B. Guiffard, L. Lebrun, A. Hajjaji, and D. Guyomar, "Oxygen vacancies effect on ionic conductivity and relaxation phenomenon in undoped and Mn doped PZN-4.5PT single crystals", *J. Phys. D Appl. Phys.,* vol. 40, no. 9, pp. 2920-2926, 2007.
[http://dx.doi.org/10.1088/0022-3727/40/9/038]

[37]   M. S. Habib, M. A. Rafiq, A. Ali, Q. K. Muhammad, A. Shuaib, and A. Shahzad, "Improved sintering and impedance studies of CuO-doped multiferroic (0.98Ba0.85Ca0.15) (Zr0.1Ti0.9)·O30.02BiFeO3 ceramics", *Applied Physics A,* vol. 128, p. 238, 2022.
[http://dx.doi.org/10.1007/s00339-022-05370-x]

[38]   M.J. Iqbal, and M.N. Ashiq, "Physical and electrical properties of Zr–Cu substituted strontium hexaferrite nanoparticles synthesized by co-precipitation method", *Chem. Eng. J.,* vol. 136, no. 2-3, pp. 383-389, 2008.
[http://dx.doi.org/10.1016/j.cej.2007.05.046]

[39]   W. Kleemann, and P. Borisov, "Multiferroic and Magnetoelectric Materials for Spintronics", In: *Smart Materials for Energy, Communications and Security* Springer, 2008, pp. 3-11.

[http://dx.doi.org/10.1007/978-1-4020-8796-7_1]

[40]   *Materials (Basel),* 2008.

[41]   V.K. Wadhawan, "Ferroic materials: A primer", *Resonance,* vol. 7, no. 7, pp. 15-24, 2002.
       [http://dx.doi.org/10.1007/BF02836749]

[42]   M.A. Rafiq, M.N. Rafiq, and K. Venkata Saravanan, "Dielectric and impedance spectroscopic studies
       of lead-free barium-calcium-zirconium-titanium oxide ceramics", *Ceram. Int.,* vol. 41, no. 9, pp.
       11436-11444, 2015.
       [http://dx.doi.org/10.1016/j.ceramint.2015.05.107]

[43]   H.-Y. Zhang, Z.-X. Zhang, X.-J. Song, X.-G. Chen, and R.-G. Xiong, "Two-Dimensional Hybrid
       Perovskite Ferroelectric Induced by Perfluorinated Substitution", *Journal of the American Chemical
       Society,* vol. 142, pp. 20208-20215, 2020.
       [http://dx.doi.org/10.1021/jacs.0c10686]

[44]   W. Eerenstein, N. Mathur, and J. F. Scott, "Multiferroic and magnetoelectric materials", *Nature,* vol.
       442, p. 759, 2006.

[45]   V.K. Wadhawan, *Smart structures: Blurring the distinction between the living and the nonliving.* vol.
       65. Oxford University Press, 2007.
       [http://dx.doi.org/10.1093/acprof:oso/9780199229178.001.0001]

[46]   A. Sathiya Priya, I.B. Shameem Banu, and S. Anwar, "Investigation of multiferroic properties of
       doped BiFeO3–BaTiO3 composite ceramics", *Mater. Lett.,* vol. 142, pp. 42-44, 2015.
       [http://dx.doi.org/10.1016/j.matlet.2014.11.111]

[47]   M. Kumar, S. Shankar, Brijmohan, S. Kumar, O.P. Thakur, and A.K. Ghosh, "Impedance
       spectroscopy and conductivity analysis of multiferroic BFO–BT solid solutions", *Phys. Lett. A,* vol.
       381, no. 4, pp. 379-386, 2017.
       [http://dx.doi.org/10.1016/j.physleta.2016.11.009]

[48]   S.W. Cheong, and M. Mostovoy, "Multiferroics: a magnetic twist for ferroelectricity", *Nat. Mater.,* vol.
       6, no. 1, pp. 13-20, 2007.
       [http://dx.doi.org/10.1038/nmat1804] [PMID: 17199121]

[49]   W. Gao, R. Brennan, Y. Hu, M. Wuttig, G. Yuan, E. Quandt, and S. Ren, "Energy transduction ferroic
       materials", *Mater. Today,* vol. 21, no. 7, pp. 771-784, 2018.
       [http://dx.doi.org/10.1016/j.mattod.2018.01.032]

[50]   C.W. Nan, "Magnetoelectric effect in composites of piezoelectric and piezomagnetic phases", *Phys.
       Rev. B Condens. Matter,* vol. 50, no. 9, pp. 6082-6088, 1994.
       [http://dx.doi.org/10.1103/PhysRevB.50.6082] [PMID: 9976980]

[51]   D. I. Lone, J. Aslam, N. R. E. Radwan, A. Bashal, A. Ajlouni, and A. Akhter, *Multiferroic ABO3
       Transition Metal Oxides: a Rare Interaction of Ferroelectricity and Magnetism* vol. 14. Nanoscale
       Research Letters, 2019.
       [http://dx.doi.org/10.1186/s11671-019-2961-7]

[52]   C.W. Nan, M.I. Bichurin, S. Dong, D. Viehland, and G. Srinivasan, "Multiferroic magnetoelectric
       composites: Historical perspective, status, and future directions", *J. Appl. Phys.,* vol. 103, no. 3, p.
       031101, 2008.
       [http://dx.doi.org/10.1063/1.2836410]

[53]   M.A. Rafiq, M.N. Rafiq, F. Ahmed, L. Ali, and M.Y. Anwar, "Structure, Dielectric and Impedance
       Studies of Li Doped (K0. 5Na0. 5) NbO3 Ceramics", *Journal of Faculty of Engineering & Technology,*
       vol. 21, pp. 167-178, 2014.

[54]   *Pharmaceutics,* 2008.

[55]   M. Mehrabian, E. N. Afshar, and O. Akhavan, "TiO2 and C60 transport nanolayers in optimized Pb-
       free CH3NH3SnI3-based perovskite solar cells", *Materials Science and Engineering: B,* vol. 287, p.

116146, 2023.
[http://dx.doi.org/10.1016/j.mseb.2022.116146]

[56]   M. Haseeb ur Rehman, and R. Nazar, "Development of PANI-TPU/MWCNTs based nanocomposites for piezoresistive strain sensing applications", *Materials Letters,* vol. 328, p. 133110, 2022.

[57]   M. Mehrabian, M. Taleb-Abbasi, and O. Akhavan, "Effects of electron transport layer type on the performance of Pb-free Cs2AgBiBr6 double perovskites: a SCAPS-1D solar simulator–based study", *Environmental Science and Pollution Research,* vol. 30, pp. 1-10, 2023.
[http://dx.doi.org/10.1007/s11356-023-30732-0]

[58]   N.A. Hill, Why are there so few magnetic ferroelectrics?*J. Phys. Chem. B* vol. 104. ACS Publications, 2000, pp. 6694-6709.
[http://dx.doi.org/10.1021/jp000114x]

[59]   M. Addamo, V. Augugliaro, A. D. Paola, E. García-López, V. Loddo, and G. Marcì, "Removal of drugs in aqueous systems by photoassisted degradation", *Journal of Applied Electrochemistry,* vol. 35, pp. 765-774, 2005.
[http://dx.doi.org/10.1007/s10800-005-1630-y]

[60]   P. Lawita, P. Siriprapa, A. Watcharapasorn, and S. Jiansirisomboon, "Effects of Nd and Co co-doping on phase, microstructure and ferromagnetic properties of bismuth ferrite ceramics", *Ceram. Int.,* vol. 39, pp. S253-S256, 2013.
[http://dx.doi.org/10.1016/j.ceramint.2012.10.072]

[61]   K. Takahashi, and M. Tonouchi, "Influence of manganese doping in multiferroic bismuth ferrite thin films", *J. Magn. Magn. Mater.,* vol. 310, no. 2, pp. 1174-1176, 2007.
[http://dx.doi.org/10.1016/j.jmmm.2006.10.280]

[62]   A. Singh, V. Singh, and K.K. Bamzai, "Structural and magnetic studies on (x)PbTiO3 – (1 – x)SrFe12O19 composite multiferroics", *Mater. Chem. Phys.,* vol. 155, pp. 92-98, 2015.
[http://dx.doi.org/10.1016/j.matchemphys.2015.02.004]

[63]   A. Kostopoulou, K. Brintakis, N.K. Nasikas, and E. Stratakis, "Perovskite nanocrystals for energy conversion and storage", *Nanophotonics,* vol. 8, no. 10, pp. 1607-1640, 2019.
[http://dx.doi.org/10.1515/nanoph-2019-0119]

[64]   C.K. Jeong, C. Baek, A.I. Kingon, K.I. Park, and S.H. Kim, "Lead-free perovskite nanowire-employed piezopolymer for highly efficient flexible nanocomposite energy harvester", *Small,* vol. 14, no. 19, p. 1704022, 2018.
[http://dx.doi.org/10.1002/smll.201704022] [PMID: 29655226]

[65]   A. Reghunadhan, and A. Ajitha, Development of perovskite nanomaterials for energy applications*Design, Fabrication, and Characterization of Multifunctional Nanomaterials* Elsevier, 2022, pp. 269-294.
[http://dx.doi.org/10.1016/B978-0-12-820558-7.00020-0]

[66]   M. Barsoum, and M. Barsoum, *Fundamentals of ceramics.* CRC press, 2002.
[http://dx.doi.org/10.1201/b21299]

[67]   H. Lu, J. Lin, W. Yang, and L. Liu, "Improved dielectric strength and loss tangent by interface modification in PI@BCZT/PVDF nano-composite films with high permittivity", *Journal of Materials Science: Materials in Electronics,* vol. 28, pp. 13360-13370, 2017.
[http://dx.doi.org/10.1007/s10854-017-7173-2]

[68]   J. Ryu, J.W. Lee, H. Yu, J. Yun, K. Lee, J. Lee, D. Hwang, J. Kang, S.K. Kim, and J. Jang, "Size effects of a graphene quantum dot modified-blocking TiO 2 layer for efficient planar perovskite solar cells", *J. Mater. Chem. A Mater. Energy Sustain.,* vol. 5, no. 32, pp. 16834-16842, 2017.
[http://dx.doi.org/10.1039/C7TA02242E]

[69]   C. Zhang, Q. Luo, J. Shi, L. Yue, Z. Wang, X. Chen, and S. Huang, "Efficient perovskite solar cells by combination use of Au nanoparticles and insulating metal oxide", *Nanoscale,* vol. 9, no. 8, pp. 2852-

2864, 2017.
[http://dx.doi.org/10.1039/C6NR09972F] [PMID: 28169383]

[70]   F. Moradi, Z. Shariatinia, N. Safari, and E. Mohajerani, "Boosted performances of mesoscopic perovskite solar cells using LaFeO3 inorganic perovskite nanomaterial", *J. Electroanal. Chem. (Lausanne),* vol. 916, p. 116376, 2022.
[http://dx.doi.org/10.1016/j.jelechem.2022.116376]

[71]   J. Cha, M.K. Kim, W. Lee, H. Jin, H. Na, D. Cung Tien Nguyen, S-H. Lee, J. Lim, and M. Kim, "Perovskite nanowires as defect passivators and charge transport networks for efficient and stable perovskite solar cells", *Chem. Eng. J.,* vol. 451, p. 138920, 2023.
[http://dx.doi.org/10.1016/j.cej.2022.138920]

[72]   F. Ziaeifar, A. Alizadeh, and Z. Shariatinia, "Dye sensitized solar cells fabricated based on nanocomposite photoanodes of TiO_2 and AlMo0.5O3 perovskite nanoparticles", *Sol. Energy,* vol. 218, pp. 435-444, 2021.
[http://dx.doi.org/10.1016/j.solener.2021.03.024]

[73]   M. Samiul Islam, K. Sobayel, A. Al-Kahtani, M.A. Islam, G. Muhammad, N. Amin, M. Shahiduzzaman, and M. Akhtaruzzaman, "Defect study and modelling of SnX3-based perovskite solar cells with SCAPS-1D", *Nanomaterials (Basel),* vol. 11, no. 5, p. 1218, 2021.
[http://dx.doi.org/10.3390/nano11051218] [PMID: 34063020]

[74]   J. Pastuszak, and P. Węgierek, "Photovoltaic Cell Generations and Current Research Directions for Their Development", *Materials (Basel),* vol. 15, no. 16, p. 5542, 2022.
[http://dx.doi.org/10.3390/ma15165542] [PMID: 36013679]

[75]   S. Rhee, K. An, and K.T. Kang, "Recent Advances and Challenges in Halide Perovskite Crystals in Optoelectronic Devices from Solar Cells to Other Applications", *Crystals (Basel),* vol. 11, no. 1, p. 39, 2020.
[http://dx.doi.org/10.3390/cryst11010039]

[76]   A. Facchetti, "π-Conjugated polymers for organic electronics and photovoltaic cell applications", *Chem. Mater.,* vol. 23, no. 3, pp. 733-758, 2011.
[http://dx.doi.org/10.1021/cm102419z]

[77]   R.W. Miles, G. Zoppi, and I. Forbes, "Inorganic photovoltaic cells", *Mater. Today,* vol. 10, no. 11, pp. 20-27, 2007.
[http://dx.doi.org/10.1016/S1369-7021(07)70275-4]

CHAPTER 3

Copper Oxide Nanoparticles in Oil and Gas Industries: Current Developments

Mhd Hazli Rosli[1], Nur Farahah Mohd Khairuddin[2], Mohamed Abdelmonem[3] and Che Azurahanim Che Abdullah[4,*]

[1] *Nanomaterial Synthesis and Characterization Lab, Institute of Nanoscience and Nanotechnology, Universiti Putra Malaysia, 43400, UPM Serdang, Selangor, Malaysia*

[2] *Centre of Foundation Studies for Agricultural Science, Universiti Putra Malaysia, 43400, UPM Serdang, Selangor, Malaysia*

[3] *Department of Physics, Faculty of Science, Universiti Putra Malaysia, 43400, UPM Serdang, Selangor, Malaysia*

[4] *Centre for Diagnostic Nuclear Imaging, Universiti Putra Malaysia, 43400, Serdang, Selangor, Malaysia*

Abstract: This chapter presents an in-depth analysis of Copper oxide nanoparticles (CuONPs) and their emerging role in the oil and gas industry. Over the past five years, nanomaterial technology, especially CuONPs, has attracted significant attention due to its diverse applications in fields like petroleum. In the context of the oil and gas industry, CuONPs have been revolutionary, particularly in enhancing oil recovery (EOR) and as innovative drilling fluids. Their application leads to more efficient extraction and reduced viscosity of trapped oil. The synthesis of CuONPs has evolved, with biological methods standing out for their cost-effectiveness, safety, and environmental friendliness. These green synthesis methods have redefined industry standards by offering a sustainable alternative to traditional physical and chemical approaches. The chapter aims to provide a comprehensive overview of the practical applications of CuONPs in the oil and gas sector, emphasizing their production through green routes. It also addresses the challenges and prospects of CuONPs, setting a foundation for further research and technological advancements in this field.

Keywords: Copper oxide, Green synthesis, Gas, Metal oxide, Nanomaterial, Oil recovery, Oil.

---

* **Corresponding author Che Azurahanim Che Abdullah:** Centre for Diagnostic Nuclear Imaging, Universiti Putra Malaysia, 43400, Serdang, Selangor, Malaysia; E-mail: azurahanim@upm.edu.my

Virat Khanna, Suneev Anil Bansal, Vishal Chaudhary and Reddicherla Umapathi (Eds.)

INTRODUCTION

In the forthcoming decades, the oil and gas sector is anticipated to face increasingly complex technical challenges [1]. As the availability of easily extractable resources rapidly depletes, the challenge of locating oil and gas reserves is escalating at a swift pace. Over the long term, it is predicted that the costs associated with exploration will persist in their upward trajectory. Concurrently, there is a consistent annual rise in worldwide energy consumption. Moreover, the exploration of new oil reserves presents significant challenges, with an estimated 30-60% of oil remaining inaccessible in presently exploited reservoirs. Consequently, both researchers and oil companies are actively seeking innovative and efficient methods to recover this residual oil from mature reservoirs. In this context, nanotechnology has emerged as a particularly promising and highly regarded technology, offering novel solutions to enhance oil recovery in these settings [2]. To address this pressing demand, the industry is in urgent need of technological advancements. To date, only limited progress has been made in addressing these issues through the application of traditional macro and micro materials [3, 4]. To mitigate these challenges, Enhanced Oil Recovery (EOR) methods have significantly advanced the efficiency of oil extraction from reservoirs, particularly after the application of primary and secondary recovery techniques. These methods utilize principles such as snap-off mechanisms and capillary pressure to extract a substantial portion, approximately two-thirds, of the original oil in place. The incremental increase in oil recovery, even by a mere 1%, through EOR techniques can yield an approximate addition of 70 billion barrels from conventional oil reserves. EOR encompasses a variety of technological processes aimed at augmenting the recovery factor from existing reservoirs. These include, but are not limited to, the injection of various fluids, and more recently, the use of microbial injections. These interventions are designed to augment the reservoir's natural energy and facilitate the effective displacement of oil towards the production wells. The interaction between the injected fluids and the reservoir's rock and oil composition creates conducive conditions for oil extraction, thereby enhancing overall production efficiency [5, 6].

Nanotechnology represents a significant breakthrough in industrial applications, particularly within the oil and gas sector, where it has instigated numerous innovative developments. The unique characteristics of nanomaterials, notably their high-volume concentration and large surface area, contribute to their distinctive properties. These nanoscale dimensions impart special magnetic, mechanical, thermal, and chemical attributes to the materials. Furthermore, nanomaterials are noteworthy for their versatility in chemical treatment, allowing for the alteration of their properties to meet specific technical requirements. This adaptability and the array of properties they exhibit have made nanotechnology a crucial element in advancing various aspects of the oil and gas industry [7, 8, 9, 10]. Green nano-biotechnology broadly refers to the process of creating nanoparticles (NPs) or nanomaterials through biological pathways, utilizing microorganisms, plants, viruses, and their derivatives like proteins and lipids, coupled with various biotechnological techniques. (NPs) synthesized via green methodologies are typically more advantageous than those produced through physical and chemical means. Key benefits of green synthesis include the elimination of costly chemicals, reduced energy consumption, and the production of environmentally friendly products and byproducts. The twelve principles of green chemistry have become a globally acknowledged framework for scientists, researchers, chemists, and chemical technologists, guiding the development of less hazardous chemical products and byproducts. These principles serve as a cornerstone for sustainable and responsible chemical research and development [11, 12].

The principal objective of this book chapter is to provide an exhaustive examination of both the triumphs and obstacles encountered in the realm of nanotechnology application. Specifically, it delves into the categorization of nanomaterials utilized within the oil and gas sector, highlighting the prominence of green synthesis methodologies, particularly those involving plant extracts and microorganism mediated synthesis, which are deemed superior to traditional chemical and physical approaches. Furthermore, this chapter offers an in-depth exploration of CuONPs, emphasizing their recent advancements and applications in the oil and gas industry. This analysis not only enhances understanding of nanotechnology's role in these sectors but also underscores the integration of green chemistry principles in the development and utilization of various nanomaterials. Fig. (1) provides the representation of NPs having a high-level surface-to-volume ratio (modified after

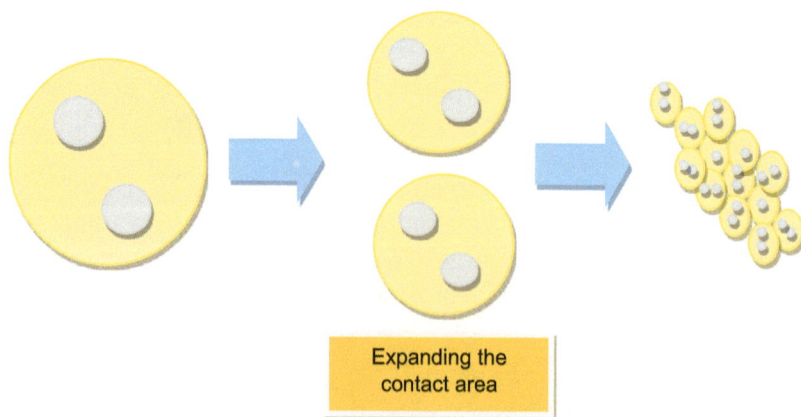

Expanding the contact area

Fig. (1). Illustration of NPs having a high-level surface-to-volume ratio (modified after [13]).

NANOMATERIALS CLASSIFICATION

Nanomaterials are composite materials with an exterior dimension smaller than 1mm at the nanoscale [14]. The shape of nanomaterials can be tubular, spherical, or irregular, and they can be single, fused, agglomerated, or aggregated. According to their structure and profile, nanomaterials are categorised into three broad categories. There are three types of nanoparticles: NPs, nanoclays, and nanoemulsions. A nanoparticle is a microscopic particle with a diameter of between 10 nm and 100 nm [15]. The two types of NPs are organic and inorganic. Inorganic NPs encompass metal and ceramic NPs. Polymeric NPs, carbon-based NPs, and lipid NPs are examples of organic NPs. The latter materials are subdivided into nanocrystals, nanoclusters, nanotubes, and supermolecules. Nanoclays are NPs that range in size from 70 nm up to 150 nm manufactured from thin layers of mineral silicate (1 nm) [16]. Saponite kaolinite and montmorillonite (MMT) are among nanoclays. Nanoemulsions contain particles ranging in size from 10 to 1000 nm and they can be either water in oil nanoemulsion, oil in water nanoemulsion, or continuous nanoemulsion [14].

Nanoparticles

The size of a nanoparticle ranges between approximately 1 - 100 nm. In nanocomposites or structures, NPs can be discovered [14]. A nanostructure is almost always composed of a smaller block. It may be zero-dimensional, one-dimensional, two-dimensional, or three-dimensional. It can also be referred to as a quantum dot [17]. They are critical components for the fabrication of nanostructured materials.

Constrained to a centralised location, we can determine it in zero-dimensional nanomaterials by their length, breadth and height. Meanwhile, in one-dimensional nanomaterials, such as nanowires and nanotubes, it has a single parameter length or breadth. Furthermore, two-dimensional nanomaterials are called two-dimensional because they have two dimensions. The material can have a length, a breadth, or a height. They are usually used for thin surface coatings. Finally, three-dimensional nanomaterials are discussed. It shares the same length, breadth, and height parameters as crystal.

The classification of NPs is based on organic or inorganic together with their material composition [13]. Precipitation of inorganic salts that are covalently and metallically coupled to molecules results in the formation of inorganic NPs. Upon synthesis, organic NPs form a three-dimensional structure. Synthetic and natural materials can be used to synthesize organic NPs, such as lipids, proteins, milk emulsions, and complex adaptive systems in nature, such as viruses.

Nanoclays

Nanoclay is a fine-grained natural or synthetic material particle with its sizes ranging between 10 and 100 nm [18]. Nanoclays have a layered silicate composition and a well-defined multi-layered geometrical structure [19]. It is composed of two-dimensional silica tetrahedral sheets in which each of the tetrahedral sheets is connected to each other and forms an octahedral alumina sheet [20]. The expansion and development of organic and inorganic NPs in polymers has gained researchers' interest [21]. Common nanoclay minerals are bentonite, saponite, montmorillonite, smectite, hectorite, phyllosilicate, mica, and kaolin. One of the most popular applications of nanoclay is as inorganic fillers in a polymer nanocomposite [22]. Clay NPs are incorporated into polymers to improve their mechanical, electrical, thermal, gas barrier, and flame resistance properties [23]. The chemical composition of nanoclays can be determined using X-ray diffraction, gravimetric analysis, Fourier Transforms Infrared Spectroscopy, and inductively coupled plasma [24].

Charge exchange capacity (CEC) ratios of layered silicates and nanoclays are extremely high. The number of interchangeable cations in the interlayer determines the water absorption capacity of nanoclays.

Nanoclays such as smectite, montmorillonite, sepiolite, cloisite, and other minerals have been reinforced into matrices such as olefins, epoxy, rubber, polystyrene, and polyamide, as well as chitosan. By incorporating nano clay, the nonflammability, self-healing, mechanical properties (modulus, stiffness, tensile strength, yield stress), and thermal stability of polymers were enhanced [19]. The inflammability of nano clay-based materials is critical in industries such as construction, automotive, aviation, electronics, and structure. For industrial applications, polymer, and nano clay nanocomposites have proved to be highly effective. Table **1** gives the significance of polymers and clay nanocomposite.

Table 1. The significance of polymers and clay nanocomposite (nanoclays).

Matrix	Nanoclays	Properties	References
Polyamide	Montmorillonite	Improvement of mechanical properties.	[25]
Styrene butadiene rubber	Montmorillonite	Increased flame retardancy.	[26]
Polystyrene	Montmorillonite	Enhanced thermal stability.	[27]
Polyamide 6	Sepiolite	Outstanding stiffness and yield stress.	[28]
Chitosan/polycaprolactone	Cloisite	Controlled drug delivery	[29]
Poly (ethylene-co-methacrylic acid)	Cloisite	Self-healing characteristics	[30]

Nanoemulsion

Nanoemulsions, are colloidal systems composed of submicron particles ranging in size from 10 to 1000 nm [31]. Nanoemulsions are created using colloidal dispersions of two immiscible liquid phases. Nanoemulsions appear to be similar to other lyotropic liquid crystalline phases, such as microemulsions and mesophases, from both a structural and a content standpoint. Increased drug loading and bioavailability are some of the advantages of nanoemulsions. It enables the creation of a repeatable plasma drug profile. It is frequently used to administer medications on a long-term or even targeted basis. The dispersion is thermodynamically stable due to its extremely poor interfacial tension and high Oil-Water (o/w) interfacial area. They share microemulsions' kinetic stability and optical transparency. Two methods for producing and manufacturing nanoemulsions were elucidated [32]. The first method is high-energy emulsification which consists of ultrasonication, high-pressure homogenisation and microfluidisers. In the second technique, phase inversion composition is combined with temperature and solvent displacement through low-energy emulsification. Table **2** lists several of the nanoemulsion active ingredients.

Table 2. Formulation of ingredients usually included in the preparation of various nanoemulsions.

Ingredients	Examples
Oils	Evening primrose oil, mineral oil, coconut oil, corn oil, olive oil, castor oil, and linseed oil.
Emulgents	Castor oil, natural lecithin from animal or plant sources, phospholipids.
Surfactants	Polysorbate20, castor oil, Polyoxy60, Sorbitan monooleate, Polysorbate80.
Cosurfactants	Polyene glycol, glycerine, PEG400, poloxamer, ethanol, PEG300.
Tonicity modifiers	Xylitol, sorbitol, glycerol.
Additives	Propylene glycol, maltose, glucose, lower alcohol (ethanol), propylene glycol, sucrose 1, 3-butylenes glycol, fructose.
Antioxidants	Tocopherol, ascorbic acid

COPPER OXIDE NANOPARTICLES (CuONPs)

Nanoparticles' technological advancements have had a significant impact on the pharmaceutical, medical, and textile industries over the last few decades. Metal nanoparticles such as Zn, Ag, and Au have long been used in medical institutions as therapeutic agents. Subsequently, numerous studies began evaluating the bioactivities of metal oxide nanoparticles such as CuONPs, which have been shown to have superior photocatalytic and biological activities to those generated by metal nanoparticles.

Plants contain a large number of reactive organic chemicals such as carboxylic acids, amino acids, phenols, and ketones, as well as proteins that act as stabilising and reducing agents, they are bio-friendly and bio-green assets available for the synthesis of nanomaterials. Fruits, roots, leaves, and flowers of plants can all be used to create nanomaterials. Citrus and berry fruits that are high in antioxidants are notable examples in the literature. Additionally, there have been studies carried out where seeds and peel extracts were also used in the synthesis of nanomaterials. Fig. (**2**) demonstrates the different biological approaches used for the synthesis of CuONPs and the Table **3** provides the Previous work on the synthesis and structural properties of CuONPs utilizing various plant based as precursors.

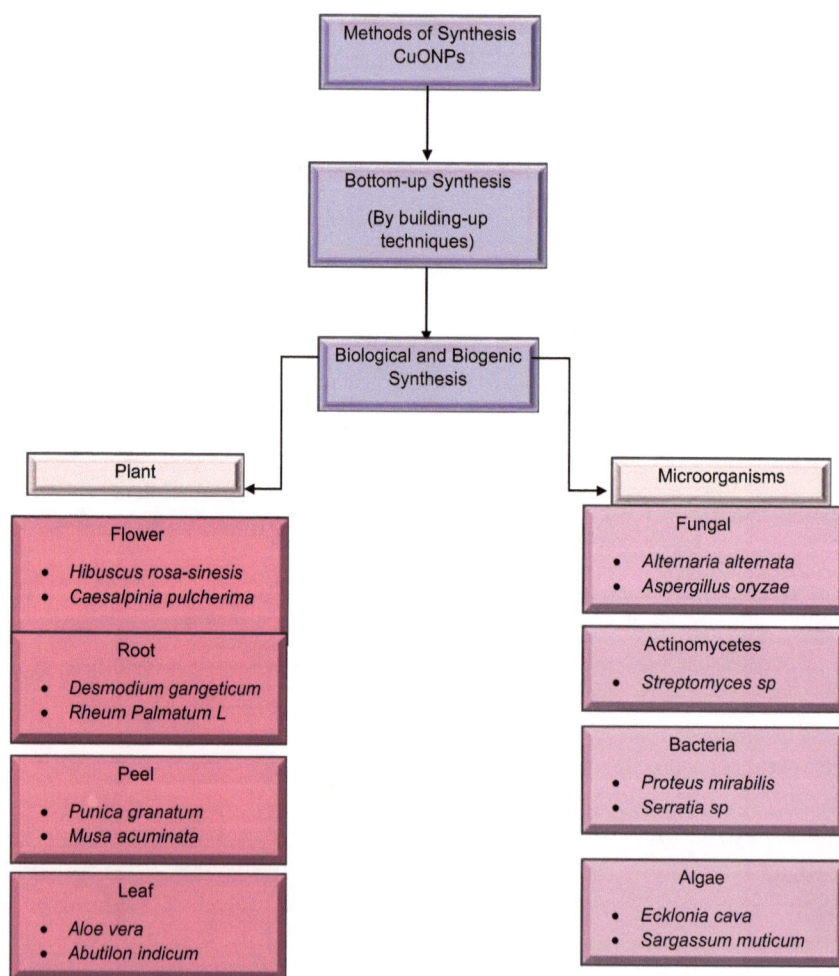

Fig. (2). demonstrates different biological approaches used for the synthesis of CuONPs.

Table 3. Previous work on the synthesis and structural properties of CuONPs utilizing various plant based as precursors [33].

Plant Name	Family	Utilised Part	Phytochemical Components	Structure	Size (nm)	characterization	References
Desmodium gangeticum	*Fabaceae*	Root extract	Flavonoids, glycosides, pterocarpenoids, lipids, glycolipids, amino glucosyl glycerolipid and alkaloids.	Spherical	12	FT- IR, SEM, TEM, TGA, UV-VIS, XRD,	[34]
Citrus medica Linn	*Rutaceae*	Juice	-	Spherical	10 - 60	FT-IR, UV-VIS XRD,	[35]
Malva sylvestris	*Malvaceae*	Leaf Extract	Alcohol, amide, phenols, flavonoids	Spherical	14	FTIR, SEM, XRD	[36]
Erzincan Cimin	*Vitaceae*	Fruit extract	Anthocyanins, phenolic acid, flavonoids, proanthocyanidins	Uniform Spherical	25 - 50	FTIR, SEM, UV–VIS, XRD,	[37]
Cordia sebestena	*Boraginaceae*	Flower extract	Alkaloids, carbohydrates, proteins/amino acids, phenols/tannin, saponins flavonoids	Spherical	20 - 40	XRD, FTIR, FESEM-EDX, SEM, TEM	[38]
Olea europaea	*Oleaceae*	Leaf extract	Hydroxytyrosol, Oleuropein	Spherical	20 - 50	UV-VIS, XRD, FTIR, TEM, and SEM	[39]

Due to their obvious efficacy and ease of access, the manufacturing of nanoparticles from algae is gaining interest at the moment [40]. Various materials with fantastic morphologies at nanoscale dimensions also were fabricated using bacteria via an extracellular and internal mechanism [33]. Bacteria have an enormous opportunity for nanoparticle synthesis [41]. In addition to their high stability and ability to generate extracellular nanoparticles, the bacteria have unique characteristics that make them ideally suited to be easily manipulated at the microscopic level, such as short organism generation times, ease of culture, easy cultivation, and benign experiment conditions. The bacteria have been reported to produce a variety of important compounds containing thiol groups when under oxidative stress [33]. This compound is a good capping agent to prevent metal oxide nanoparticles from oxidising. Table **4** shows bacteria-mediated synthesis of CuONPs. Fig. (**3**) provides some main characterization techniques for as synthesized nanoparticles.

Table 4. Example of previous studies related to bacteria-mediated synthesis of CuONPs [33].

Strain of Bacteria	Gram +/ Gram -	Size (nm)	Shape	characterization
Escherichia coli	Gram -	10.00 – 40.00	Polydispersed and Quasi-spherical	FT-IR, SAED, SEM, TEM, XRD
Serratia sp	Gram -	10.00 – 30.00	Polydispersed	FT-IR, TEM, UV-VIS, XPS, XRD
Halomonas elongata	Gram -	57.00 – 79.00	Rectangular	FESEM, FT-IR, UV-VIS, XRD
Streptomyces cyaneus	Gram +	29.80	Spherical	DLS, FT-IR, TEM, UV-VIS, XRD
Lactobacillus casei	Gram +	200.00	Spherical	EDX, FESEM, FT-IR, TEM, XRD

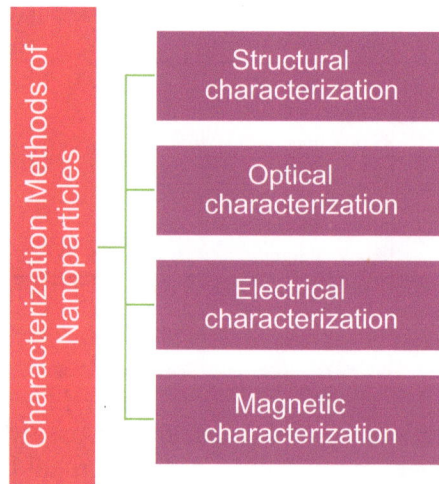

Fig. (3). Main characterization techniques for as synthesized nanoparticles.

Nanomaterials Characterization

Characterization is critical for the advancement of nanotechnology because it sheds light on the particles' shape, synthesis, structural, and size components. The following techniques are used to characterize synthesized nanoparticles: Atomic Force Microscopy (AFM), Dynamic Light Scattering (DLS), Mass Spectroscopy (MS), Scanning Electron Microscopy (SEM), Fourier Transform Infrared

Spectroscopy (FTIR), Fluorescence Correlation Spectroscopy (FCS), Nuclear Magnetic Resonance (NMR), X-ray Diffraction (XRD), Raman Scattering (RS), X-ray Photoelectron Spectroscopy (XPS), Zeta potential, and Transmission Electron Microscopy (TEM). These techniques are used to characterize material properties, surface characteristics, shape, chemical compositions, size, thermal analysis, microscopic structure, and the orientation and sequence of functional groups associated with nanoparticles. The methods listed above can be classified into structural, optical, electrical, and magnetic, all of which are discussed in greater depth below (Fig. **4**).

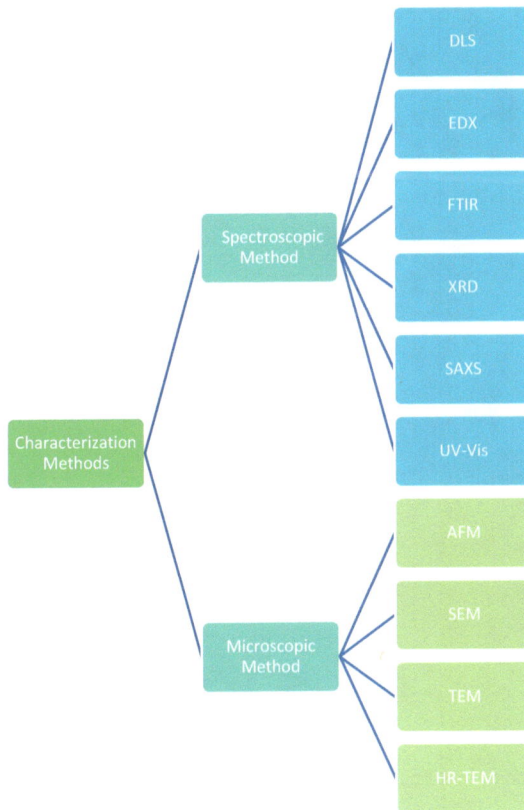

Fig. (4). Spectroscopy and microscopy approaches in the characterization of nanomaterials.

Additionally, nanoparticle characterization can be classified into two categories: spectroscopic and microscopic methods. There are six primary analyses based on the spectroscopic method (Fig. **4**), and four primary analyses based on the microscopic method (Figs. **4**, **5** and **6**) emphasize the major characterization and its subdivision.

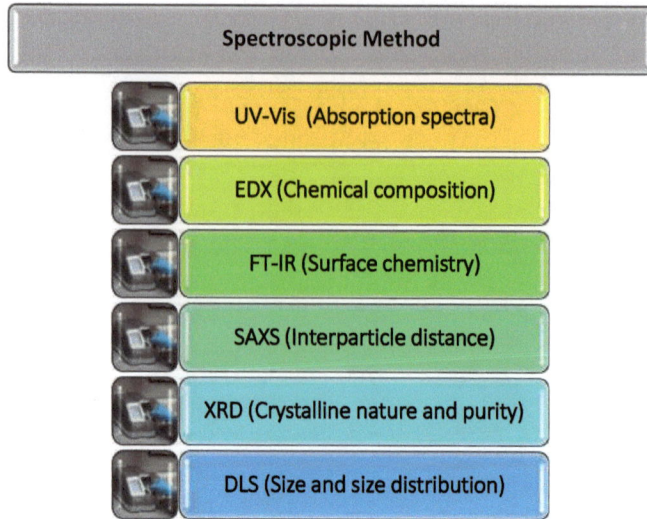

Fig. (5). Spectroscopic methods are mainly used in the characterization of nanomaterials.

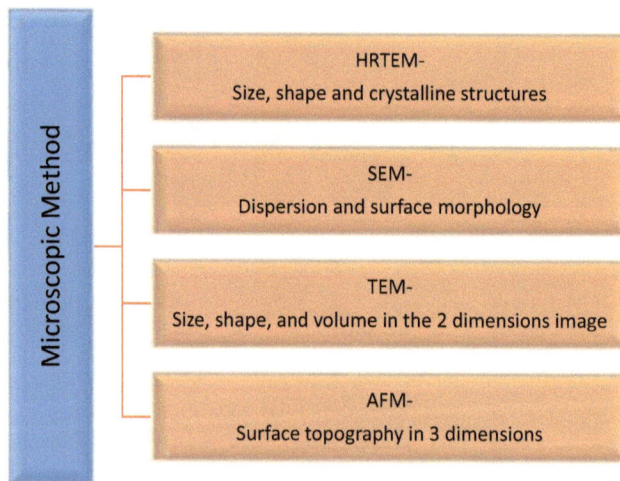

Fig. (6). Microscopy approach of nanomaterials characterization.

CuONPs APPLICATION IN OIL AND GAS

Here is a thorough overview of some of the newest applications of nanoparticles in the oil and gas industry, including drilling fluid, enhanced oil recovery (EOR), and H_2S gas sensor. A tabular summary of the employed CuONPs, their efficacy in terms of enhancing the desired parameters and the examined parameters are included in the discussion.

Drilling Fluid

As a subset of nanotechnology, nanoparticles offer enormous potential for improving drilling fluids. Despite this, the involvement of NPs in this domain is still in its infancy, and as a result, it has garnered much more attention in recent years. Research aims to optimize the thermal and physical-mechanical properties of drilling fluids and managed to solve drilling fluid issues using NPs with noteworthy features such as high thermal conductivity and a large surface area [42].

Drilling fluid is an example of a petrochemical sector core component. They are used mainly for controlling or suppressing formation pore pressure and transporting rock fragments from the bottom of the hole to the top for reprocessing. Apart from these functions, the drilling fluid performs a variety of additional functions due to its appropriate rheological characteristics, which include gel strength, yield point, apparent and plastic viscosity. Drilling fluids with particular gravities greater than formation pressure prevent fluids from entering the well by generating a hydrostatic head greater than formation pressure. To prevent sagging during the hole cleaning process, the weighing agents must have a high viscosity and yield point [43]. In a study, the addition of 0.8 wt% CuO nanoparticles to water-based drilling fluids resulted in a 30.2% reduction in filtration loss and a 27.6% decrease in mud cake thickness under high-pressure and high-temperature conditions, highlighting their effectiveness in enhancing fluid properties [44]. Moreover, in another study, CuO and ZnO nanoparticle-based nanofluids (<50 nm), at concentrations of 0.1, 0.3, and 0.5 wt%, were added to xanthan gum-based drilling muds at 1% volume to create nanofluid-enhanced water-based drilling muds (NWBM). These NWBMs demonstrated a 35% improvement in thermal and electrical properties over standard water-based muds (WBM). CuO-based NWBMs exhibited superior thermal properties and high-pressure, high-temperature (HPHT) resistance compared to ZnO-based NWBMs. Rheological stability was enhanced at temperatures of 25, 70, 90, and 110 °C and pressures of 0.1 and 10 MPa. The Herschel-Bulkley model was identified as the best fit for describing the rheological behavior of NWBMs [45].

The loss of fluid from the manufacturing zone degraded the reservoir by blocking the pore throats in the well bore, affecting wettability, and decreasing the relative permeability of the sand phase area to oil. A stimulation job appears to become unavoidable as a result of declining productivity levels, incurring additional costs. Due to this problem, non-damaging drilling fluids (NDDF) were developed, which basically are solid-free drilling fluids [46].

According to the review, high NPs concentrations are not recommended due to the insignificant performance enhancement observed when low NPs concentrations are compared to high NPs concentrations [47, 48]. Additionally, high concentrations may increase the friction coefficient of the particle, affecting lubricity and hole cleaning efficiency [49]. Certain NPs may have a detrimental effect on the filtration properties at elevated temperatures. In terms of rheological properties, CuONPs are more thermally stable than MgONPs or, more precisely, Al_2O_3 [47]. Additionally, it was discovered that SiO_2 NPs may increase pressure losses within bentonite-based water-based mud, resulting in an increase in frictional forces [50].

Enhanced Oil Recovery (EOR)

The oil recovery process consists of three distinct stages. Primary production occurs when natural energy is transferred within a reservoir, which includes gas-cap drive, solution-gas drive, and natural-water drive [51]. Gas injection and waterflooding are two secondary recovery methods. Upon completion of the secondary oil recovery procedure, tertiary oil recovery processes use chemicals and miscible gases with or without thermal energy to relocate additional oil [13]. Primary and secondary processes have differing mechanisms of action. The injections of fluids used in secondary processes supplement the reservoir's inherent energy for oil extraction. The recovery efficiency is largely determined by the pressure maintenance technique used. Tertiary processes are thus affected by injected fluids interacting with reservoir rocks or oil systems. The interactions described above may result in decreased viscosity, oil swelling, altered wettability, or favourable phase behaviour and interfacial tensions (IFT). In certain circumstances, tertiary processes can be used as secondary operations. As a result, the term "tertiary recovery" became unpopular in the literature, while the term "EOR" gained popularity. Given the finite nature of oil reserves and the growing energy market, researchers are seeking high-efficiency EOR techniques or ways to improve the efficiency of currently proposed EOR.There are relatively few instances of this nanoparticle being used in research as an EOR agent. For example, in a study, a nanofluid formulated from copper oxide (PHPA) and nanoclay (NF2) was analyzed for oil recovery efficiency. The rheological analysis showed that this nanofluid exhibits pseudoplastic behavior and significant shear resistance. Its viscosity was found to reduce from 51 mPa.s to 8.5 mPa.s, indicating enhanced fluid properties suitable for oil recovery applications [52].Researchers have identified the use of metal oxide nanoparticles to inhibit EOR, in addition to their advantages over silica nanoparticles. A novel EOR technique known as electrokinetic (EK)-assisted nanofluid flooding has also been developed. This technique employed EK (2 V/cm direct current voltage) to enhance oil recovery from nanofluid flooding. They assessed the effectiveness of various NPs against EK following waterflooding in Abu Dhabi carbonate cores. They discovered that CuONPs has outperformed NiONPs and FeONPs due to its higher density and electrical conductivity. Fig. (7) below provided the application of nanotechnology in EOR while the Table **5** shows classifications of readily accessible EOR advancements.

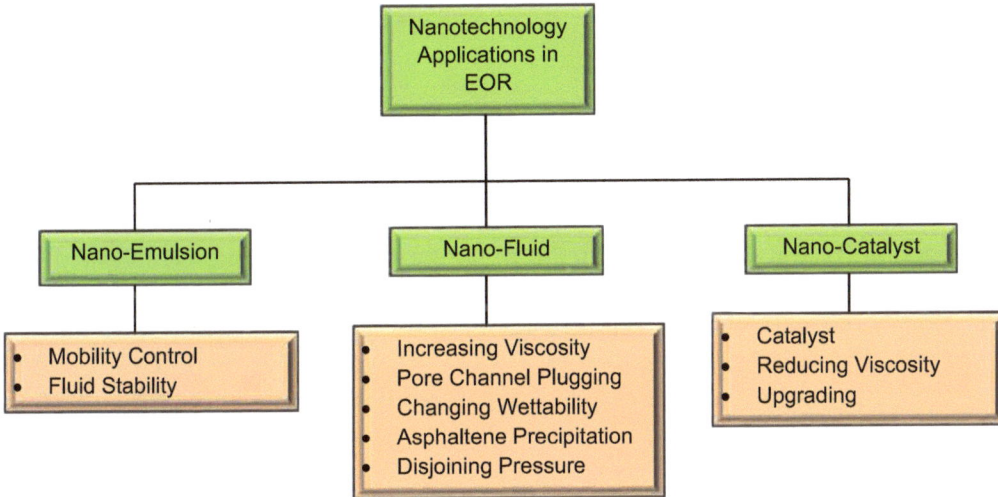

Fig. (7). Application of nanotechnology in EOR (modified from [53]).

Table 5. shows the classifications of readily accessible EOR advancements (modified based on [13]).

	Thermal Techniques	**Chemical Techniques**	**Gas Techniques**
Detailed Techniques	• Steam Flooding • SAGD • Electrical Heating • In-situ Combustion • CSS	• Surfactant Flooding • Alkaline Flooding • ASP Flooding • Polymer Flooding • Micellar Flooding	• CO_2 Injection • Air Injection • N_2 Injection • WAG Injection • Hydrocarbon Gas Injection
EOR process	• Oil Expansion • Gravity Drainage • IFT Reduction • Viscosity Reduction • Steam Distillation	• Emulsification • IFT Reduction • Mobility Control • Wettability Alteration	• Miscibility • Oil Expansion • Pressure Maintenance • Viscosity Reduction

	Thermal Techniques	Chemical Techniques	Gas Techniques
Provocation	• Thermal deterioration is ineffective. • Heat transfers out from the heat generator towards the reservoir • High priced energy • Low thermal conductivity of rock and fluids • Heat leakage to the undesired layers	• Decelerate diffusion rate in the pore structure. • Unsatisfactory mobility ratio. • The excessive expense as a result of the additional amount required. • Ineffectiveness on IFT and viscosity changes. • Incompatibility results in damage.	• Asphaltene deposition occurs • CO_2 corrosion • Fingering and early gas breakthrough • Miscible flooding needs high MMP • Supersede of gravitation

H_2S Gas Sensor

Hydrogen Sulfide (H_2S) is a significant pollutant that is encountered frequently in the petrochemical and sewage treatment industries [54]. H_2S is a caustic, poisonous, corrosive, and flammable gas that poses numerous industrial challenges. H_2S concentrations greater than 10 ppm should be avoided for human health reasons, as they can cause significant eye irritation. As much as 750 ppm can cause respiratory paralysis within minutes, resulting in death. Almost any concentration, between 4.5 - 45% above this limit, can cause an outburst. Various stages of the oil and gas industry are afflicted by H_2S corrosion, including extraction, transportation, and processing, resulting in significant financial and economic losses justified that the involvement of H_2S gas in the atmosphere stimulates the formation of CuS [55]. The alternative CuS production pathway involves a chemical reaction that is complemented by an electrical potential shift [56]. Equation (1) and (2) depict the spontaneous chemical process.

$$CuO + H_2S \longrightarrow CuS + H_2O \tag{1}$$

$$2CuS + 3O_2 \longrightarrow 2CuO + 2SO_2 \tag{2}$$

As H_2S is formed during the good drilling phase, mixed with oil and other gases, as well as being highly flammable, it is important that sensors for this gas are isolated from either the measurement site or the wellhead. Numerous sensing technologies, including catalytic-based sensors, electrolyte solutions, optical sensors, and electrochemical sensors are used in industry and have been recommended in the literature to determine the presence and intensity of H_2S. In order for electrochemical sensors to function, they must be in direct contact with H_2S at the measurement location in order to sustain an electrical resistance change. This implies that the sensor must be powered by an electrical source [57]. Despite this, the sensors are commonly operated at elevated temperatures, making them risky to use in combustible environments. Temperatures exceeding those needed to operate the sensor are generally not acceptable. Additionally, a dependable electrical source must be available at the measurement site, in case of an electrical failure. As a result, reliable H_2S sensing via electrochemical sensors is limited, as it is site-dependent for a solution, a fibre optic-based sensor is the most suitable since it relies on a waveguide that can be supervised remotely from either the sensing point or the sensitive zone for H_2S detection. The fibre optic-based sensors can be operated at room temperature (RT).

In one of the recent studies, CuONPs-coated plastic optical fiber (POF) sensors showed high sensitivity in detecting 200 ppm H_2S gas. However, these CuONPs sensors exhibited a non-reversible chemisorption response to H2S, marking a limitation despite their heightened sensitivity compared to iron oxide nanoparticle-coated sensors [58]. Furthermore, in another study, a U-shaped POF sensor coated with CuONPs effectively detected H2S gas. The CuO-coated sensor responded within 1 minute to 200 ppm H2S and could detect concentrations as low as 10 ppm. Notably, this sensor, functioning at room temperature (25°C) and without electrical charging or heating, demonstrated high selectivity and a logarithmic response to varying H2S concentrations [59]. In different research works, high crystalline CuONPs, synthesized via a hydrothermal method, demonstrated enhanced sensitivity and selectivity for H2S gas sensing at low temperatures. The study identified two mechanisms in H2S detection: H2S oxidation and copper sulphide (CuS) formation, both enhancing sensor response. However, CuS formation at low temperatures inhibits sensor recovery, a process that weakens with increasing operating temperatures from 40 °C to 250 °C, as confirmed by XPS and micro-Raman spectral analysis [60].SEM images demonstrated that morphology CuONPs is look-liked of 'grapes' structures. The CuONP particles size is approximately around 20 nm. The incredible structure of CuONPs is an advantage for its gas sensing property. The H_2S gas can be adsorbed not only by particles outside, but also by particles inside the spherical shape before being adsorbed by particles inside. Table **6** gives the Comparison of CuO-based nanoparticles H2S gas sensing performance.

Table **6**. Comparison of CuO-based nanoparticles H_2S gas sensing performance [61].

Materials	Concentration (ppm)	Temperature	Response	Time to Response (s)	Time to Recovery (s)
CuONPs	5	40 °C	4.9 ± 0.43	297.5±9.2	54 ± 7.1
Cu$_2$O/CuO	0.1	95 °C	2.5	76	75
CuO thin film	5	25 °C	3.5	> 100	> 4000
Cu$_2$O	0.1	25 °C	1.7	100	N/A

In the case of H_2S, the sensitive chemical could be an SMOX. Apart from SnO_2, copper and nickel oxide have demonstrated a high sensitivity to H_2S. This is due to nickel and copper's inherent affinity for sulphur. When CuO is exposed to H_2S, it partially converts to copper sulphide (covellite), a moderate conductor. Additionally, several studies have demonstrated that when CuO thin films are exposed to H_2S, their electrical resistance dramatically decreases [62].

To conclude, CuONPs have been commonly used to produce high-performance gas sensing devices for detecting the most prevalent dangerous gases and contaminants in the environment. In several situations, exceptional sensitivity involving detection limits as low as ppb-level concentrations or lower was achievable [63].

Heavy Oil Exploration Catalyst

The world is awash in heavy oil reserves, but the vast majority of them remain unexplored. Increasing crude oil production makes heavy oil an important alternative energy source [64]. Due to its high viscosity and complicated chemical composition, heavy oil is difficult to extract from deep reserves [65]. Heavy oil has a high viscosity, density, and fluidity, making it unsuitable for transmission and processing via long-distance pipelines. Thus, an energy-intensive pipeline heat exchanger is needed to ensure efficient pipeline operation [66]. Clearly, reducing the viscosity and flow resistance of heavy oil is critical for pipeline transportation. Methods of reducing heavy oil viscosity include mixing light oils with heavy oils, heating them, using catalytic methods, and mixing surfactants with heavy oils [67].

Chemical experts have studied a variety of different types of catalysts in the laboratory and determined how they affect heavy oil [68]. Catalysts for reducing the viscosity of heavy oils can be grouped into four categories as shown in Table 7. CuONPs have played a major role as a catalyst for a wide range of catalytic reactions, including reduction, oxygenation, hydrogenation, and hydrocarbon combustion [70]. For example, in a recent study conducted on well-dispersed CuONPs were synthesized in situ as catalysts for reducing the viscosity of Shengli heavy oil. Using a $Cu(OH)_2$-contained microemulsion, the catalytic process under aquathermolysis conditions (240 °C, 2.5 MPa of N_2) resulted in a significant viscosity reduction of 94.6% in the oil, while also converting 22.4% of asphaltenes into lighter components. The research highlights the efficacy of CuO nanoparticles in enhancing heavy oil processing with minimal nanoparticle agglomeration [71]. In another study, hydrophobic CuONPs with a mean diameter of 84.3 nm were synthesized and utilized to catalyze the low-temperature oxidation (LTO) of heavy oil. The addition of CuO nanoparticles resulted in a

1.4-fold increase in the average oxygen consumption rate and reduced the residue oxygen content in the tail gas from 8.76% to 3.9% compared to the experiment without nanoparticles. Additionally, the nanoparticles facilitated the cleavage reaction during LTO, offering potential improvements for the safety of air flooding technology in heavy oil extraction [72]. Furthermore, in a study exploring thermocatalytic upgrading of heavy oil, copper oxide nanoparticles synthesized via a non-hydrothermal method acted as an effective catalyst. The application of 0.2 wt% of this catalyst at 350°C for 40 minutes resulted in an 85.75% reduction in heavy oil viscosity. Additionally, the presence of a hydrogen-donor in the reaction, catalyzed by the copper oxide nanoparticles, was found to enhance the chemical composition of resin and asphaltene in heavy oil [73]. Table **8** below shows the Optimum thermocatalytic activity of CuONPs in the heavy oil.

Table 7. Lists of available catalyst used in reducing heavy oil viscosity [69].

Catalysts
Water-soluble
Mineral
Oil-soluble
Dispersed

Table 8. Optimum thermocatalytic activity of CuONPs in the heavy oil [65].

Chemical Substances	Cu (NO$_3$)$_2$. 3H$_2$O and CO(NH$_2$)$_2$
Temperature	350 °C
Time	40 minutes
Catalyst Content	0.2 wt%
Reduced Viscosity of Heavy Oil	16000 mPa.s to 2280 mPa.s
Ratio of Reduction	85.75%
Further Improvement	6.88% with adding 3.0 wt% THF hydrogen donor in the catalytic reaction
Average CuONPs Catalyst	56 nm increasing to 21.12%

CHALLENGES AND OPPORTUNITIES FOR FUTURE RESEARCH

Perspective on Enhanced Oil Recovery (EOR)

Even though NPs have been shown and demonstrated to be prospective agents in a variety of EOR processes, the majority of them are limited to a series of experiments and are therefore unsuitable for field-scale implementations. In order for nano assisted EOR systems to be deployed in practical situations, a number of obstacles and barriers need to be removed.

1. There are numerous technical difficulties associated with the preparation of nanofluids. NPs generally congregate in difficult reservoir conditions as a result of the significant Vander Waals interactions [74]. As a result, the primary technological challenge is to create an NPs suspension that is homogenous.
2. Currently, nanofluid flooding still seems to be the major focus of research. Other nano assisted EOR approaches are not well understood, as the interaction between NPs, reservoir fluid and rock properties are not well considered.
3. While a limited number of experiments have been conducted to examine the effectiveness of metal nanoparticles as EOR agents, research has found some discrepancies regarding their potential benefits over $SiO2$ nanoparticles and CuONPs. The review of literature on mixed nanofluids is also in its infancy. In addition, a lack of experimental research on nanofluid mixtures limits their utility in EOR.
4. It remains challenging to gain a fundamental understanding of nano assisted EOR techniques that require precise calculations and detailed modelling, due to the complexity and unknown mechanisms involved in nano assisted EOR methods.

Perspectives on H_2S Gas Sensor

A CuONP-based gas sensor is among the technologies being considered for detecting harmful gases and pollutants. Despite some promising studies on nanoparticle decoration and doping, CuONPs-based devices have shown limited gas selectivity, preventing their practical use in real-world applications. However, it may be possible to achieve exceptional sensitivity with lower detection limits.

1. Taking a step forward would be beneficial in pursuing greater advancements in the development of smart sensor systems that would utilise a variety of sensor devices and would associate pattern recognition, machine learning, artificial intelligence, and neural networks, which could be beneficial for further advancement of smart sensor systems.
2. An unsupervised principal component analysis was successfully used to analyze the resistance transients of CuONPs thin film devices for the purpose of distinguishing acetone, methanol, ethanol, and 2-propanol. Gas-selective pre-filters may also be utilized to mitigate the effect of either H_2S or SO_2 tolerance [75].
3. Notwithstanding from further technological advancements, a few fundamental scientific issues must be addressed, most notably those relating to the gas

interactions and underlying mechanisms of humidity target with CuONPs surfaces, as well as the role of additives. In order to ensure sensor functionality, experimental approaches that enable the characterization of sensor material characterization under ambient conditions to those encountered during sensor operation, particularly infrared spectroscopy performed in operation [76] will be critical to achieve this goal. Transmission electron microscopes (TEM) have the potential to shed new light on environment changes. Exposure to reactive gas species at elevated temperatures alters the chemical composition of nanostructures [77].

Perspectives on Heavy Oil Exploration Catalyst

The use of nanoparticles or nanofluids to reduce heavy oil viscosity has attracted a lot of attention in recent decades. Nevertheless, this innovation is still in its infancy and cannot be used in a practical setting. Even though it is energy efficient, as well as environmentally beneficial, the main limitations are the nanoparticles' self-association and inherent specificity for certain oils. Many issues remain unresolved. It has been demonstrated that oil viscosity reducers are frequently particular to the structure and properties of heavy oils. Although some papers have suggested some potential viscosity-reducing methods, direct solid evidence to demonstrate the interaction between nanoparticles, transporting fluids, and oil at the molecular level is insufficient. Through this perspective, we propose three concepts for resolving the issues:

1. Use current instruments such as Transmission Electron Microscope (TEM), Scanning Electron Microscope (SEM), and X-ray Photoelectron Spectroscopy (XPS) to identify structural alterations of heavy oil through the existence of nanoparticles to determine the properties that enhance nanoparticles' performance and reduce oil viscosity.
2. After analysing the structure of specific oil molecules, use molecular dynamics simulation methods to design and select oil viscosity reducers.
3. Additional research should focus on designing and selecting targeted materials based on heavy oil properties. The most effective heavy oil viscosity reducers have only been documented in laboratory.

If we want to improve the flow of oil and the recovery of oil, we need to build a new pipeline as well as recommend effective viscosity reducers for use in core flooding tests and field applications.

CONCLUSION

This section explores the applications of CuONPs in the O&G industries. Nowadays, most of the current research is mainly focused on the production of CuONPs by developing and producing them using either chemical or physical approaches. Green approaches to CuONPs offer various advantages and have gathered a lot of attention in recent years, especially for a wide range of applications in O&G industries and manufacturing. A critical component to consider is the ideal properties of green biosynthesized CuONPs. Future work will relate to improving the limitation of CuONPs of biogenic synthesis, through data collection on their biological properties and limiting properties. Moreover, biosynthesized green CuONPs have potential for petrochemical utilisation, but there is a need to optimize the reaction parameters, as well as the method for achieving large-scale and bulk production of these materials.

REFERENCES

[1]   Z. Zhe, and A. Yuxiu, "Nanotechnology for the oil and gas industry – an overview of recent progress", *Nanotechnol. Rev.,* vol. 7, no. 4, pp. 341-353, 2018.
[http://dx.doi.org/10.1515/ntrev-2018-0061]

[2]   J.A. Ali, K. Kolo, A.K. Manshad, and K.D. Stephen, "Potential application of low-salinity polymeric-nanofluid in carbonate oil reservoirs: IFT reduction, wettability alteration, rheology and emulsification characteristics", *J. Mol. Liq.,* vol. 284, pp. 735-747, 2019.
[http://dx.doi.org/10.1016/j.molliq.2019.04.053]

[3]   Md. Amanullah, and A.M. Al-Tahini, "Nano-Technology - Its Significance in Smart Fluid Development for Oil and Gas Field Application". *All Days, SPE,* 2009 Al-Khobar, Saudi Arabia
[http://dx.doi.org/10.2118/126102-MS]

[4]   X. Kong, and M.M. Ohadi, Applications of Micro and Nano Technologies in the Oil and Gas Industry-An Overview of the Recent Progress.*All Days.* SPE, 2010.
[http://dx.doi.org/10.2118/138241-MS]

[5]   L. Lake, R.T. Johns, W.R. Rossen, and G.A. Pope, "Fundamentals of Enhanced Oil Recovery", *Society of Petroleum Engineers,* 2014.
[http://dx.doi.org/10.2118/9781613993286]

[6]   S. Thomas, "Enhanced Oil Recovery - An Overview", *Oil Gas Sci. Technol.,* vol. 63, no. 1, pp. 9-19, 2008.
[http://dx.doi.org/10.2516/ogst:2007060]

[7]   B. Peng, J. Tang, J. Luo, P. Wang, B. Ding, and K.C. Tam, "Applications of nanotechnology in oil and gas industry: Progress and perspective", *Can. J. Chem. Eng.,* vol. 96, no. 1, pp. 91-100, 2018.
[http://dx.doi.org/10.1002/cjce.23042]

[8]   S. Mokhatab, M.A. Fresky, and M.R. Islam, "Applications of Nanotechnology in Oil and Gas E&P", *J. Pet. Technol.,* vol. 58, no. 4, pp. 48-51, 2006.
[http://dx.doi.org/10.2118/0406-0048-JPT]

[9]   B. Bhushan, Ed., *Springer Handbook of Nanotechnology.* Springer Berlin Heidelberg: Berlin, Heidelberg, 2017.
[http://dx.doi.org/10.1007/978-3-662-54357-3]

[10]   "Library", *MRS Bull.,* vol. 30, no. 5, pp. 394-395, 2005.
[http://dx.doi.org/10.1557/mrs2005.114]

[11]   O.V. Kharissova, H.V.R. Dias, B.I. Kharisov, B.O. Pérez, and V.M.J. Pérez, "The greener synthesis of nanoparticles", *Trends Biotechnol.*, vol. 31, no. 4, pp. 240-248, 2013.
[http://dx.doi.org/10.1016/j.tibtech.2013.01.003] [PMID: 23434153]

[12]   B.A. de Marco, B.S. Rechelo, E.G. Tótoli, A.C. Kogawa, and H.R.N. Salgado, "Evolution of green chemistry and its multidimensional impacts: A review", *Saudi Pharm. J.*, vol. 27, no. 1, pp. 1-8, 2019.
[http://dx.doi.org/10.1016/j.jsps.2018.07.011] [PMID: 30627046]

[13]   X. Sun, Y. Zhang, G. Chen, and Z. Gai, "Application of nanoparticles in enhanced oil recovery: A critical review of recent progress", *Energies*, vol. 10, no. 3, p. 345, 2017.
[http://dx.doi.org/10.3390/en10030345]

[14]   S. Ranjan, N. Dasgupta, and E. Lichtfouse, *Nanoscience in Food and Agriculture 3* vol. 3. Springer, 2016.
[http://dx.doi.org/10.1007/978-3-319-48009-1]

[15]   R.S. Bhosale, K.Y. Hajare, and B. Mulay, "Biosynthesis, Characterization and Study of Antimicrobial Effect of Silver Nanoparticles by Actinomycetes spp", *Int. J. Curr. Microbiol. Appl. Sci.*, vol. 2, no. 2, pp. 144-151, 2015.

[16]   A. Patel, F. Patra, N. Shah, and C. Khedkar, *Application of Nanotechnology in the Food Industry: Present Status and Future Prospects* Impact of Nanoscience in the Food Industry, 2018, pp. 1-27.
[http://dx.doi.org/10.1016/B978-0-12-811441-4.00001-7]

[17]   S. Maiti, "Tailored bio-polysaccharide nanomicelles for targeted drug delivery", In: *Nanoparticles' Promises and Risks.* Springer, Cham., 2015.
[http://dx.doi.org/10.1007/978-3-319-11728-7_16]

[18]   F. Bergaya, and G. Lagaly, "General Introduction: Clays, Clay Minerals, and Clay Science", *Dev. Clay Sci.*, vol. 1, no. C, pp. 1-18, 2006.
[http://dx.doi.org/10.1016/S1572-4352(05)01001-9]

[19]   A. Kausar, "A review of fundamental principles and applications of polymer nanocomposites filled with both nanoclay and nano-sized carbon allotropes – Graphene and carbon nanotubes", *J. Plast. Film Sheeting*, vol. 36, no. 2, pp. 209-228, 2020.
[http://dx.doi.org/10.1177/8756087919884607]

[20]   H.H. Murray, "Applied clay mineralogy today and tomorrow", *Clay Miner.*, vol. 34, no. 1, pp. 39-49, 1999.
[http://dx.doi.org/10.1180/000985599546055]

[21]   L.B. de Paiva, A.R. Morales, and F.R. Valenzuela Díaz, "Organoclays: Properties, preparation and applications", *Appl. Clay Sci.*, vol. 42, no. 1-2, pp. 8-24, 2008.
[http://dx.doi.org/10.1016/j.clay.2008.02.006]

[22]   M. Galimberti, *Rubber Clay Nanocomposites* IntechOpen, 2012.
[http://dx.doi.org/10.5772/51410]

[23]   M.M.A.B. Abdullah, L.Y. Ming, H.C. Yong, and M.F.M. Tahir, *Clay-Based Materials in Geopolymer Technology.* Cement Based Materials, 2018.
[http://dx.doi.org/10.5772/intechopen.74438]

[24]   A. Sircar, K. Rayavarapu, N. Bist, K. Yadav, and S. Singh, "Applications of nanoparticles in enhanced oil recovery", *Petroleum Research*, vol. 7, no. 1, pp. 77-90, 2022.
[http://dx.doi.org/10.1016/j.ptlrs.2021.08.004]

[25]   N. Hasegawa, H. Okamoto, M. Kato, A. Usuki, and N. Sato, "Nylon 6/Na–montmorillonite nanocomposites prepared by compounding Nylon 6 with Na–montmorillonite slurry", *Polymer (Guild f.)*, vol. 44, no. 10, pp. 2933-2937, 2003.
[http://dx.doi.org/10.1016/S0032-3861(03)00215-5]

[26]   H. Zhang, Y. Wang, Y. Wu, L. Zhang, and J. Yang, "Study on flammability of

montmorillonite/styrene-butadiene rubber (SBR) nanocomposites", *J. Appl. Polym. Sci.,* vol. 97, no. 3, pp. 844-849, 2005.
[http://dx.doi.org/10.1002/app.21797]

[27] Z. Yang, H. Peng, W. Wang, and T. Liu, "Crystallization behavior of poly(ε-caprolactone)/layered double hydroxide nanocomposites", *J. Appl. Polym. Sci.,* vol. 116, no. 5, pp. 2658-2667, 2010.
[http://dx.doi.org/10.1002/app.31787]

[28] E. Bilotti, R. Zhang, H. Deng, F. Quero, H.R. Fischer, and T. Peijs, "Sepiolite needle-like clay for PA6 nanocomposites: An alternative to layered silicates?", *Compos. Sci. Technol.,* vol. 69, no. 15-16, pp. 2587-2595, 2009.
[http://dx.doi.org/10.1016/j.compscitech.2009.07.016]

[29] S. Sahoo, A. Sasmal, D. Sahoo, and P. Nayak, "Synthesis and characterization of chitosan-polycaprolactone blended with organoclay for control release of doxycycline", *J. Appl. Polym. Sci.,* vol. 118, no. 6, pp. 3167-3175, 2010.
[http://dx.doi.org/10.1002/app.32474]

[30] J. Asadi, N. Golshan Ebrahimi, and M. Razzaghi-Kashani, "Self-healing property of epoxy/nanoclay nanocomposite using poly(ethylene-co-methacrylic acid) agent", *Compos., Part A Appl. Sci. Manuf.,* vol. 68, pp. 56-61, 2015.
[http://dx.doi.org/10.1016/j.compositesa.2014.09.017]

[31] M. Agista, K. Guo, and Z. Yu, "A State-of-the-Art Review of Nanoparticles Application in Petroleum with a Focus on Enhanced Oil Recovery", *Appl. Sci. (Basel),* vol. 8, no. 6, p. 871, 2018.
[http://dx.doi.org/10.3390/app8060871]

[32] S.N. Kale, and S.L. Deore, "Emulsion Micro Emulsion and Nano Emulsion: A Review", *Syst. Rev. Pharm.,* vol. 8, no. 1, pp. 39-47, 2016.
[http://dx.doi.org/10.5530/srp.2017.1.8]

[33] A. Waris, M. Din, A. Ali, M. Ali, S. Afridi, A. Baset, and A. Ullah Khan, "A comprehensive review of green synthesis of copper oxide nanoparticles and their diverse biomedical applications", *Inorg. Chem. Commun.,* vol. 123, p. 108369, 2021.
[http://dx.doi.org/10.1016/j.inoche.2020.108369]

[34] R. Guin, A. Shakila Banu, and G.A. Kurian, "Synthesis of Copper oxide nanoparticles using Desmodium gangeticum aqueous root extract", *Int. J. Pharm. Pharm. Sci.,* vol. 7, no. January, pp. 60-65, 2015.

[35] S. Shende, A.P. Ingle, A. Gade, and M. Rai, "Green synthesis of copper nanoparticles by Citrus medica Linn. (Idilimbu) juice and its antimicrobial activity", *World J. Microbiol. Biotechnol.,* vol. 31, no. 6, pp. 865-873, 2015.
[http://dx.doi.org/10.1007/s11274-015-1840-3] [PMID: 25761857]

[36] A.M. Awwad, and M.W. Amer, "Biosynthesis of copper oxide nanoparticles using Ailanthus altissima leaf extract and antibacterial activity", *Chem. Int.,* vol. 6, no. 4, pp. 210-2017, 2020.

[37] D.D.E.M.I.R.C.I.G.U.L.T.E.K.I.N. Demet Gultekin, H. Nadaroglu, A. Alayli Gungor, and N. Horasan Kishali, "Biosynthesis and Characterization of Copper Oxide Nanoparticles using Cimin Grape (Vitis vinifera cv.) Extract", *International Journal of Secondary Metabolite,* no. December, pp. 77-84, 2017.
[http://dx.doi.org/10.21448/ijsm.362672]

[38] S. Prakash, N. Elavarasan, A. Venkatesan, K. Subashini, M. Sowndharya, and V. Sujatha, "Green synthesis of copper oxide nanoparticles and its effective applications in Biginelli reaction, BTB photodegradation and antibacterial activity", *Adv. Powder Technol.,* vol. 29, no. 12, pp. 3315-3326, 2018.
[http://dx.doi.org/10.1016/j.apt.2018.09.009]

[39] G. M. Sulaiman, A. T. Tawfeeq, and M. D. Jaaffer, "Biogenic synthesis of copper oxide nanoparticles using Olea europaea leaf extract and evaluation of their toxicity activities : An in vivo and in vitro study", *Biotechnol Prog.,* vol. 34, no. 1, pp. 218-230, 2018.

[40]   M. Shah, D. Fawcett, S. Sharma, S. Tripathy, and G. Poinern, "Green synthesis of metallic nanoparticles via biological entities", *Materials (Basel),* vol. 8, no. 11, pp. 7278-7308, 2015. [http://dx.doi.org/10.3390/ma8115377] [PMID: 28793638]

[41]   P. Mukherjee, A. Ahmad, D. Mandal, S. Senapati, S.R. Sainkar, M.I. Khan, R. Ramani, R. Parischa, P.V. Ajayakumar, M. Alam, M. Sastry, and R. Kumar, "Bioreduction of AuCl4− Ions by the Fungus, Verticillium sp. and Surface Trapping of the Gold Nanoparticles Formed", *Angew. Chem. Int. Ed.,* vol. 40, no. 19, pp. 3585-3588, 2001. [http://dx.doi.org/10.1002/1521-3773(20011001)40:19<3585::AID-ANIE3585>3.0.CO;2-K] [PMID: 11592189]

[42]   G. Cheraghian, "Nanoparticles in drilling fluid: A review of the state-of-the-art", *J. Mater. Res. Technol.,* vol. 13, pp. 737-753, 2021. [http://dx.doi.org/10.1016/j.jmrt.2021.04.089]

[43]   P.O. Ogbeide, and S.A. Igbinere, "The effect of additives on rheological properties of drilling fluid in highly deviated wells", *Futo Journal Series,* vol. 2, no. 2, pp. 68-82, 2016.

[44]   P. Dejtaradon, H. Hamidi, M.H. Chuks, D. Wilkinson, and R. Rafati, "Impact of ZnO and CuO nanoparticles on the rheological and filtration properties of water-based drilling fluid", *Colloids Surf. A Physicochem. Eng. Asp.,* vol. 570, pp. 354-367, 2019. [http://dx.doi.org/10.1016/j.colsurfa.2019.03.050]

[45]   J.K.M. William, S. Ponmani, R. Samuel, R. Nagarajan, and J.S. Sangwai, "Effect of CuO and ZnO nanofluids in xanthan gum on thermal, electrical and high pressure rheology of water-based drilling fluids", *J. Petrol. Sci. Eng.,* vol. 117, pp. 15-27, 2014. [http://dx.doi.org/10.1016/j.petrol.2014.03.005]

[46]   M. Kuma, B.M. Das, and P. Talukdar, "The effect of salts and haematite on carboxymethyl cellulose–bentonite and partially hydrolyzed polyacrylamide–bentonite muds for an effective drilling in shale formations", *J. Pet. Explor. Prod. Technol.,* vol. 10, no. 2, pp. 395-405, 2020. [http://dx.doi.org/10.1007/s13202-019-0722-x]

[47]   O. Contreras, M. Alsaba, G. Hareland, M. Husein, and R. Nygaard, "Effect on Fracture Pressure by Adding Iron-Based and Calcium-Based Nanoparticles to a Nonaqueous Drilling Fluid for Permeable Formations", *J. Energy Resour. Technol.,* vol. 138, no. 3, p. 032906, 2016. [http://dx.doi.org/10.1115/1.4032542]

[48]   O. Mahmoud, H.A. Nasr-El-Din, Z. Vryzas, and V.C. Kelessidis, "Characterization of filter cake generated by nanoparticle-based drilling fluid for HP/HT applications", *Proceedings - SPE International Symposium on Oilfield Chemistry,* 2017pp. 74-92 Montgomery, Texas, USA [http://dx.doi.org/10.2118/184572-MS]

[49]   M. A. A. Alvi, M. Belayneh, and A. Saasen, "The effect of micro-sized boron nitride BN and iron trioxide Fe_2O_3 nanoparticles on the properties of laboratory bentonite drilling fluid", *Society of Petroleum Engineers - SPE Norway One Day Seminar,* 2018pp. 478-491 Bergen, Norway [http://dx.doi.org/10.2118/191307-MS]

[50]   "The effect of nanoparticles additives in the drilling fluid on pressure loss and cutting transport efficiency in the verticle boreholes", In: *Journal of Petroleum Science and Engineering* vol. 171. , 2018, pp. 1-24.

[51]   T. Ahmed, *Oil Recovery Mechanisms and the Material Balance Equation.* Academia.edu, 2010. [http://dx.doi.org/10.1016/B978-1-85617-803-7.50019-5]

[52]   T. K. Dora, G. Kumar, V. Chaudhary, S. K. Govindarajan, and S. R. Devarapu, "Effects of Nanoparticles in Polymer-Based Enhanced Oil Recovery Technique", *Journal of Nano- and Electronic Physics,* vol. 14, no. 6, pp. 06009-1-06009–4, 2022. [http://dx.doi.org/10.21272/jnep.14(6).06009]

[53]   J. Ali, A.K. Manshad, I. Imani, S.M. Sajadi, and A. Keshavarz, "Greenly Synthesized

Magnetite@SiO2@Xanthan Nanocomposites and Its Application in Enhanced Oil Recovery: IFT Reduction and Wettability Alteration", *Arab. J. Sci. Eng.,* vol. 45, no. 9, pp. 7751-7761, 2020.
[http://dx.doi.org/10.1007/s13369-020-04377-x]

[54]   A. K. Torghabeh, A. Kalantariasl, M. Kamali, and M. G. Akbarifard, "Reservoir gas isotope fingerprinting and mechanism for increased H2S: An example from Middle East Shanul gas field", *J Pet Sci Eng,* vol. 199, p. 108325, 2021.
[http://dx.doi.org/10.1016/j.petrol.2020.108325]

[55]   P. Han, C. Chen, H. Yu, Y. Xu, and Y. Zheng, "Study of pitting corrosion of L245 steel in H 2 S environments induced by imidazoline quaternary ammonium salts", *Corros. Sci.,* vol. 112, pp. 128--, 2016.
[http://dx.doi.org/10.1016/j.corsci.2016.07.006]

[56]   Y. Nie, P. Deng, Y. Zhao, P. Wang, L. Xing, Y. Zhang, and X. Xue, "The conversion of PN-junction influencing the piezoelectric output of a CuO/ZnO nanoarray nanogenerator and its application as a room-temperature self-powered active H 2 S sensor", *Nanotechnology,* vol. 25, no. 26, p. 265501, 2014.
[http://dx.doi.org/10.1088/0957-4484/25/26/265501] [PMID: 24916033]

[57]   J. López-Molino, and P. Amo-Ochoa, "Gas Sensors Based on Copper-Containing Metal-Organic Frameworks, Coordination Polymers, and Complexes", *ChemPlusChem,* vol. 85, no. 7, pp. 1564-1579, 2020.
[http://dx.doi.org/10.1002/cplu.202000428] [PMID: 32725963]

[58]   J.D. Lopez, M. Keley, A. Dante, and M.M. Werneck, "Optical fiber sensor coated with copper and iron oxide nanoparticles for hydrogen sulfide sensing", *Opt. Fiber Technol.,* vol. 67, p. 102731, 2021.
[http://dx.doi.org/10.1016/j.yofte.2021.102731]

[59]   M.M. Keley, F.F. Borghi, R.C. Allil, A. Mello, and M.M. Werneck, "Cu2-O-functionalized plastic optical fiber for H2S sensing", *Opt. Fiber Technol.,* vol. 62, p. 102469, 2021.
[http://dx.doi.org/10.1016/j.yofte.2021.102469]

[60]   F. Peng, Y. Sun, Y. Lu, W. Yu, M. Ge, J. Shi, R. Cong, J. Hao, and N. Dai, "Studies on Sensing Properties and Mechanism of CuO Nanoparticles to H2S Gas", *Nanomaterials (Basel),* vol. 10, no. 4, p. 774, 2020.
[http://dx.doi.org/10.3390/nano10040774] [PMID: 32316393]

[61]   F. Peng, Y. Sun, Y. Lu, W. Yu, M. Ge, J. Shi, R. Cong, J. Hao, and N. Dai, "Studies on Sensing Properties and Mechanism of CuO Nanoparticles to H2S Gas", *Nanomaterials (Basel),* vol. 10, no. 4, p. 774, 2020.
[http://dx.doi.org/10.3390/nano10040774] [PMID: 32316393]

[62]   K. Unioersity, "Enhanced H2S Sensing Properties of Porous SnO2 Nanofibers Modified with CuO", *Sens. Actuators B Chem.,* vol. 15, no. May, pp. 575-578, 2003.

[63]   S. Steinhauer, J. Zhao, V. Singh, T. Pavloudis, J. Kioseoglou, K. Nordlund, F. Djurabekova, P. Grammatikopoulos, and M. Sowwan, "Thermal Oxidation of Size-Selected Pd Nanoparticles Supported on CuO Nanowires: The Role of the CuO–Pd Interface", *Chem. Mater.,* vol. 29, no. 14, pp. 6153-6160, 2017.
[http://dx.doi.org/10.1021/acs.chemmater.7b02242]

[64]   R. Shah, A. Pathan, H. Vaghela, S.C. Ameta, and K. Parmar, "Green Synthesis and Characterization of Copper Nanoparticles Using Mixture (Zingiber officinale, Piper nigrum and Piper longum) Extract and its Antimicrobial Activity", *Chem. Sci. Trans.,* vol. 8, no. 1, pp. 63-69, 2019.
[http://dx.doi.org/10.7598/cst2019.1517]

[65]   Y. Zhong, X. Tang, J. Li, T. Zhou, and C. Deng, *Thermocatalytic upgrading and viscosity reduction of heavy oil using copper oxide nanoparticles.* Pet Sci Technol, 2020, pp. 1-13.
[http://dx.doi.org/10.1080/10916466.2020.1788079]

[66]   N.H. Abdurahman, Y.M. Rosli, N.H. Azhari, and B.A. Hayder, "Pipeline transportation of viscous

crudes as concentrated oil-in-water emulsions", *J. Petrol. Sci. Eng.,* vol. 90-91, pp. 139-144, 2012.
[http://dx.doi.org/10.1016/j.petrol.2012.04.025]

[67]   H.A. Faris, N.A. Sami, A.A. Abdulrazak, and J.S. Sangwai, "The performance of toluene and naphtha as viscosity and drag reducing solvents for the pipeline transportation of heavy crude oil", *Petrol. Sci. Technol.,* vol. 33, no. 8, pp. 952-960, 2015.
[http://dx.doi.org/10.1080/10916466.2015.1030079]

[68]   Y. Chen, J. He, Y. Wang, and P. Li, "GC-MS used in study on the mechanism of the viscosity reduction of heavy oil through aquathermolysis catalyzed by aromatic sulfonic H3PMo12O40", *Energy,* vol. 35, no. 8, pp. 3454-3460, 2010.
[http://dx.doi.org/10.1016/j.energy.2010.04.041]

[69]   K. Chao, Y. Chen, H. Liu, X. Zhang, and J. Li, "Laboratory experiments and field test of a difunctional catalyst for catalytic aquathermolysis of heavy oil", *Energy Fuels,* vol. 26, no. 2, pp. 1152-1159, 2012.
[http://dx.doi.org/10.1021/ef2018385]

[70]   A. Yoosefi Booshehri, R. Wang, and R. Xu, "Simple method of deposition of CuO nanoparticles on a cellulose paper and its antibacterial activity", *Chem. Eng. J.,* vol. 262, pp. 999-1008, 2015.
[http://dx.doi.org/10.1016/j.cej.2014.09.096]

[71]   M. Chen, C. Li, G.R. Li, Y.L. Chen, and C.G. Zhou, "In situ preparation of well-dispersed CuO nanocatalysts in heavy oil for catalytic aquathermolysis", *Petrol. Sci.,* vol. 16, no. 2, pp. 439-446, 2019.
[http://dx.doi.org/10.1007/s12182-019-0300-3]

[72]   X. Tang, G. Liang, J. Li, Y. Wei, and T. Dang, "Catalytic effect of in-situ preparation of copper oxide nanoparticles on the heavy oil low-temperature oxidation process in air injection recovery", *Petrol. Sci. Technol.,* vol. 35, no. 13, pp. 1321-1326, 2017.
[http://dx.doi.org/10.1080/10916466.2017.1318145]

[73]   Y.T. Zhong, X.D. Tang, J.J. Li, T.D. Zhou, and C.L. Deng, "Thermocatalytic upgrading and viscosity reduction of heavy oil using copper oxide nanoparticles", *Petrol. Sci. Technol.,* vol. 38, no. 18, pp. 891-903, 2020.
[http://dx.doi.org/10.1080/10916466.2020.1788079]

[74]   H. Ehtesabi, M.M. Ahadian, and V. Taghikhani, Enhanced heavy oil recovery using TiO_2 nanoparticles: Investigation of deposition during transport in core plug., *Energy Fuels,* vol. 29, no. 1, pp. 1-8, 2015.
[http://dx.doi.org/10.1021/ef5015605]

[75]   A. Ghosh, A. Maity, R. Banerjee, and S.B. Majumder, "Volatile organic compound sensing using copper oxide thin films: Addressing the cross sensitivity issue", *J. Alloys Compd.,* vol. 692, pp. 108-118, 2017.
[http://dx.doi.org/10.1016/j.jallcom.2016.09.001]

[76]   D. Degler, U. Weimar, and N. Barsan, "Current Understanding of the Fundamental Mechanisms of Doped and Loaded Semiconducting Metal-Oxide-Based Gas Sensing Materials", *ACS Sens.,* vol. 4, no. 9, pp. 2228-2249, 2019.
[http://dx.doi.org/10.1021/acssensors.9b00975] [PMID: 31365820]

[77]   S. Steinhauer, "Gas Sensors Based on Copper Oxide Nanomaterials: A Review", *Chemosensors (Basel),* vol. 9, no. 3, p. 51, 2021.
[http://dx.doi.org/10.3390/chemosensors9030051]

CHAPTER 4

Combating Hot Corrosion of Metallic Substrate by Nano-Coating

Santosh Kumar[1,*]

[1] *Department of Mechanical Engineering, Chandigarh Group of College, Landran, Mohali, Punjab, India*

Abstract: Corrosion of metallic materials poses a serious threat to the efficiency of the manufacturing and construction industries. To overcome this, various surface modification techniques are employed. But, surface protection by nano-coating is gaining great potential owing to its numerous benefits. These include surface hardness, high-resistance against hot corrosion, high wear resistance, and adhesive strength. Additionally, nano-coatings can be deposited in thinner and smoother thicknesses, allowing for increased efficiency, more flexible equipment design, smaller carbon footprints, and lower operating and maintenance costs. Hence, the aim of this chapter is to provide an overview of the corrosion performance of ceramic, metallic, and nanocomposite coatings on the surface of the metallic substrate. In addition, the role of nanocoating to combat corrosion of metallic substrate is explored. Finally, the diverse applications of nano-coating in different fields including aircraft, automobile, marine, defense, electronic, and medical industries are discussed.

Keywords: Coating materials, Coating deposition techniques, Corrosion, Nano-coating, Surface protection.

INTRODUCTION

The battle against hot corrosion of metallic substrates has been a long and ongoing one, with engineers and scientists constantly seeking new and improved solutions [1]. The emergence of nano-coatings in recent decades has offered a revolutionary approach, paving the way for dramatic advancements in combating this destructive phenomenon [2]. The historical development of nano-coating is mentioned below:

Early Efforts (Pre-1980s): Traditional methods relied on thick, sacrificial coatings like chromium or aluminum, which offered limited protection at high temperatures and suffered from diffusion limitations. The research focused on

---

[*] **Corresponding author Santosh Kumar:** Department of Mechanical Engineering, Chandigarh Group of Colleges, Landran, Mohali, Punjab, India; E-mail: santoshdgc@gmail.com

Virat Khanna, Suneev Anil Bansal, Vishal Chaudhary and Reddicherla Umapathi (Eds.)

understanding the complex mechanisms of hot corrosion, involving oxidation, sulfidation, and molten salt attack. Further, the development of new alloy compositions with improved inherent high-temperature corrosion resistance, albeit limited in their effectiveness [3 - 5].

Rise of Thin Film Technologies (1980s-2000s): The introduction of physical vapor deposition (PVD) and chemical vapor deposition (CVD) techniques enabled the creation of thin film coatings. However ceramic and metallic coatings like alumina, yttria-stabilized zirconia (YSZ), and MCrAlY (where M represents transition metals) offered superior high-temperature oxidation resistance. Further, multi-layered coatings were developed with tailored compositions and microstructures for enhanced protection against specific corrosion mechanisms [6 - 8].

Nano-Coating Revolution (2000s-Present): Advances in nanotechnology paved the way for the development of ultra-thin nano-coatings with unique properties, although nano-structured coatings offered improved adhesion, reduced crack formation, and enhanced diffusion barrier properties. Moreover, the research shifted towards exploring various material compositions, including nano-composite coatings, self-healing coatings, and bio-inspired designs [9, 10].

One of the main issues that the nation's economy faces is corrosion, which results from a metal's contact with its environment [11]. In terms of economic elements, corrosion damages include maintenance and repair expenses, material losses, equipment damage, decreased productivity, and equipment damage. Corrosion damages also have additional negative societal repercussions, such as resource depletion, personal injury, pollution from contaminated hazardous items, and safety implications (cause of fire, explosions, release of poisonous chemicals) [12]. According to the "National Association of Corrosion Engineers (NACE)" research, the cost of corrosion worldwide is projected to be $255 billion USD or 3.4% of the world's gross domestic product (GDP) [13]. Corrosion is predicted to cost the U.S. economy $552 billion per year in direct and indirect expenditures, or 6% of GDP [14]. The expense of maintaining appliances, highway bridges, cars, aeroplanes, and industrial facilities such as petrochemical, desalination, pharmaceutical, and energy production and distribution systems are only a few examples of the direct consequences of corrosion. Indirect corrosion costs might include taxes, overhead corrosion expenses, and productivity losses from delays, failures, or outages, among other things. These costs are just as significant as direct corrosion costs. As illustrated in Fig. (1) [14, 15], the corrosion costs for the economic sector for five distinct locations were gathered.

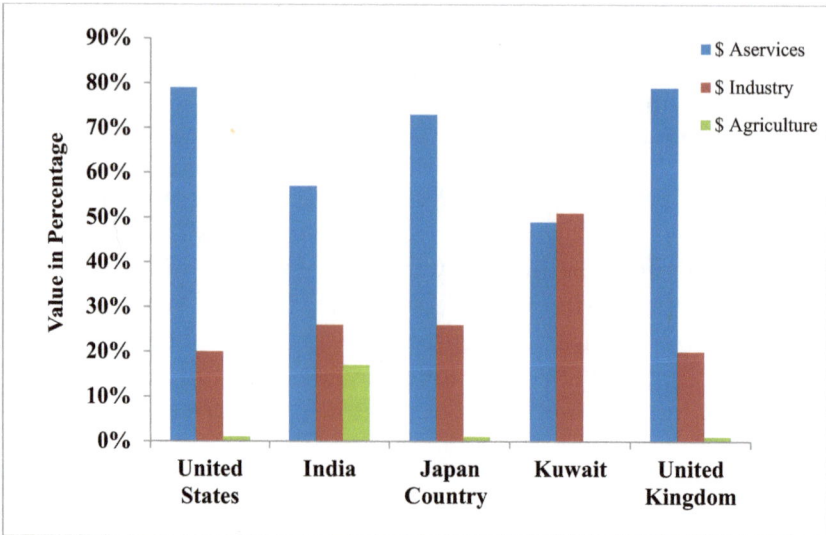

Fig. (1). According to the Worldwide actions of anticipation, application, and economics of corrosion technology research, a NACE international publication, the cost of corrosion varies amongst economic sectors in five different nations.

Installing a layer of coating is one of the simplest ways to stop metals from corroding [11]. Corrosion experts have used a variety of coatings, most of which are based on electrochemical principles, to prevent corrosion. Fortunately, the metal and metallurgy sectors frequently use four different coating kinds based on cost and efficacy. Utilizing barrier coatings like plastic, powder, and paint is one of the most affordable and efficient ways to avoid corrosion. Therefore, nanocomposite coatings applied by electrochemical deposition provide superior metal surface corrosion and abrasion resistance [16]. Instead of replacing the components, these coatings might be utilised to restore them, which would save maintenance costs and disruption [11, 17].

The word "nano" is derived from the Greek word for "dwarf." It relates to a dimension scale between 10^{-9} and 10^{-8} metres. An interdisciplinary field of study known as nanotechnology and nanoscience uses the principles of physics, biology, chemistry, and engineering to investigate and manage matter at the molecular and atomic scales. Nano-materials and nano-particles frequently have high surface region-to-volume ratios when compared to bulk materials, which has a variety of intriguing impacts on the subsequent behaviour of these materials [18 - 24]. Different strategies are used to prevent corrosion, and selecting the best one requires balancing method cost, technique performance, and corrosion impacts. Thus, the corrosion can be avoided by: (a) Choosing materials that are either relatively inert in the -galvanic series or capable of forming a passivating oxide layer in a specific atmosphere; (b) Altering the atm parameters, such as by employing inhibitors [25, 26], and modifying the pH and temperature of the atm, diminishing the flow velocity, removing sand and other debris, and so on; (c) Surface modifications, which involve reducing fissures and cracks by using physical barriers like films and coatings [27 - 30]; (d) Cathodic protection, (*i.e.* corrosive current is stifled and made to flow to the metal that has to be protected). It can be done by applying a power source or by covering the structure that has to be protected with a more active (anodic) substance [31]. "The best protection strategy to use relies on the operational conditions' needs because each protection approach has advantages and downsides of its own. The coating may be applied to various surfaces in distinct temperature ranges, and it even offers the added merits of smoother surfaces, which improves the effectiveness of the interface and flow on surfaces" [32]. Although depositing a coating may be expensive, it is seen to be more practical for long-term applications since it offers significant savings in terms of repair costs, natural resource use, safety, equipment life, *etc* [33]. Coatings often prevent corrosion by either active or passive protection [34 - 36]. When chemicals are introduced to harsh settings to stop or reduce corrosion, active protection is obtained. Inhibitors reduce the rate of corrosion by interacting with the corrosive component in the aqueous medium [37]. In recent years, one of the active and quickly growing fields has been nano-composite coatings, which uses a variety of methods and facets of nano-technology and nanoscience to design and create various kinds of surface coatings with higher performance and increased durability at a reduced cost. Additionally, nano-composite coatings allow for the incorporation of novel and varied capabilities, opening the door to the potential of creating qualities through the manufacture of clever multifunctional coatings [11]. Recently, a lot of research has been conducted on nano-coating and corrosion (Fig. **2**).

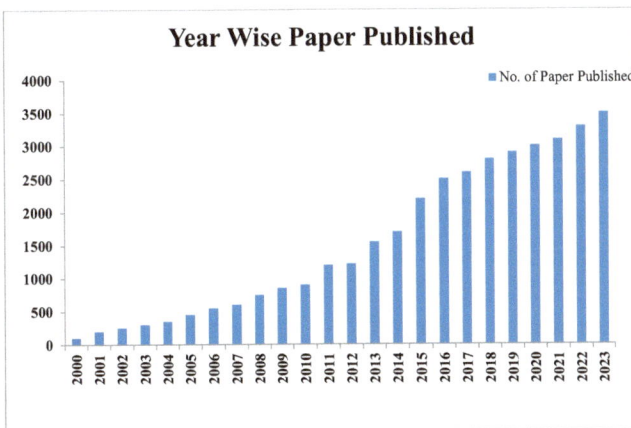

Fig. (2). Year wise (2000-2022) research paper published in the area of corrosion and nano-coating.

However, the share of published articles on corrosion and nano-coating among distinct journals is represented in Fig. (3).

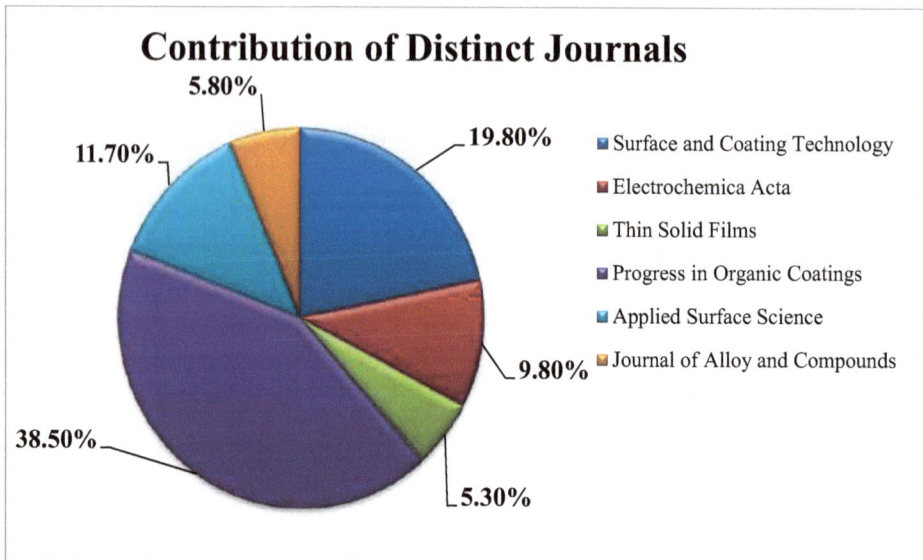

Contribution of Distinct Journals

- Surface and Coating Technology
- Electrochemica Acta
- Thin Solid Films
- Progress in Organic Coatings
- Applied Surface Science
- Journal of Alloy and Compounds

5.80%
19.80%
11.70%
9.80%
38.50%
5.30%

Fig. (3). Share of published research articles in corrosion and nano-coating among distinct journals.

NANO-COATING AND ITS ROLE IN CORROSION PREVENTION

Due to the importance of metallic materials in the construction and industrial sectors, it is now crucial to prevent corrosion failures and the resulting financial loss. In addition, due to the widespread usage in heavy construction, such as the maritime and chemical industries, as well as vehicles, home appliances, commercial machinery, and other items, steel has become a valuable commodity in our daily lives. Therefore, in order to keep metallic equipment functioning while being exposed to potentially hazardous environments, it is required to modify its surface properties. The coating's barrier makes it possible for coated metallic surfaces to withstand more than just their atmosphere, but, also to exhibit individualised performance in line with the chosen surface treatment [38]. A shielding surface coating has been created using a number of methods, such as sole-gel approaches, PVD, *etc.* To stop corrosion, polymeric coatings, and cathodic protection are frequently applied to metal buildings. Because it produces superior wear and corrosion-resistant coatings like Ni, Cu, Co, and Pd, electroless plating has become more popular for coating creation [39 - 42]. Single metal or alloy coatings can be created via electroplating. When it comes to preventing the corrosion of Fe and other materials, pure and composite Zn coatings are thought to be the most favourable [43]. Due to their superior hardness, wear resistance, Co and associated alloys have also received significant attention as a potential replacement for the electro-deposited Cr coatings [44, 45]. Under specific test settings, CoeW coatings seemed to have wear resistance comparable to hard Cr [46].

Nanoparticles are frequently added to metal and metal alloy coatings to improve their mechanical and physical qualities. Co or Ni-based coatings with the addition of alumina are examples of coatings with metal nanoparticles [47, 48].

One of the most popular techniques in the coatings business is electrodeposition (particularly in terms of direct current), which can be used on a massive scale while still producing coatings of excellent quality and adaptability. Due to the increased deposition rate, pulsed electro-deposition has also garnered interest in creating coatings with enhanced structure and characteristics. For instance, Tao (2006) [49] observed that Cu coatings electrodeposited using pulsed current modes had higher hardness, a lower coefficient of friction, and a slower rate of wear than coatings electrodeposited using direct current modes. Regarding electrodeposited Co-Mo alloy coatings, similar results were obtained [50].

MATERIALS EMPLOYED IN NANO-COATINGS FOR ANTICORROSION

Metallic Coating Material

The anti-corrosion qualities of materials that are often exposed to corrosive environments are aided by a metal and alloy coating. Metallic nanocoating serves two main functions: cathodic protection and barrier effect [51]. A metallic coating should ideally be self-passivating and have a lower potential than the work piece so that it can have a long lifespan and a high level of protection. The following [52 - 54] discusses certain metals and their alloys employed as anti-corrosive coatings.

A) Nickel: In the majority of harsh environments, alloys based on nickel exhibit stronger corrosion resistance than stainless steels, while having upper initial costs that are offset by longer lifespans and less equipment downtime. The formability of nickel-based alloys is good, and they can be welded into more intricate, bigger industrial components [55].

B) Titanium: When Ti and Ti-based alloys are exposed as a nano-coating on the workpiece surface in contact with the environment, durable protective oxide coatings spontaneously develop. In both industrial and maritime contexts, TiO_2 on the surface provides good resistance to air corrosion [56].

C) Aluminum: Al's nanocoated surface creates a thin, colourless, durable oxide sheet that serves as a primary barrier against oxidation and chemical reactions. Al flake is a pigment that resists corrosion. Water can travel down this flake's convoluted channel at the matrix interface, and ions nearby produce oxides that can block the coating's retained pores [57].

D) Zinc: Steel is frequently coated with zinc nano coating to prevent corrosion. In saltwater or chloride-containing maritime settings, zinc-coated steel performs well. In a maritime environment, Zn has a superior cathodic protective effect, but an inadequate barrier influence. Steel substrates need cathodic protection to last a long time, and thermally sprayed ZneNi coating can provide that protection [58].

Polymeric Coating Material

In order to prevent and regulate corrosion on particular metals, ceramics, and synthetic materials, organic nanocoatings are frequently used. By creating an impermeable layer on the surface. Polymer-based paint can offer high adhesion, corrosion resistance, and protection against harmful environmental conditions [59]. The only drawback of polymer nanocoatings is poor wear resistance.

Composite Coating Material

Utilizing the necessary combination qualities in the coating materials is permitted by composite materials. Better corrosion characteristics towards the protected iron substrate are produced by composite layers in multilayer systems, which is crucial for using the iron substrate under more aggressive corrosion circumstances. Good adhesive qualities, strain endurance, self-healing qualities, and heat conductivity are a few of the traits. MMCs produce good results for the creation of thin coatings [60]. Aluminum doping can significantly change the microstructure of the metallic alloys resulting in the formation of materials showing high resistance against corrosion [61]. Adding aluminum to metallic alloy may significantly alter the microstructure (grain refinement, precipitation development, passivation layer development, *etc.*). In addition, finer grains provide fewer pathways for corrosive agents to penetrate the material, leading to improved resistance to pitting and intergranular corrosion. In some cases, aluminum itself can contribute to the formation of a stable and protective oxide layer, further enhancing corrosion resistance. A few examples are:

Aluminum in stainless steel: Adding aluminum to stainless steel improves its high-temperature oxidation resistance and resistance to specific corrosive environments like seawater.

Aluminum in magnesium alloys: Aluminum additions refine the microstructure, strengthen the alloy, and promote the formation of a protective oxide layer, enhancing overall corrosion resistance.

Aluminum in aluminum alloys: Controlled aluminum additions in aluminum-magnesium or aluminum-copper alloys can further refine the microstructure and improve resistance to specific types of corrosion [61, 62].

Ceramic Coating Material

Ceramics are used in the creation of coatings that prevent corrosion. A resistive coating is applied to the surfaces of many pieces of automotive equipment, including turbine engines, heat exchangers, and IC engines. In extremely corrosive settings, several researchers have heavily utilised erosion-resistant ceramic nanocoatings like TiN and CrN [61, 62].

New and Advanced Coating Materials

The coatings business is very interested in smart and nanomaterials among new and developing materials [63, 64].

Smart Coating Material

Any coating that alters a material's characteristics in response to an external stimulation is known as a smart coating, and the variations are most often made in a reversible way. Smart coating in this context refers to coated materials that are sensitive, adaptable, or active towards the stimuli or environment of touch. According to the presumptions, as there are numerous stimuli and reactions, there are also prospective applications that may be created [65].

Graphene and CNT Coating

"Massachusetts Institute of Technology" researchers have discovered that chemical vapor-deposited, ultrathin graphene coatings are more resistant to chemicals than conventional functional hydrophobic coatings. Graphene is a planar sheet of closely packed sp2-connected carbon atoms, which is one atom thick and has a honeycomb crystal structure [66].

CORROSION PROTECTION TECHNIQUES

The rate of corrosion of materials in service can also be considerably reduced by eliminating poor design, surface pollution, galvanic couples, and inadequate or improperly implemented surface protection systems. Coating the substrate material is one of the best ways to stop corrosion. By creating a barrier b/w the metal and its surroundings and/or by including corrosion-inhibiting chemicals, coatings can protect a substrate. There are several conceivable coating systems, each with unique benefits and drawbacks. This review's objective is to present an accurate picture of the current state of technology, with a focus on nanocoatings in particular [67].

Standard Anti-Corrosion Coating Techniques

For the purpose of preventing corrosion, several coating processes are available. This comprises vapor-phase procedures, anodizing, conversion coatings, and electrochemical plating. In the sections that follow, a brief description of each of them will be provided.

Electrochemical Plating

In order to increase a workpiece's resistance to corrosion and wear, electrical conductivity, or ornamental appeal, it is frequently desired to change the surface qualities of the workpiece. This can be achieved by coating the component with a metal that possesses the requisite qualities for a particular use. Electrochemical plating is one of the most affordable and straightforward methods for applying a metallic coating to a workpiece. In both situations, a metal salt in sol^n is transformed onto the workpiece's surface into its metallic form. A good plating process must overcome a variety of obstacles, particularly for a material like Mg alloys. Due to its strong reactivity, magnesium is categorised as a metal that is challenging to the plate. This implies that Mg very hurriedly generates a passive oxide layer in the presence of air, which must be detached before metal plating. Because the oxide layer forms quickly, the surface must be properly prepped in order to create a surface layer that inhibits oxidation while being simple to remove during the plating process [68 - 71].

Anodizing

"A thick, stable oxide layer is created on metals and alloys by the electrolytic method of anodizing" [72]. These films can be used as a passivation action, a key for dyeing, or to increase paint adherence to metal [73]. The processing phases include [72]; (1) Mechanical pretreatment, (2) Washing, pickling, and degreasing; electro brightening or polishing; anodizing. However, at the metal-coating interface of the films, there is a thin barrier layer, followed by a layer with a cellular formation [73].

Conversion Type of Coatings

These coatings are made by chemically or electrochemically treating a metal surface to create a thin layer of chemically bound workpiece metal oxides, and other chemicals. Conversion coatings operate as an insulating barrier with low solubility b/w the metal surface and the atm. and/or include corrosion-inhibiting chemicals to protect the workpiece from corrosion [74 - 76].

Gas Phase Deposition Technique

The gas phase can be used to create protective coatings. Although, they can also be organic coatings. They are primarily metallic coatings. All of these procedures have the benefit of having minimally detrimental environmental effects. However, these approaches often come at a hefty capital expense.

Thermal Sprayed Coatings

The coating materials are put into a torch or cannon, then heated to a point where it melts or almost melts. The resultant droplets are absorbed by tiny lamellar particles and stick to the substrate after being propelled onto it by a gas stream [77]. This category covers a variety of coating methods depicted in Fig. (**4**) [78, 79].

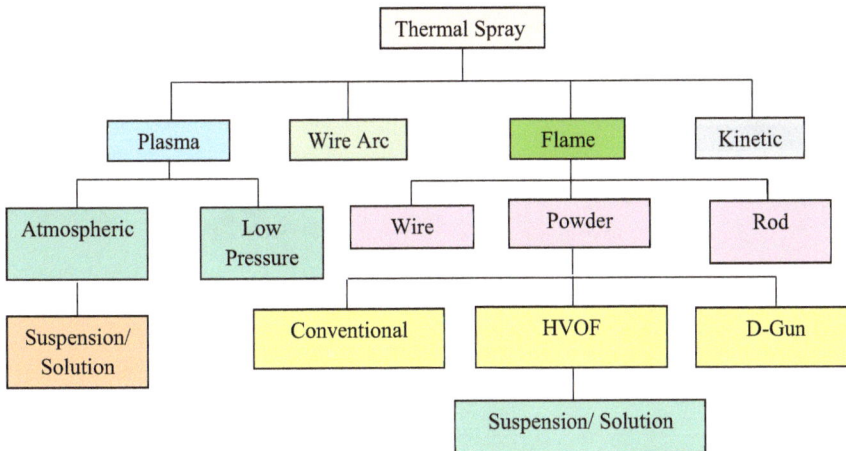

Fig. (4). Types of TS coating processes.

Classification According to the Energy Obtained from Kinetic Energy of Compressed Gases

Cold Gas Dynamic Spray (CS): Using a de Laval nozzle, solid powder particles are sprayed at the substrate using the thermal spray technique known as CS [80, 81]. The schematic representation of CS system is shown in Fig. (**5**).

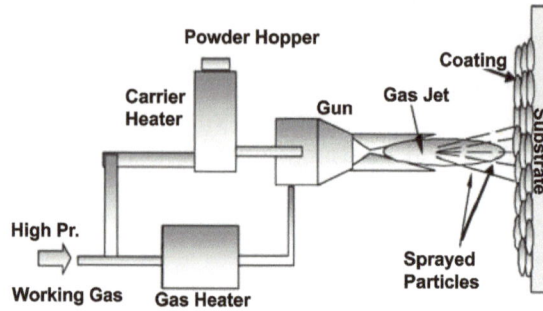

Fig. (5). Schematic representation of CS process.

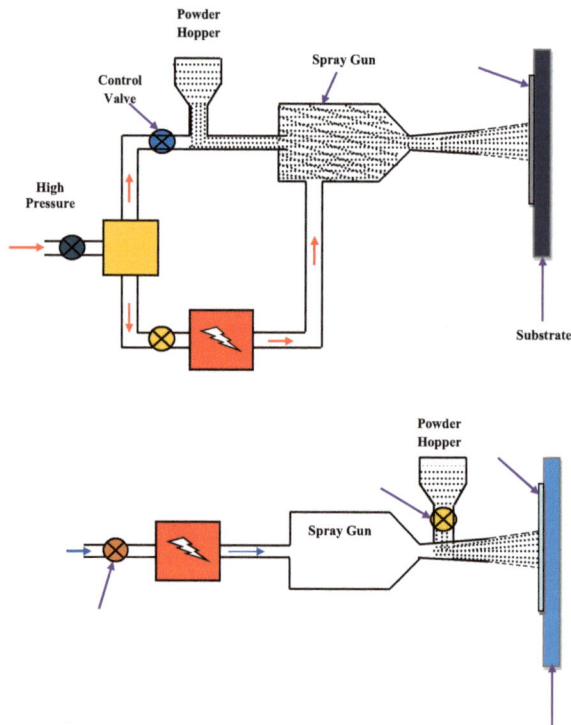

Fig. (6). Cold spray system (**a**) High pressure (**b**) Low pressure.

During HPCS, the powder particles are injected before the spray nozzle throat through HP compressed air. Due to high weight and size of the system, these systems are stationary and utilize gases at high pressure and produce high velocity of powder particles (800 to 1400 m/s). Helium and nitrogen gases are used as propellant gases during the coating process [82].

However, particles of powder are fed into the de-laval nozzle's diverging portion during LPCS. This system uses compressed air supply with low pressure (LP). In addition, these systems are more compact, portable, use air or nitrogen as propellant gases and produce low particle velocities (300m/s to 600 m/s) [83-86].

Classification According to the Energy Obtained from Combustion of Gases

D- Gun Spray: Detonation spraying is one of the various thermal spraying methods used to apply a protective coating to a material at supersonic speeds in order to alter its surface properties [64, 65]. The schematic illustration of D-Gun process is shown in Fig. (**7**) [87, 88].

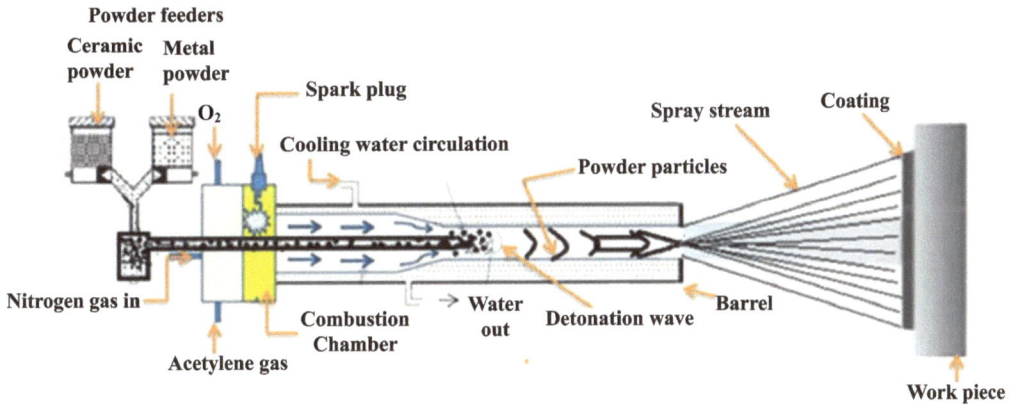

Fig. (7). Schematics illustration of D-Gun.

Flame Spray: It is a 1st generation of thermal spray method that makes use of feedstock materials (wire, rod, or powder). Acetylene (C_2H_2) and oxygen are utilized in this process to melt the feedstock material as shown in Fig. (**8**). C_2H_2 can be substituted with distinct gases such as hydrogen, propane, and methyl-acetylene-propane [89].

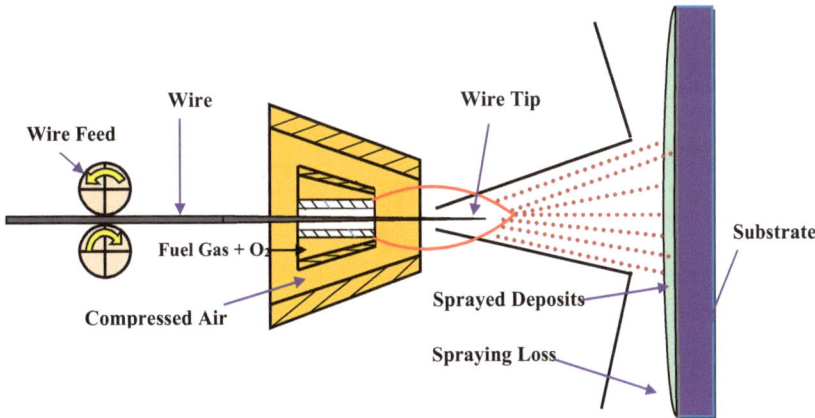

Fig. (8). Schematic diagram of Flame spraying.

Powder flame spray: Using the powder feed method, a hopper is used to feed the coating material to the cannon. The powder is transported across the flame via gas aspiration. In the flame's path, the powder particles melt and deposit on the substrate [90] as depicted in Fig. (**9**).

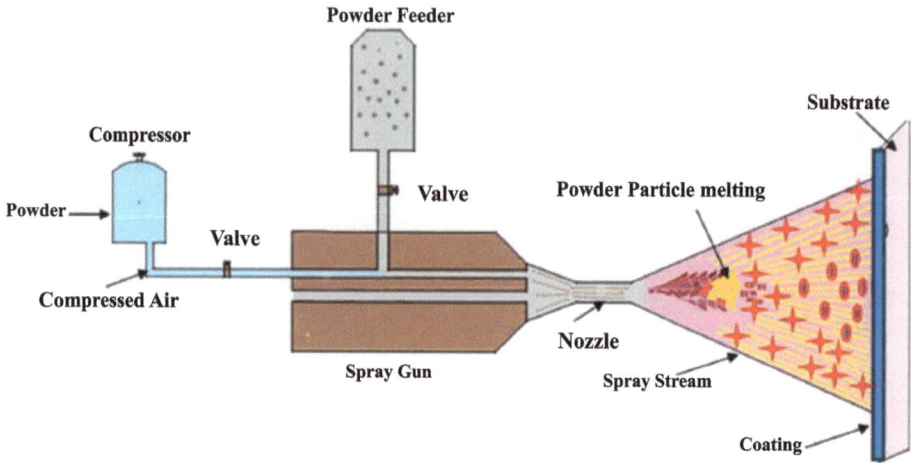

Fig. (9). Schematic diagram of powder FS.

HVOF: HVOF spraying is a technique used to apply functional anti-corrosion coatings with different thicknesses on surfaces of certain sizes and forms. The HVOF cannon is fed with these coating materials, which come in the form of metals, alloys, ceramics, *etc.*, in a variety of shapes (powder/wire) and fine-sized particles. The burning of gas or liquid fuel is utilized as energy for High-velocity oxy-fuel spraying as depicted in Fig. (**10**) [91 - 93].

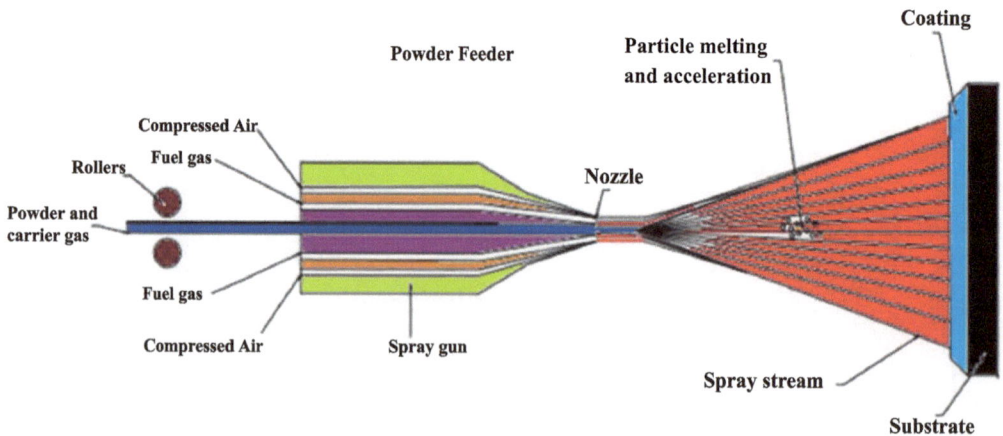

Fig. (10). Schematic illustration of HVOF.

Classification According to the Energy Obtained from Electric or Gas Discharge

Plasma Spray: An inert gas (Ar), heat source, and a high temperature are used in this procedure to form coating with high quality [94] as shown in Fig. (**11**)

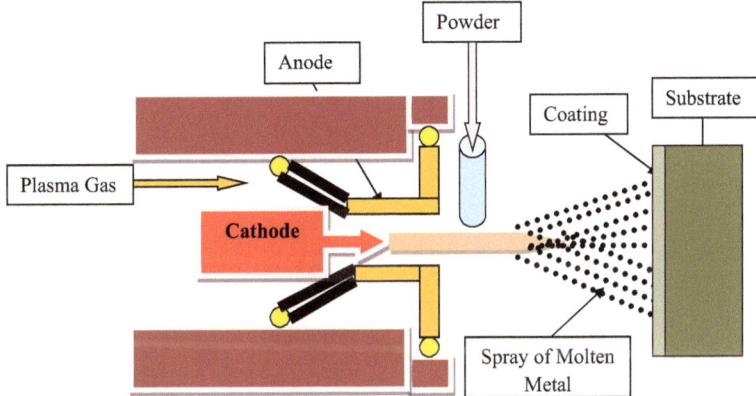

Fig. (11). Schematic diagram of plasma spray.

Electric Arc Spray Coating Process: Two metallic wires are electrically charged throughout this procedure, and they are fed into the arc cannon at regulated rates. The wire tips melt as a result of the opposing charges on the wires. The molten metal is then atomized and accelerated onto the workpiece to produce a coating after being supplied with compressed HP air [95 - 100]. Recently, Ni-20Cr and Ni-5Al coating proved well corrosion resistant especially for boiler applications by using this process [101] (Fig. **12**).

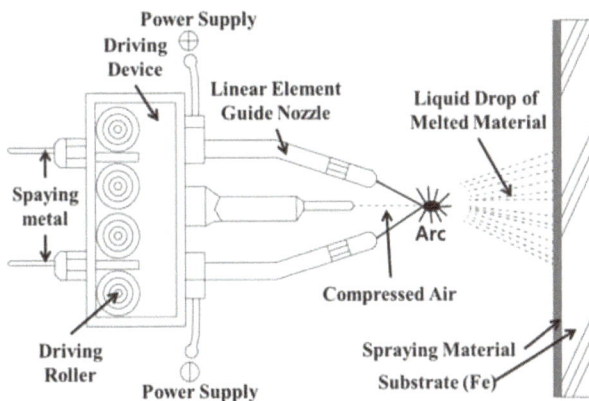

Fig. (12). Schematic representation of wire arc spraying.

COMPARISON OF DISTINCT THERMAL SPRAY TECHNIQUES:

Different thermal spray methods such as cold spray, HVOF, D-Gun, plasma and wire arc spray are compared on the basis of their coating properties as depicted [102 - 105] in Fig. (**13A**).

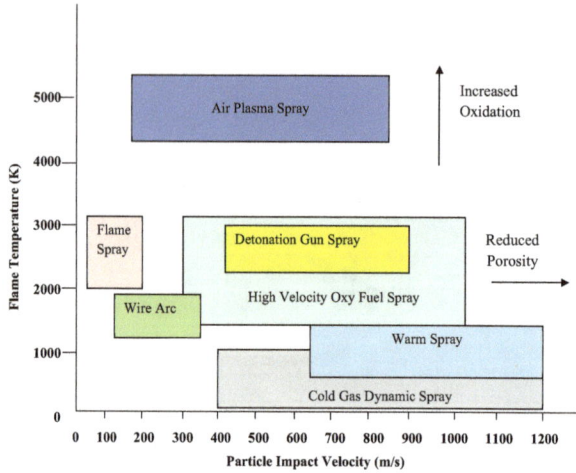

Fig. (13A). Comparison chart of distinct thermal spray process.

Physical Vapor Deposition Processes

Atoms/molecules from the vapour phase are deposited onto a metal workpiece during PVD. The operation's high operating costs and the small sample sizes under coatings are drawbacks of this method [106, 107] (Fig. **13B**).

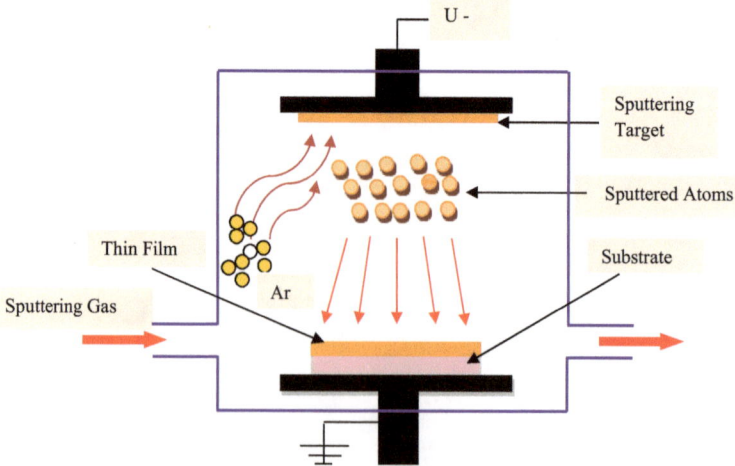

Fig. (13B). Schematics diagram of the PVD process.

Chemical Vapor Deposition

CVD is the process of depositing a solid by a gas-phase chemical reaction on a heated surface. The deposition of high temp materials far below their melting temperatures, attainment of almost control over particle size and position, and processing at atm. pressure and strong adhesion are all benefits of this method [108].

Laser -Surface Alloying

Rapid solidification processing is used in this procedure, although only the surface area is altered. This method has the potential to handle complicated geometries, up to millimeters of treatment depth, cheaper operation costs, and more control over the changed layer concentration [109 - 111].

POTENTIAL APPLICATION OF NANOCOATING IN ANTICORROSION

Aircraft Industries

The phenomenon of corrosion is the deterioration of materials or their surfaces as a result of environmental factors. Organic thin-film coatings and surface passivation appear to be the most effective methods of shielding materials from corrosive attack. Long-term corrosion attacks, however, can still occur and pose serious risks to safety, the environment, and the economy. According to estimates, an industrialized country's gross national product (GNP) is spent on environmental preservation, maintenance, and replacing damaged parts. Considering the prices in 2003, this translates to a cost to the American economy of around $280 billion annually [112 - 117]. Vehicle strength and durability have increased because of innovative building materials. Their integration has benefitted the aviation sector for more than 20 years in a variety of ways, including structural lightness, aerial sensory surveillance, *etc.* This industry must step up its efforts in response to the COVID-19 pandemic surge and collaborate with cutting-edge surface protection technologies to construct pathogen surveillance. In the high-stakes space and military industries, engineered nano-materials (ENMs) and coatings have been instrumental in improving the surface wear and friction qualities of parts. But now, necessity has replaced luxury as the top concern [118, 119].

Automobile Industries

Since humans rely on cars more frequently than either air or water transportation, the automotive industry is where much nanotechnology-based research and development are concentrated. Automobile interiors, electrics and electronics, engines, drive trains, chassis and tyres, and emissions are all affected by nanotechnology. Nanotechnology shapes the vital components of autos [120].

Marine Industries

The main method of preventing corrosion in maritime engineering structures is coating. Due to their exceptional qualities, nanomaterials have emerged as the best materials for preventing corrosion. The study and use of novel nanocoating materials have received a lot of interest lately [121 - 126].

Defense Industries

A roadmap for nanotechnology was created in order to stay up with technological advancements and enable the military sector to get ready for the future. In general, military personnel are stationed in unfavourable weather conditions, which can range from high humidity in coastal or marine areas to high temperatures, salt spray, and caustic desert regions, where wind-blown sand erodes surface materials and gets into every crack. The effects of corrosion on several pieces of defensive technology are substantial [127, 128].

Electronic Industries

The popularity of smaller, lighter, and more streamlined gadgets among consumers has led to an increase in the usage of portable electronics and communication tools like tablets and smartphones. The increasing use of these gadgets throughout the world is also being facilitated by technological developments in hardware and downsizing. Even while these devices can meet customer expectations, one of their biggest downsides is that they are extremely vulnerable to damage from dust, moisture, and humidity. The gadgets become damaged as a result, and their functionality is lost. To shield these gadgets from harm from the outside and moisture, a variety of protective coatings, including hydrophobic and superhydrophobic coatings, have been created [129].

Medical Industries

Numerous medical gadgets for use in healthcare are being produced as a result of recent discoveries fusing nanotechnology with electronics and computers. Developments have been achieved in the use of coatings to improve the performance of stents, hip prostheses, dental implants, and other devices. Recently, high entropy alloy coatings also played a significant role in the medical sector [130]. The use of coatings in the healthcare sector has a huge and mostly unexplored potential, despite several publications [131 - 140].

Thermal Power Plant

Nanocoatings are revolutionizing various aspects of thermal power plants, offering innovative solutions for protecting components, enhancing efficiency, and reducing maintenance costs. Turbine Blades: Nano coatings protect blades from high-temperature oxidation and hot gas corrosion, extending their lifespan and reducing thermal efficiency losses. Recently, Nanocoating has found wide applications in boiler tubes, burners, and heat transfer equipment (condensers, heat exchangers, pipelines) to prevent internal corrosion and erosion [141 - 145].

RESULTS AND DISCUSSION

While the potential of nano-coatings for combating hot corrosion is undeniably exciting, understanding their impact requires quantifiable data and analysis. Here's a breakdown of key aspects:

Improved Performance

Increased Lifespan: Nanocoatings can extend the lifespan of metallic substrates in high-temperature environments by 2-5 times compared to traditional coatings.

This translates to significant cost savings due to reduced component replacements and downtime.

Enhanced Efficiency: By minimizing oxidation and corrosion, nanocoatings can improve the efficiency of turbines, boilers, and heat exchangers by 5-10%. This translates to increased power generation and fuel savings.

Reduced Emissions: Improved combustion efficiency and reduction of hot corrosion-related particulates in flue gas can lead to reductions in NOx and SOx emissions by 5-10% [146].

Material Advancements

Nano-YSZ: Studies show nano-YSZ coatings offer up to 95% reduction in oxidation rate compared to bare nickel at 800°C.

Nanocomposite Coatings: Incorporating metallic nanoparticles within ceramic matrices can improve oxidation and sulfidation resistance by 20-30% compared to single-phase coatings.

Self-Healing Coatings: Embedded nanoparticles or stimuli-responsive polymers can autonomously repair minor damage, extending the protective lifespan of nanocoatings by 30-50% [147].

Economic Considerations

Cost of Nanocoatings: While more expensive than traditional coatings initially, the extended lifespan and improved efficiency of components lead to a net cost-saving within 2-5 years depending on the application.

Scaling Up Production: The development of cost-effective deposition techniques like spray coating is crucial for wider adoption, potentially reducing costs by 20-30%.

Maintenance Savings: Reduced component replacements and downtime due to the enhanced durability of nanocoatings can lead to significant maintenance cost savings, estimated at 10-20% per year [147].

Environmental Impact

Reduced Emissions: Lower NOx and SOx emissions contribute to cleaner air and improved environmental quality.

Increased Efficiency: Improved fuel utilization lowers CO_2 emissions, contributing to the fight against climate change.

Extended Equipment Lifespan: Minimizes resource consumption associated with frequent component replacements [146 - 150].

CONCLUSION

From the extensive literature, it is clear that nanocoating plays a significant role in corrosion. Distinct types of nano-coating material deposition techniques have been discussed. Further, this chapter explores the application of distinct materials including metallic, polymeric material, ceramic, and composite material coating, which are an efficient medium for anti-corrosion. Finally, the diverse applications of nano-coatings are outlined clearly to help us comprehend how corrosion loss affects our daily lives.

REFERENCES

[1]   G.L. Song, and A. Atrens, "Recently deepened insights regarding Mg corrosion and advanced engineering applications of Mg alloys", *Journal of Magnesium and Alloys,* vol. 11, no. 11, pp. 3948-3991, 2023.
[http://dx.doi.org/10.1016/j.jma.2023.08.012]

[2]   S. Malik, K. Muhammad, and Y. Waheed, "Nanotechnology: A Revolution in Modern Industry", *Molecules,* vol. 28, no. 2, p. 661, 2023.
[http://dx.doi.org/10.3390/molecules28020661] [PMID: 36677717]

[3]   Y. Gu, K. Xia, D. Wu, J. Mou, and S. Zheng, "Technical Characteristics and Wear-Resistant Mechanism of Nano Coatings: A Review", *Coatings,* vol. 10, no. 3, p. 233, 2020.

[http://dx.doi.org/10.3390/coatings10030233]

[4]   X.Y. Ren, "Developing status of nanocoaters and coating technology", *Nonferrous Metals.,* vol. 56, pp. 31-34, 2004.

[5]   L. Ajdelsztajn, J.A. Picas, G.E. Kim, F.L. Bastian, J. Schoenung, and V. Provenzano, "Oxidation behavior of HVOF sprayed nanocrystalline NiCrAlY powder", *Mater. Sci. Eng. A,* vol. 338, no. 1-2, pp. 33-43, 2002.
[http://dx.doi.org/10.1016/S0921-5093(02)00008-4]

[6]   Q. He, J.M. Paiva, J. Kohlscheen, B.D. Beake, and S.C. Veldhuis, "An integrative approach to coating/carbide substrate design of CVD and PVD coated cutting tools during the machining of austenitic stainless steel", *Ceram. Int.,* vol. 46, no. 4, pp. 5149-5158, 2020.
[http://dx.doi.org/10.1016/j.ceramint.2019.10.259]

[7]   T. Ishigaki, S. Tatsuoka, K. Sato, K. Yanagisawa, K. Yamaguchi, and S. Nishida, "Influence of the Al content on mechanical properties of CVD aluminum titanium nitride coatings", *Int. J. Refract. Hard Met.,* vol. 71, pp. 227-231, 2018.
[http://dx.doi.org/10.1016/j.ijrmhm.2017.11.028]

[8]   Q. He, J.M. Paiva, J. Kohlscheen, B.D. Beake, and S.C. Veldhuis, "An integrative approach to coating/carbide substrate design of CVD and PVD coated cutting tools during the machining of austenitic stainless steel", *Ceram. Int.,* vol. 46, no. 4, pp. 5149-5158, 2020.
[http://dx.doi.org/10.1016/j.ceramint.2019.10.259]

[9]   A. Chauhan, M. Kumar, and S. Kumar, "Fabrication of polymer hybrid composites for automobile leaf spring application", *Mater. Today Proc.,* vol. 48, no. 5, pp. 1371-1377, 2022.
[http://dx.doi.org/10.1016/j.matpr.2021.09.114]

[10]   I. Ielo, F. Giacobello, S. Sfameni, G. Rando, M. Galletta, V. Trovato, G. Rosace, and M.R. Plutino, "Nanostructured Surface Finishing and Coatings: Functional Properties and Applications", *Materials (Basel),* vol. 14, no. 11, p. 2733, 2021.
[http://dx.doi.org/10.3390/ma14112733] [PMID: 34067241]

[11]   T. Brindha, R. Rathinam, S. Dheenadhayalan, and R. Sivakumar, "Nanocomposite Coatings in Corrosion Protection Applications: An Overview", *Orient. J. Chem.,* vol. 37, no. 5, pp. 1062-1067, 2021. Available from: https://bit.ly/3Bo9M5x

[12]   K. Kumar, S. Kumar, and H.S. Gill, "Role of Surface Modification Techniques to Prevent Failure of Components Subjected to the Fireside of Boilers", *J. Fail. Anal. Prev.,* vol. 23, pp. 1-15, 2022.
[http://dx.doi.org/10.1007/s11668-022-01556-w]

[13]   G. Koch, J. Varney, N. Thompson, O. Moghissi, M. Gould, and J. Payer, *International Measures of Prevention, Application, and Economics of Corrosion Technologies Study.* NACE International: Houston, TX, USA, 2016.

[14]   G.H. Koc, M.P.H. Brongers, N.G. Thompson, Y.P. Virmani, and J.H. Payer, *Corrosion Costs and Preventive Strategies in the United States.* NACE International: McLean, VA, USA, 2002.

[15]   D. Abdeen, M. El Hachach, M. Koc, and M. Atieh, "A Review on the Corrosion Behaviour of Nanocoatings on Metallic Substrates", *Materials (Basel),* vol. 12, no. 2, p. 210, 2019.
[http://dx.doi.org/10.3390/ma12020210] [PMID: 30634551]

[16]   Phuong Nguyen-Tri, Tuan Anh Nguyen, Pascal Carriere, and Cuong Ngo Xuan, "Nanocomposite Coatings: Preparation, Characterization, Properties, and Applications", *International Journal of Corrosion,* p. 19, 2018.
[http://dx.doi.org/10.1155/2018/4749501]

[17]   A. Meldrum, L.A. Boatner, and C.W. White, "Nanocomposites formed by ion implantation: Recent developments and future opportunities", *Nucl. Instrum. Methods Phys. Res. B,* vol. 178, no. 1-4, pp. 7-16, 2001.
[http://dx.doi.org/10.1016/S0168-583X(00)00501-2]

[18]   Y. Nedal, "Recent Advances in Nanocomposite Coatings for Corrosion Protection Applications", Handbook of Nanoceramic and Nanocomposite Coatings and Materials: Butterworth-Heinemann, 2015, pp. 515-549.
[http://dx.doi.org/10.1016/B978-0-12-799947-0.00024-9]

[19]   T.S. Bedi, S. Kumar, and R. Kumar, "Corrosion performance of hydroxyapaite and hydroxyapaite/titania bond coating for biomedical applications", *Mater. Res. Express,* vol. 7, no. 1, p. 015402, 2020.
[http://dx.doi.org/10.1088/2053-1591/ab5cc5]

[20]   S. Kumar, M. Kumar, and A. Handa, "Combating hot corrosion of boiler tubes – A study", *Eng. Fail. Anal.,* vol. 94, pp. 379-395, 2018.
[http://dx.doi.org/10.1016/j.engfailanal.2018.08.004]

[21]   X. Zhang, P. Lan, Y. Lu, J. Li, H. Xu, J. Zhang, Y. Lee, J.Y. Rhee, K.L. Choy, and W. Song, "Multifunctional antireflection coatings based on novel hollow silica-silica nanocomposites", *ACS Appl. Mater. Interfaces,* vol. 6, no. 3, pp. 1415-1423, 2014.
[http://dx.doi.org/10.1021/am405258d] [PMID: 24443948]

[22]   U. Sultan, J. Kumar, S. Dadra, and S. Kumar, "Experimental investigations on the tribological behaviour of advanced aluminium metal matrix composites using grey relational analysis", *Materials Today: Proceedings,* 2022.
[http://dx.doi.org/10.1016/j.matpr.2022.12.171]

[23]   T. Mishra, A.K. Mohanty, and S.K. Tiwari, "Recent development in clay based functional coating for corrosion protection", *Key Eng. Mater.,* vol. 571, pp. 93-109, 2013.
[http://dx.doi.org/10.4028/www.scientific.net/KEM.571.93]

[24]   M. Omastová, and M. Mičušík, "Polypyrrole coating of inorganic and organic materials by chemical oxidative polymerisation", *Chem. Pap.,* vol. 66, no. 5, pp. 392-414, 2012.
[http://dx.doi.org/10.2478/s11696-011-0120-4]

[25]   S.V. Lamaka, M.L. Zheludkevich, K.A. Yasakau, M.F. Montemor, and M.G.S. Ferreira, "High effective organic corrosion inhibitors for 2024 aluminium alloy", *Electrochim. Acta,* vol. 52, no. 25, pp. 7231-7247, 2007.
[http://dx.doi.org/10.1016/j.electacta.2007.05.058]

[26]   K. Rahmani, R. Jadidian, and S. Haghtalab, "Evaluation of inhibitors and biocides on the corrosion, scaling and biofouling control of carbon steel and copper–nickel alloys in a power plant cooling water system", *Desalination,* vol. 393, pp. 174-185, 2016.
[http://dx.doi.org/10.1016/j.desal.2015.07.026]

[27]   C.P. Cho, O.S. Kwon, and Y.J. Lee, "Effects of the sulfur content of liquefied petroleum gas on regulated and unregulated emissions from liquefied petroleum gas vehicle", *Fuel,* vol. 137, pp. 328-334, 2014.
[http://dx.doi.org/10.1016/j.fuel.2014.07.090]

[28]   D. Du, K. Chen, H. Lu, L. Zhang, X. Shi, X. Xu, and P.L. Andresen, "Effects of chloride and oxygen on stress corrosion cracking of cold worked 316/316L austenitic stainless steel in high temperature water", *Eval. Program Plann.,* vol. 110, pp. 134-142, 2016.

[29]   J.A. Calderón, J.P. Jiménez, and A.A. Zuleta, "Improvement of the erosion-corrosion resistance of magnesium by electroless Ni-P/Ni(OH)2-ceramic nanoparticle composite coatings", *Surf. Coat. Tech.,* vol. 304, pp. 167-178, 2016.
[http://dx.doi.org/10.1016/j.surfcoat.2016.04.063]

[30]   J. Telegdi, T. Szabó, L. Románszki, and M. Pávai, The use of nano-/microlayers, self-healing and slow-release coatings to prevent corrosion and biofouling.*Handbook of Smart Coatings for Materials Protection.* Woodhead Publishing: Cambridge, UK, 2014, pp. 135-182.
[http://dx.doi.org/10.1533/9780857096883.2.135]

[31]   S. Wang, Z. Ma, Z. Liao, J. Song, K. Yang, and W. Liu, "Study on improved tribological properties by alloying copper to CP-Ti and Ti–6Al–4V alloy", *Mater. Sci. Eng. C,* vol. 57, pp. 123-132, 2015.
[http://dx.doi.org/10.1016/j.msec.2015.07.046] [PMID: 26354247]

[32]   R. Singh, Coating for Corrosion Prevention.*Corrosion Control for Offshore Structures: Cathodic Protection and High Efficiency Coating.* Gulf Professional Publishing: Waltham, MA, USA, 2014, pp. 115-129.
[http://dx.doi.org/10.1016/B978-0-12-404615-3.00008-5]

[33]   A. Samimiã, and S. Zarinabadi, "An Analysis of Polyethylene Coating Corrosion in Oil and Gas Pipelines", *J. Am. Sci.,* vol. 7, pp. 1032-1036, 2011.

[34]   N. Van Velson, and M. Flannery, "Performance Life Testing of a Nanoscale Coating for Erosion and Corrosion Protection in Copper Microchannel Coolers", *Proceedings of the 15th IEEE Intersociety Conference on Thermal and Thermomechanical Phenomena in Electronic Systems (ITherm),* pp. 662-669 Las Vegas, NV, USA
[http://dx.doi.org/10.1109/ITHERM.2016.7517612]

[35]   V.S. Saji, The impact of nanotechnology on reducing corrosion cost.*Corrosion Protection and Control Using Nanomaterials.,* V.S. Saji, R. Cook, Eds., Woodhead Publishing Limited: Philadelphia, PA, USA, 2012, pp. 3-15.
[http://dx.doi.org/10.1533/9780857095800.1.3]

[36]   M. Yao, Y. He, Y. Zhang, and Q. Yang, "Al2O3-Y2O3 Nano- and Micro-Composite Coatings on Fe-9Cr-Mo Alloy", *J. Rare Earths,* vol. 24, no. 5, pp. 587-590, 2006.
[http://dx.doi.org/10.1016/S1002-0721(06)60169-5]

[37]   C.G. Dariva, and A.F. Galio, "Corrosion Inhibitors—Principles, Mechanisms and Applications", In: *In Developments in Corrosion Protection* IntechOpen Limited: London, UK, 2014, p. 16.

[38]   A. Pruna, "Nanocoatings for protection against steel corrosion, Nanotechnology in Eco-efficient Construction", In: *Materials* Second Edition. Processes and Applications Woodhead Publishing Series in Civil and Structural Engineering, 2019, pp. 337-359.
[http://dx.doi.org/10.1016/B978-0-08-102641-0.00015-3]

[39]   S. Haag, M. Burgard, and B. Ernst, "Pure nickel coating on a mesoporous alumina membrane: preparation by electroless plating and characterization", *Surface and Coatings Technology,* vol. 201, no. 6, pp. 2166-2173, 2006.
[http://dx.doi.org/10.1016/j.surfcoat.2006.03.023]

[40]   H. Zhao, Z. Huang, and J. Cui, "Electroless plating of copper on AZ31 magnesium alloy substrates", *Microelectronic Engineering,* vol. 85, pp. 253-e258, 2008.
[http://dx.doi.org/10.1016/j.mee.2007.05.068]

[41]   M. Kitiwan, and D. Atong, "Effects of porous alumina support and plating time on electroless plating of palladium membrane", *Journal of Materials Science & Technology,* vol. 26, pp. 1148-e1152, 2010.
[http://dx.doi.org/10.1016/S1005-0302(11)60016-9]

[42]   S.L. Cheng, T.L. Hsu, T. Lee, S.W. Lee, J.C. Hu, and L.T. Chen, "Characterization and kinetic investigation of electroless deposition of pure cobalt thin films on silicon substrates", *Applied Surface Science,* vol. 264, pp. 732-e736, 2013.
[http://dx.doi.org/10.1016/j.apsusc.2012.10.111]

[43]   L. Benea, and E. Danaila, "Nucleation and growth mechanism of Ni/TiO$_2$ nanoparticles electro-codeposition", *Journal of the Electrochemical Society,* vol. 163, no. 13, pp. D655-eD662, 2016.

[44]   F. Su, and P. Huang, "Microstructure and tribological property of nanocrystalline Co–W alloy coating produced by dual-pulse electrodeposition", *Mater. Chem. Phys.,* vol. 134, no. 1, pp. 350-359, 2012.
[http://dx.doi.org/10.1016/j.matchemphys.2012.03.001]

[45]   J. Vazquez-Arenas, T. Treeratanaphitak, and M. Pritzker, "Formation of Co–Ni alloy coatings under direct current, pulse current and pulse-reverse plating conditions", *Electrochim. Acta,* vol. 62, pp. 63-

72, 2012.
[http://dx.doi.org/10.1016/j.electacta.2011.11.085]

[46]   H. Capel, P.H. Shipway, and S.J. Harris, "Sliding wear behaviour of electrodeposited cobalt–tungsten and cobalt–tungsten–iron alloys", *Wear,* vol. 255, no. 7-12, pp. 917-923, 2003.
[http://dx.doi.org/10.1016/S0043-1648(03)00241-2]

[47]   H. Gül, F. Kılıç, S. Aslan, A. Alp, and H. Akbulut, "Characteristics of electro-co-deposited Ni–Al$_2$O$_3$ nano-particle reinforced metal matrix composite (MMC) coatings", *Wear,* vol. 267, no. 5-8, pp. 976-990, 2009.
[http://dx.doi.org/10.1016/j.wear.2008.12.022]

[48]   F. Hou, W. Wang, and H. Guo, "Effect of the dispersibility of ZrO$_2$ nanoparticles in Ni-ZrO$_2$ electroplated nanocomposite coatings on the mechanical properties of nano-composite coatings", *Applied Surface Science,* vol. 252, pp. 3812-e3817, 2006.

[49]   S. Tao, and D.Y. Li, "Tribological, mechanical and electrochemical properties of nanocrystalline copper deposits produced by pulse electrodeposition", *Nanotechnology,* vol. 17, no. 1, pp. 65-78, 2006.
[http://dx.doi.org/10.1088/0957-4484/17/1/012]

[50]   E. Pellicer, E. Gómez, and E. Vallés, "Use of the reverse pulse plating method to improve the properties of cobalt–molybdenum electrodeposits", *Surf. Coat. Tech.,* vol. 201, no. 6, pp. 2351-2357, 2006.
[http://dx.doi.org/10.1016/j.surfcoat.2006.04.011]

[51]   Q. Li, D. Zeng, and M. An, "An, Elevating the photo-generated cathodic protection of corrosion product layers on electrogalvanized steel through nano-electrodeposition", *Chem. Phys. Lett.,* vol. 722, pp. 1-e5, 2019.

[52]   S. Kumar, and M. Kumar, "Tribological and Mechanical Performance of Coatings on Piston to Avoid Failure—A Review", *J. Fail. Anal. Prev.,* vol. 22, no. 4, pp. 1346-1369, 2022.
[http://dx.doi.org/10.1007/s11668-022-01436-3]

[53]   Z. Bai, and B. Zhang, "Fabrication of superhydrophobic reduced-graphene oxide/nickel coating with mechanical durability, self-cleaning and anticorrosion performance", *Nano Mater. Sci.,* vol. 2, no. 2, pp. 151-158, 2020. in press

[54]   S. Teng, Y. Gao, F. Cao, D. Kong, and L. Zhi, "Zinc-reduced graphene oxide for enhanced corrosion protection of zincrich epoxy coatings", *Prog. Org. Coat.,* vol. 123, pp. 185-e189, 2018.

[55]   G. Genchi, A. Carocci, G. Lauria, M.S. Sinicropi, and A. Catalano, "Nickel: Human health and environmental toxicology", *Int. J. Environ. Res. Public Health,* vol. 17, no. 3, p. 679, 2020.
[http://dx.doi.org/10.3390/ijerph17030679] [PMID: 31973020]

[56]   A. Behera, P. Mallick, and S.S. Mohapatra, "Nanocoatings for anticorrosion: An introduction", In: *Corrosion protection at the nanoscale.* Elsevier, 2020, pp. 227-243.
[http://dx.doi.org/10.1016/B978-0-12-819359-4.00013-1]

[57]   D. A. Roberts, *"Magnetron sputtering and corrosion of Ti-Al-C and Cr-Al-C coatings for Zr-alloy nuclear fuel cladding",* Thesis: University of Tennessee - Knoxville, 2016.

[58]   S. Kumar, and R. Kumar, "Influence of processing conditions on the properties of thermal sprayed coating: a review", *Surf. Eng.,* vol. 37, no. 11, pp. 1339-1372, 2021.
[http://dx.doi.org/10.1080/02670844.2021.1967024]

[59]   H. Kaur, J. Sharma, D. Jindal, R.K. Arya, and S.B. Arya, "Crosslinked polymer doped binary coatings for corrosion protection", *Prog. Org. Coat.,* vol. 125, pp. 32-e39, 2018.

[60]   W. Shen, K. Cai, Z. Yang, Y. Yan, and P. Liu, "Improved endothelialization of NiTi alloy by VEGF functionalized nanocoating", *Colloids Surf., B Biointerfaces,* vol. 94, pp. 347-e353, 2012.

[61]   B. Shakeri, E. Heidari, and S.M.A. Boutorabi, "Effect of Isothermal Heat Treatment Time on the Microstructure and Properties of 4.3% Al Austempered Ductile Iron", *Int. J. Met. Cast.,* vol. 17, no. 4,

pp. 3005-3018, 2023.
[http://dx.doi.org/10.1007/s40962-023-00980-4]

[62]   P. Wang, L. Ma, X. Cheng, and X. Li, "Influence of grain refinement on the corrosion behavior of metallic materials: A review", *Int. J. Miner. Metall. Mater.,* vol. 28, no. 7, pp. 1112-1126, 2021.
[http://dx.doi.org/10.1007/s12613-021-2308-0]

[63]   J.O. Carneiro, V. Teixeira, P. Carvalho, S. Azevedo, and N. Manninen, *Self-cleaning smart nanocoatings* Technologies and Applications Woodhead Publishing Series in Metals and Surface Engineering, 2011, pp. 397-413.
[http://dx.doi.org/10.1533/9780857094902.2.397]

[64]   Q. Ren, J. Chen, F. Chu, J. Li, and J. Fang, "Graphene/star polymer nanocoating", *Prog. Org. Coat.,* vol. 103, pp. 15-e22, 2017.

[65]   R. Kumar, R. Singh, and S. Kumar, "Erosion and Hot Corrosion Phenomena in Thermal Power Plant and their Preventive Methods: A Study", *Asian Review of Mechanical Engineering,* vol. 7, no. 1, pp. 38-45, 2018.
[http://dx.doi.org/10.51983/arme-2018.7.1.2436]

[66]   P. Ajit Behera, "Nanocoatings for anticorrosion: An introduction", In: *Micro and Nano Technologies, Corrosion Protection at the Nanoscale* Elsevier, 2020, pp. 227-243.
[http://dx.doi.org/10.1016/B978-0-12-819359-4.00013-1]

[67]   A. Salam Hamdy, "Corrosion Protection Performance via Nano-Coatings Technologies", *Recent Pat. Mater. Sci.,* vol. 3, no. 3, pp. 258-267, 2010.
[http://dx.doi.org/10.2174/1874464811003030258]

[68]   W. Innes, "Electroplating and Electroless Plating on Magnesium and Magnesium Alloys", In: *Modern Electroplating* vol. 601. Wiley-Interscience: New York, 1974, p. 17.

[69]   J. Hajdu, E. Yarkosky, P. Cacciatore, and M. Suplicki, "Electroless nickel processes for memory disks", *Symposium on Magnetic Materials, Processes and Devices,* vol. 90, pp. 685-91, 1990.

[70]   H. Singh, J. Singh, and S. Kumar, "Effect of processing conditions and electrode materials on the surface roughness of EDM-processed hybrid metal matrix composites", *International Journal of Lightweight Materials and Manufacture,* vol. 7, no. 3, pp. 480-493, 2023.
[http://dx.doi.org/10.1016/j.ijlmm.2023.12.001]

[71]   J.L. Luo, and N. Cui, "Effects of microencapsulation on the electrode behavior of Mg2Ni-based hydrogen storage alloy in alkaline solution", *J. Alloys Compd.,* vol. 264, no. 1-2, pp. 299-305, 1998.
[http://dx.doi.org/10.1016/S0925-8388(97)00277-6]

[72]   L.S. Kasten, J.T. Grant, N. Grebasch, N. Voevodin, F.E. Arnold, and M.S. Donley, "An XPS study of cerium dopants in sol–gel coatings for aluminum 2024-T3", *Surf. Coat. Tech.,* vol. 140, no. 1, pp. 11-15, 2001.
[http://dx.doi.org/10.1016/S0257-8972(01)01004-0]

[73]   C.K. Mittal, "Transactions of the Metal Finishers Association of India", *Academic Resource Index,* vol. 4, p. 227, 1995.

[74]   F.A. Lowenheim, *Modern electroplating.* Wiley: New York, 1974, pp. 486-551.

[75]   J.B. Mohler, "Plating cells", *Met. Finish.,* vol. 93, no. 1, pp. 539-543, 1995.
[http://dx.doi.org/10.1016/0026-0576(95)93402-N]

[76]   P.L. Hagans, and C.M. Haas, Chromate conversion coatings*ASM Int* vol. 5. ASM handbook, surface engineering, 1994, p. 405.

[77]   R.H. Unger, Thermal spray coatings*Corrosion* vol. 13. ASM Handbook, 1987, p. 459.

[78]   T. Yoshinori, *"Method for plating magnesium alloy",* JP60024383

[79]   P. Vuoristo, *Thermal Spray Coating Processes.* Comprehensive Materials Processing, 2014, pp. 229-

276.
[http://dx.doi.org/10.1016/B978-0-08-096532-1.00407-6]

[80]   A. Papyrin, *Adv. Mater. Process.,* pp. 43-59, 2001.

[81]   M.R. Rokni, S.R. Nutt, C.A. Widener, V.K. Champagne, and R.H. Hrabe, "Review of Relationship Between Particle Deformation, Coating Microstructure, and Properties in High-Pressure Cold Spray", *Journal of Thermal Spray Technology,* vol. 26, no. 6, pp. 1308-1355, 2017.
[http://dx.doi.org/10.1007/s11666-017-0575-0]

[82]   M. Grujicic, C.L. Zhao, W.S. DeRosset, and D. Helfritch, "Adiabatic shear instability based mechanism for particles/substrate bonding in the cold-gas dynamic-spray process", *Mater. Des.,* vol. 25, no. 8, pp. 681-688, 2004.
[http://dx.doi.org/10.1016/j.matdes.2004.03.008]

[83]   R. Maev, and V. Leshchynsky, "Air Gas Dynamic Spraying of Powder Mixtures: Theory and Application", *J. Therm. Spray. Technol.,* vol. 7, pp. 205-212, 1998.

[84]   S. Kumar, "Influence of processing conditions on the mechanical, tribological and fatigue performance of cold spray coating: a review", *Surf. Eng.,* vol. 38, no. 4, pp. 324-365, 2022.
[http://dx.doi.org/10.1080/02670844.2022.2073424]

[85]   S. Singh, S. Kumar, and V. Khanna, *A review on surface modification techniques* Material Today Proceedings, 2023, pp. 1-10.
[http://dx.doi.org/10.1016/j.matpr.2023.01.010]

[86]   S. Kumar, M. Kumar, and N. Jindal, "Overview of Cold Spray Coatings Applications and Comparisons: A Critical Review", In: *World Journal of Engineering* vol. 17. , 2020, no. 1, pp. 27-51.
[http://dx.doi.org/10.1016/j.matpr.2023.01.010]

[87]   S. Amin, and H. Panchal, *International Journal of Current Trends in Engineering & Research* vol. 2. Scientific J. Impact Factor, 2016, pp. 556-563. [IJCTER].

[88]   V.A.D. Souza, and A. Neville, "Aspects of microstructure on the synergy and overall material loss of thermal spray coatings in erosion–corrosion environments", *Wear,* vol. 263, no. 1-6, pp. 339-346, 2007.
[http://dx.doi.org/10.1016/j.wear.2007.01.071]

[89]   M.P. Groover, *Fundamentals of Modern Manufacturing: Materials, Processes, and Systems.* Wiley, 2002, pp. 1-1008.

[90]   S Amin, and H Panchal, "International", *Journal of Current Trends in Engineering & search,* vol. 2, pp. 556-563, 2016.

[91]   S. Kumar, R. Kumar, S. Singh, H. Singh, and A. Handa, "The Role of Thermal Spray Coating to Combat Hot corrosion of Boiler Tubes: A Study", *Journal of Xidian University,* vol. 14, no. 5, pp. 229-239, 2020.
[http://dx.doi.org/10.37896/jxu14.5/024]

[92]   R. Kumar Sr, R. Kumar Sr, and S. Kumar Sr, "Erosion Corrosion Study of HVOF Sprayed Thermal Sprayed Coatings on Boiler Tubes:A Review", *International Journal of Science and Management Studies (IJSMS),* vol. 1, no. 3, p. ijsms-v1i3p101, 2018.
[http://dx.doi.org/10.51386/25815946/ijsms-v1i3p101]

[93]   R. Yadaw, S.K. Singh, S. Chattopadhyaya, and S. Kumar, *Int. J. Eng. Technol.,* vol. 7, pp. 1656-1663, 2018.

[94]   K.G. Budinski, *Prentice Hall*New Jersey, USA, 1988.

[95]   R. Kumar, and S. Kumar, "Thermal Spray Coating Process: A Study", *Int. J. Eng. Sci. Res. Technol.,* vol. 7, no. 3, pp. 610-617, 2018.
[http://dx.doi.org/10.5281/zenodo.1207005]

[96]   K. Rakesh, and K. Santosh, "Comparative Parabolic Rate Constant and Coating Properties of Nickel, Cobalt, Iron and Metal Oxide Based Coating: A Review", *i-manager's Journal on Material Science,*

vol. 6, no. 1, pp. 45-56, 2018.
[http://dx.doi.org/10.26634/jms.6.1.14379]

[97]   S. Kumar, A. Handa, and R. Kumar, "Overview of Wire Arc Spray Process: A Review", *A Journal of Composition Theory,* vol. 12, no. 7, pp. 900-907, 2019.

[98]   V. Sharma, S. Kumar, M. Kumar, and D. Deepak, "High temperature oxidation performance of Ni-C--Ti and Ni-5Al coatings", *Mater. Today Proc.,* vol. 26, no. 3, pp. 3397-3406, 2020.
[http://dx.doi.org/10.1016/j.matpr.2019.11.048]

[99]   M. Kumar, S. Kant, and S. Kumar, "Corrosion behavior of wire arc sprayed Ni-based coatings in extreme environment", *Mater. Res. Express,* vol. 6, no. 10, p. 106427, 2019.
[http://dx.doi.org/10.1088/2053-1591/ab3bd8]

[100]   S. Kumar, M. Kumar, and A. Handa, "High temperature oxidation and erosion-corrosion behaviour of wire arc sprayed Ni-Cr coating on boiler steel", *Mater. Res. Express,* vol. 6, no. 12, p. 125533, 2020.
[http://dx.doi.org/10.1088/2053-1591/ab5fae]

[101]   S. Kumar, M. Kumar, and A. Handa, "Comparative Study of High Temperature Oxidation Behavior of Wire Arc Sprayed Ni-Cr and Ni-Al Coatings", *Eng. Fail. Anal.,* vol. 106, pp. 104173-104189, 2019.
[http://dx.doi.org/10.1016/j.engfailanal.2019.104173]

[102]   G. Kasi, T. Sabo, T. Goyal, and T.S. Sidhu, *National Conference on Advancements and Futuristic Trends in Mechanical and Materials Engineering,* vol. 3, pp. 77-88, 2010.

[103]   S. Kumar, A. Handa, V. Chawla, N.K. Grover, and R. Kumar, "Performance of thermal-sprayed coatings to combat hot corrosion of coal-fired boiler tube and effect of process parameters and post-coating heat treatment on coating performance: a review", *Surf. Eng.,* vol. 37, no. 7, pp. 833-860, 2021.
[http://dx.doi.org/10.1080/02670844.2021.1924506]

[104]   S. Kumar, M. Kumar, and A. Handa, "Erosion corrosion behaviour and mechanical properties of wire arc sprayed Ni-Cr and Ni-Al coating on boiler steels in a real boiler environment", *Mater. High Temp.,* vol. 37, no. 6, pp. 370-384, 2020.
[http://dx.doi.org/10.1080/09603409.2020.1810922]

[105]   S. Harsimran, K. Santosh, and K. Rakesh, "Overview of Corrosion and its Control: A Critical Review", *Proceedings on Engineering Sciences,* vol. 3, no. 1, pp. 13-24, 2021.
[http://dx.doi.org/10.24874/PES03.01.002]

[106]   M. Sharma, H. Jindal, D. Kumar, S. Kumar, and R. Kumar, "Overview on Corrosion, Classification and Control Measure: A Study", *I-manager's Journal on Future Engineering & Technology,* vol. 17, no. 2, pp. 26-36, 2022. Available from: https://imanagerpublications.com/article/18501/2

[107]   R. Feurer, N. Bui, R. Morancho, M. Larhrafi, and R. Calsou, "Electrochemical behaviour of molybdenum based coatings obtained by organometallic chemical vapour deposition", *Br. Corros. J.,* vol. 24, no. 2, pp. 126-130, 1989.
[http://dx.doi.org/10.1179/000705989798270216]

[108]   J.O. Carlsson, and P.M. Martin, Chemical vapor deposition.*Handbook of Deposition Technologies for films and coatings.* William Andrew Publishing, 2010, pp. 314-363.
[http://dx.doi.org/10.1016/B978-0-8155-2031-3.00007-7]

[109]   H.O. Pierson, CVD/PVD Coatings*Corrosion* vol. 13. ASM Handbook, 1987, p. 456.

[110]   I. Nakatsugawa, *Surface modification technology for magnesium products.* Int Magnesium Assoc, 1996, p. 24.

[111]   D.S. Gnanamuthu, *"Metal surface modification",* US4401726

[112]   P.R. Roberge, *Handbook of Corrosion Engineering.* McGraw-Hill: New York, 2000.

[113]   P.A. Schweitzer, *Atmospheric Degradation and Corrosion Control.* Marcel Dekker: New York, 1999.

[114]   D. Jones, *Principles and Prevention of Corrosion.* 2nd ed. Prentice Hall: Upper Saddle River, NJ,

1996.

[115] C. Corfias, N. Pebere, and C. Lacabanne, "Characterization of a thin protective coating on galvanized steel by electrochemical impedance spectroscopy and a thermostimulated current method", *Corros. Sci.,* vol. 41, no. 8, pp. 1539-1555, 1999.
[http://dx.doi.org/10.1016/S0010-938X(98)00203-0]

[116] C. Corfias, N. Pébère, and C. Lacabanne, "Characterization of protective coatings by electrochemical impedance spectroscopy and a thermostimulated current method: influence of the polymer binder", *Corros. Sci.,* vol. 42, no. 8, pp. 1337-1350, 2000.
[http://dx.doi.org/10.1016/S0010-938X(00)00005-6]

[117] J.O. Iroh, and W. Su, "Corrosion performance of polypyrrole coating applied to low carbon steel by an electrochemical process", *Electrochim. Acta,* vol. 46, no. 1, pp. 15-24, 2000.
[http://dx.doi.org/10.1016/S0013-4686(00)00519-3]

[118] R. Asmatulu, R.O. Claus, J.B. Mecham, and S.G. Corcoran, "Nanotechnology-associated coatings for aircrafts", *Mater. Sci.,* vol. 43, no. 3, pp. 415-422, 2007.
[http://dx.doi.org/10.1007/s11003-007-0047-7]

[119] S. Pathak, G.C. Saha, M.B. Abdul Hadi, and N.K. Jain, "Engineered Nanomaterials for Aviation Industry in COVID-19 Context: A Time-Sensitive Review", *Coatings,* vol. 11, no. 4, p. 382, 2021.
[http://dx.doi.org/10.3390/coatings11040382]

[120] J. Mathew, J. Joy, and S.C. George, "Potential applications of nanotechnology in transportation: A review", *Journal of King Saud University - Science,* vol. 31, no. 4, pp. 586-594, 2019.
[http://dx.doi.org/10.1016/j.jksus.2018.03.015]

[121] Y. Zhao, Z. Shen, X. Zhang, and X. Zhao, "Research progress on application of two-dimensional nanomaterials in corrosion protection", *China Powder Sci. Technol.,* vol. 27, no. 01, pp. 11-21, 2021.
[http://dx.doi.org/10.1016/j.powtec.2021.04.092]

[122] "Research and application of nano-spraying materials", *Inner Mongolia Electric Power.,* vol. 38, no. 04, p. 3, 2020.

[123] Y. Zhou, X. Liu, J. Kang, W. Yue, W. Qin, G. Ma, Z. Fu, L. Zhu, D. She, H. Wang, J. Liang, W. Weng, and C. Wang, "Corrosion behavior of HVOF sprayed WC-10Co4Cr coatings in the simulated seawater drilling fluid under the high pressure", *Eng. Fail. Anal.,* vol. 109, p. 104338, 2020.
[http://dx.doi.org/10.1016/j.engfailanal.2019.104338]

[124] W. Wang, H. Wang, J. Zhao, X. Wang, C. Xiong, L. Song, R. Ding, P. Han, and W. Li, "Self-healing performance and corrosion resistance of graphene oxide–mesoporous silicon layer–nanosphere structure coating under marine alternating hydrostatic pressure", *Chem. Eng. J.,* vol. 361, pp. 792-804, 2019.
[http://dx.doi.org/10.1016/j.cej.2018.12.124]

[125] X. Mu, Y. Hu, T. Jin, C. Liu, and H. Wang, "Corrosion Behavior of New Nano-coating in accelerated corrosion Environment", *Equip. Environ. Eng.,* vol. 17, no. 02, pp. 31-40, 2020.

[126] E. Han, J. Chen, and Y. Su, *Corrosion and protection for marine offshore and coastal structures.* Chemical Industry Press Co., Ltd., 2016, pp. 169-171.

[127] M. Sharon, A.S. Rodriguez, C. Sharon, and P.S. Gallardo, *Nanotechnology in the Defense Industry: Advances, Innovation, and Practical Applications.* John Wiley & Sons, 2019.
[http://dx.doi.org/10.1002/9781119460503]

[128] A. Aydogdu, S. Burmaoglu, O. Saritas, and S. Cakir, "A nanotechnology roadmapping study for the Turkish defense industry", *foresight,* vol. 19, no. 4, pp. 354-375, 2017.

[129] N. Verniquet, "Innovative nanocoatings for electronic devices", *Advanced Coatings & Surface Technology,* vol. 26, no. 9, pp. 3-4, 2013.

[130] S. Kumar, "Comprehensive review on high entropy alloy-based coating", *Surface and Coatings*

Technology, vol. 477, p. 130327, 2023.
[http://dx.doi.org/10.1016/j.surfcoat.2023.130327]

[131] W. Ahmed, M. AlHannan, S. Yusuf, and M.J. Jackson, Nanocoatings for Medical Devices.*Surgical Tools and Medical Devices.,* W. Ahmed, M. Jackson, Eds., Springer: Cham, 2016.
[http://dx.doi.org/10.1007/978-3-319-33489-9_16]

[132] K. Singh, V. Khanna, S. Sonu, S. Singh, S.A. Bansal, V. Chaudhary, and A. Khosla, "Paradigm of state-of-the-art CNT reinforced copper metal matrix composites: processing, characterizations, and applications", *J. Mater. Res. Technol.,* vol. 24, pp. 8572-8605, 2023.
[http://dx.doi.org/10.1016/j.jmrt.2023.05.083]

[133] V. Khanna, V. Kumar, S.A. Bansal, C. Prakash, M. Ubaidullah, S.F. Shaikh, A. Pramanik, A. Basak, and S. Shankar, "Fabrication of efficient aluminium/graphene nanosheets (Al-GNP) composite by powder metallurgy for strength applications", *J. Mater. Res. Technol.,* vol. 22, pp. 3402-3412, 2023.
[http://dx.doi.org/10.1016/j.jmrt.2022.12.161]

[134] M. Dahiya, V. Khanna, and S. Anil Bansal, "Effect of graphene size variation on mechanical properties of aluminium graphene nanocomposites: A modeling analysis", *Mater. Today Proc.,* vol. 73, no. Part 2, pp. 249-254, 2023.
[http://dx.doi.org/10.1016/j.matpr.2022.07.259]

[135] P. Gupta, N. Ahamad, D. Kumar, N. Gupta, V. Chaudhary, S. Gupta, V. Khanna, and V. Chaudhary, "Synergetic Effect of CeO 2 Doping on Structural and Tribological Behavior of Fe-Al 2 O 3 Metal Matrix Nanocomposites", *ECS J. Solid State Sci. Technol.,* vol. 11, no. 11, p. 117001, 2022.
[http://dx.doi.org/10.1149/2162-8777/ac9c92]

[136] K. Singh, S.A. Bansal, V. Khanna, and S. Singh, "Effects of Performance Measures of Non-conventional Joining Processes on Mechanical Properties of Metal Matrix Composites", *Metal Matrix Composites,* no. Aug, pp. 135-165, 2022.
[http://dx.doi.org/10.1201/9781003194897-7]

[137] "Design and Investment of High Voltage NanoDielectrics", In: *IGI Global* Publisher of Timely Knowledge, 2020, p. 363.
[http://dx.doi.org/10.4018/978-1-7998-3829-6]

[138] "Emerging Nanotechnology Applications in Electrical Engineering", In: *IGI Global* Publisher of Timely Knowledge, 2021, p. 318.
[http://dx.doi.org/10.4018/978-1-7998-8536-8]

[139] A.T. Mohamed, and K.E.A. Ahmed, "Controlling on attraction forces of water droplets on surfaces of polypropylene nanocomposites coatings", *Transactions on Electrical and Electronic Materials,* vol. 19, no. 5, pp. 387-395, 2018.
[http://dx.doi.org/10.1007/s42341-018-0054-4]

[140] A. Thabet, M. Allam, and S.A. Shaaban, "Investigation on enhancing breakdown voltages of transformer oil nanofluids using multi☐nanoparticles technique", *IET Gener. Transm. Distrib.,* vol. 12, no. 5, pp. 1171-1176, 2018.
[http://dx.doi.org/10.1049/iet-gtd.2017.1183]

[141] A. Thabet, "Synthesis and Measurement of Optical Light Characterization for Modern Cost-fewer Polyvinyl chloride Nanocomposites Thin Films", In: *Transactions on Electrical and Electronic Materials Journal* vol. 24. Nature Springer, 2023, no. 6, pp. 98-109.

[142] V. Khanna, K. Singh, S. Kumar, S.A. Bansal, M. Channegowda, I. Kong, M. Khalid, and V. Chaudhary, "Engineering electrical and thermal attributes of two-dimensional graphene reinforced copper/aluminium metal matrix composites for smart electronics", *ECS J. Solid State Sci. Technol.,* vol. 11, no. 12, p. 127001, 2022.
[http://dx.doi.org/10.1149/2162-8777/aca933]

[143] R. Mohan, N.V. Saxena, and S. Kumar, "Performance Optimization and Numerical Analysis of Boiler at Husk Fuel Based Thermal Power Plant", *E3S Web of Conferences,* vol. 405, no. 02010, pp. 1-12,

2023.
[http://dx.doi.org/10.1051/e3sconf/202340502010]

[144]   R. Kumar, M. Kumar, J.S. Chohan, and S. Kumar, "Overview on metamaterial: History, types and applications", *Mater. Today Proc.,* vol. 56, no. 5, pp. 3016-3024, 2022.
[http://dx.doi.org/10.1016/j.matpr.2021.11.423]

[145]   R. Kumar, H. Thakur, M. Kumar, G. Luthra, and S. Kumar, "Corrosion and Wear Behaviour of Metal Matrix Composites", *i-manager's Journal on Future Engineering & Technology,* vol. 18, no. 3, pp. 1-16, 2023.
[http://dx.doi.org/10.26634/jfet.18.3.19400]

[146]   D. Abdeen, M. El Hachach, M. Koc, and M. Atieh, "A Review on the Corrosion Behaviour of Nanocoatings on Metallic Substrates", *Materials (Basel),* vol. 12, no. 2, p. 210, 2019.
[http://dx.doi.org/10.3390/ma12020210] [PMID: 30634551]

[147]   A. Thabet, M. Allam, and S.A. Shaaban, "Assessment of Individual and Multiple Nanoparticles on Electrical Insulation of Power Transformers Nanofluids", *Electr. Power Compon. Syst.,* vol. 47, no. 4-5, pp. 420-430, 2019.
[http://dx.doi.org/10.1080/15325008.2019.1609624]

[148]   A. Kumar, J. Moledina, Y. Liu, K. Chen, and P.C. Patnaik, "Nano-Micro-Structured 6%–8% YSZ Thermal Barrier Coatings: A Comprehensive Review of Comparative Performance Analysis", *Coatings,* vol. 11, no. 12, p. 1474, 2021.
[http://dx.doi.org/10.3390/coatings11121474]

[149]   A. Thabet, and N. Salem, "Experimental Progress in Electrical Properties and Dielectric Strength of Polyvinyl Chloride Thin Films Under Thermal Conditions", *Transactions on Electrical and Electronic Materials,* vol. 21, no. 2, pp. 165-174, 2020.
[http://dx.doi.org/10.1007/s42341-019-00163-1]

[150]   I.S. Seddiek, and M.M. Elgohary, "Eco-friendly selection of ship emissions reduction strategies with emphasis on SOx and NOx emissions", *Int. J. Nav. Archit. Ocean Eng.,* vol. 6, no. 3, pp. 737-748, 2014.
[http://dx.doi.org/10.2478/IJNAOE-2013-0209]

CHAPTER 5

Agro-Nanotechnology: A Way Towards Sustainable Agriculture

Aquib Khan[1] and Faria Fatima[2,*]

[1] *Department of Polytechnic, Integral University, Kursi road Lucknow, Uttar Pradesh, India*

[2] *Department of Agriculture, IIAST, Integral University, Kursi road Lucknow, Uttar Pradesh, India*

Abstract: Addressing the global population's dietary needs is crucial amid crop damage issues like insect infestations and adverse weather affecting one-third of conventionally farmed crops. Nanotechnology, recognized for its efficacy and environmental benefits, has gained attention in the past decade. While it has transformed medicine, its applications in agriculture are underexplored. Current research investigates the use of nanomaterials in agriculture for targeted delivery of genes, insecticides, fertilizers, and growth regulators. Nanotechnology shows promise in mitigating abiotic stress in plants by mimicking antioxidative enzymes. This chapter assesses nanoparticles' roles in plant research, highlighting their effectiveness as growth regulators, nanopesticides, nanofertilizers, antimicrobial agents, and targeted transporters. Understanding plant-nanomaterial interactions opens new avenues for enhancing agricultural practices, improving disease resistance, and crop productivity, and optimizing fertilizer use.

Keywords: Agriculture, Nanofertilizer, Nanopesticide, Production, Sustainable.

INTRODUCTION

Agriculture, the cornerstone of a thriving economy, plays a pivotal role in providing sustenance for an ever-growing global population. However, the agricultural sector faces challenges such as climate uncertainties, soil contamination from fertilizers and pesticides, and an escalating demand for food [1]. The imperative to boost food production by over 50% to meet the needs of the burgeoning population underscores the urgency of finding innovative solutions [2]. Industrialization, while contributing to economic growth, poses a significant threat to natural resources vital for sustenance, including forests and seas. Preserving ecological diversity is paramount for maintaining the delicate balance

---

* **Corresponding author Faria Fatima:** Department of Agriculture, IIAST, Integral University, Kursi road Lucknow, Uttar Pradesh, India E-mail: fatimafaria45@gmail.com

Virat Khanna, Suneev Anil Bansal, Vishal Chaudhary and Reddicherla Umapathi (Eds.)

between food production and the environment, enhancing agriculture's resilience to environmental stresses that could lead to crop failure. In response to these challenges, the adoption of nanomaterials (NMs) represents a progressive step in revolutionizing current farming techniques [3]. Nanomaterials, with their remarkable reactivity owing to a large surface area to volume ratio and distinctive physicochemical characteristics, can be easily tailored to meet the increasing demands in agriculture. Nanoparticles (NPs) are already making strides in various consumer products, drawing keen attention from the food and agricultural industries. These NMs exhibit unique surface chemical, electrical, and optical properties, enhancing sensitivity, detection limits, and response times [4, 5].

Fertilizers, vital for increasing crop output, often disrupt soil fertility by upsetting mineral balances and can contribute to environmental pollution. Nanoparticles offer a solution by reducing production costs, minimizing the need for plant protection agents, and mitigating nutrient losses to enhance yields. Innovative applications, such as intelligent agrochemical delivery systems utilizing NMs as carriers for active chemicals, are continually evolving [5]. Moreover, agricultural waste materials, like soy hulls and grain straws, can be transformed into advanced bio-nanocomposites with enhanced mechanical and physical qualities for industrial applications. This chapter emphasizes the superior potential of nanomaterials in enhancing agriculture and food systems [6]. Despite these advancements, challenges remain, requiring improvements in procedure design, risk assessment of nanopesticides and nanofertilizers, and regulatory frameworks for the commercialization of nano-agroproducts, to fully realize the benefits of nanotechnology in agriculture.

APPLICATIONS

Agriculture is the mainstay of the developing economy, providing people with food and a better standard of living. A wide range of problems are currently affecting agriculture, including unforeseen climate change, soil contamination from harmful environmental contaminants like pesticides and fertilizers, and dramatically rising the demand for food due to an expanding world population [7]. On the other side, industrialization has an alarmingly quick detrimental effect on the ecological diversity, woods, and seas which promote the population's way of life. In order to meet the requirements of the population's unrelenting growth, it is urgent to increase agricultural output by more than 50%. It encourages the necessity of increased agricultural output and enhanced food security as a result. Environmental biodiversity strengthens agriculture's resistance to environmental pressures that could result in crop failure, which is essential for maintaining the delicate equilibrium between food production and the environment [8]. New techniques and strategies are constantly emerging to solve these significant issues.

Given this, one action that has been taken is the creation of products based on nanomaterials (NM) that will revolutionize the way that agriculture is now done. They also offer the advantage of being easily adjusted to satisfy growing demand.

Additionally, regulation of the excessive cost of production of these fertilizers and herbicides is also necessary. The use of NMs in agriculture seeks to decrease nutrient losses to boost yields, decrease the quantities of plant protection products, and reduce production costs to optimize output [9].

The creation of products based on nanotechnology is ongoing, including sophisticated nanopackaging products, nanopesticides, agrochemical delivery products, *etc*. Fig. (**1**). It is possible to create novel bio-nanocomposites from agricultural wastes like soy hulls and wheat straw that have improved mechanical and physical properties that can be employed in industrial applications. It is also necessary to test the level of toxicity of NM-based products before releasing them onto the market Table **1**.

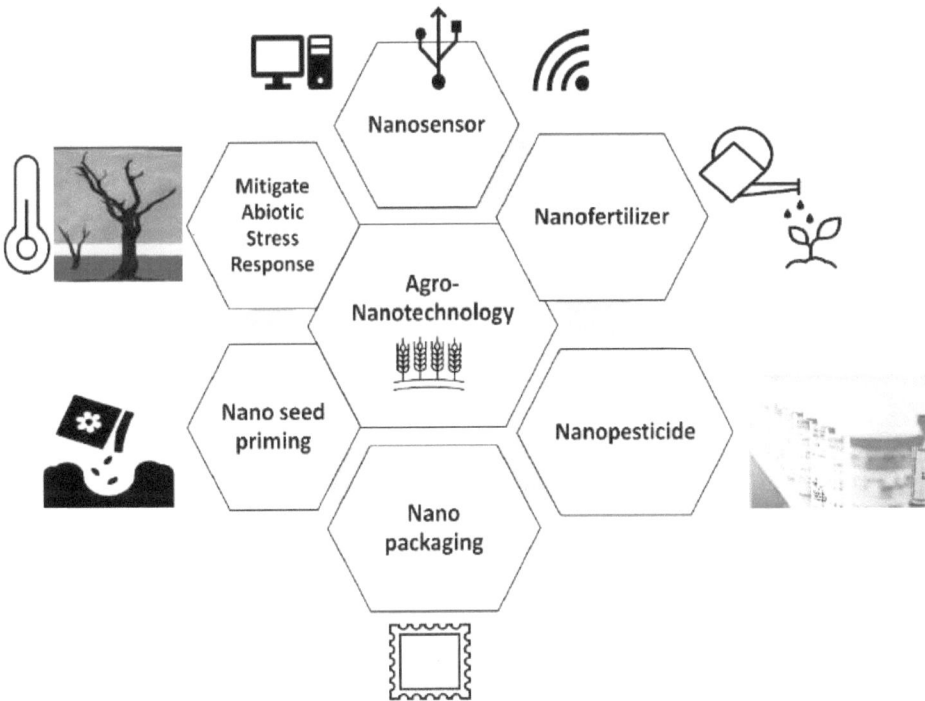

Fig. (1). Application of Nanotechnology in Agriculture.

Table 1. Commercially available nanofertilizers.

S. No	Nanofertilizer	References
1.	Nano Calcium (Magic Green)	[37]
2.	Nano Green and PPC Nano	[39]
3.	Nano Micronutrient (EcoStar)	[40]

NANOFERTILIZER

Nanofertilizers are fertilizers that are formulated with nanoparticles and are used to enhance crop growth and productivity. The tiny size of the particles allows for better and more efficient nutrient delivery to the plants, leading to increased yields and better crop health [10]. They are designed to be environmentally friendly and can improve soil health and fertility over time. Nanoscale metallic fertilizers: These are metallic fertilizers that are made into nanoparticles and are formulated to provide essential micronutrients to the plants.

1. **Nanoscale organic fertilizers**: These are organic fertilizers that are formulated with nanoparticles to improve the efficiency and bioavailability of the nutrients.

2. **Nanoscale nutrient delivery systems**: These are systems that use nanoparticles to deliver essential nutrients to plants in a controlled manner, providing plants with the right amount of nutrients at the right time.

3. **Nanoscale agrochemical delivery systems**: These are systems that use nanoparticles to deliver agrochemicals like pesticides and herbicides to plants in a more targeted and efficient manner.

Overall, nano fertilizers aim to improve the efficiency and effectiveness of nutrient delivery to plants, leading to enhanced crop growth and productivity. The advantages of nanofertilizers include:

1. **Increased crop productivity and yield:** Nanofertilizers can improve the efficiency of nutrients' delivery to plants, leading to enhanced crop growth and higher yields.

2. **Improved soil health**: Nanofertilizers can improve the fertility and health of the soil over time, making it easier for crops to grow and thrive.

3. **Better nutrient uptake**: Nanoparticles in nanofertilizers can penetrate plant tissues more easily and efficiently, leading to better nutrient uptake by the plants.

4. **Increased efficiency and cost savings**: Nanofertilizers can reduce the amount of fertilizer required to achieve the same results, leading to increased efficiency and cost savings for farmers.

5. **Environmentally friendly**: Nanofertilizers are designed to be environmentally friendly, reducing the need for chemical fertilizers and pesticides and helping to improve the overall health of the environment.

Nanotechnology has made it possible to create nanoformulations for sustainable agriculture that are more effective and contain fewer contaminants. Plants and nanoparticle contact can result in both beneficial and detrimental outcomes [11]. The ability of nanoparticles in nanofertilizers to enter plant tissues results efficiently and quickly in increased nutrient absorption by the plants and improved growth and production Fig. (**2**). Additionally, nanoparticles can help shield plants from disease and environmental stress, improving crop health. The interaction of nanoparticles with plants can also result in the production of reactive oxygen species, which can harm plant cells and induce oxidative stress [12]. Furthermore, nanoparticles can interact with the soil and the environment in ways that are yet not completely understood, which might have a negative effect on the ecosystem and the health of the soil.

Role of Nanofertilizers:

- Balanced amount of nutrients
- Increased crop production
- Quality augmentation
- Reduced production costs due to the control of nutrient migration to the environment
- Stress resistant crops
- Increase soil and microorganism activity
- Water retaining capacity

Fig. (2). Effect of nanofertilizers on plants.

To produce and deploy nanofertilizers responsibly and sustainably, it is crucial to thoroughly assess how nanoparticles interact with plants and the environment. Research is still needed to improve our understanding of how nanoparticles interact with plants and to create new, safer, and more efficient nanofertilizers.

For the controlled distribution of active compounds, it is also critical to understand how these NMs interact, favourably or unfavourably, with plants [13]. Thus, they might offer up new avenues for developing superior nanomaterial-based products. Furthermore, it is believed that natural NM concentrations are far lower than those considered to be harmful. There are still certain gaps, nevertheless, that need careful safety studies.

NANO SEED PRIMING

Before sowing, seeds are given a nanoparticle treatment known as "nano seed priming". Nano seed priming aims to boost seed germination and growth, which will increase agricultural production and output. In comparison to all previous seed priming techniques, nano-priming is a far more successful strategy. The key functions of NPs within seed priming that boost the surface response and electron exchange capacities are linked to diverse plant cells and tissues [13]. The nanoparticles employed in nano seed priming might be organic or inorganic, and they may contain important micronutrients or other substances that foster development. The microscopic size of the nanoparticles makes it possible for the nutrients to be delivered to the seeds more effectively, improving germination and development. Numerous crops, including cereals, vegetables, and legumes, have proved to thrive and produce more effectively when treated with nano seed priming [Pereira *et al.*, 2021].

To better understand the long-term effects of nano seed priming on plant development, soil health, and the environment, further study is required because the use of nanotechnology in agriculture is still a relatively new topic. The activation of ROS/antioxidant pathways in seeds, the creation of nanopores in the shoot, the generation of hydroxyl radicals to loosen cell walls, and the fast hydrolysis of starch are all induced by nano-priming. Furthermore, it facilitates the distribution of H_2O_2, or ROS, across biological membranes and promotes the activation of aquaporin genes, which are involved in water intake. By stimulating amylase, nano-priming causes starch breakdown, which in turn stimulates seed germination [15]. A modest ROS is produced by nano-priming, which serves as a major signalling cue for a number of signalling cascade events involved in the generation of secondary metabolites and stress tolerance. In this article, we discussed the potential mechanisms of seed germination, breaking seed dormancy, and how these factors affect the production of primary and secondary metabolites Fig. (3). Given their existing position, it was also investigated how well nano-based fertilizers and insecticides may be used for nano-priming and the development of plant growth.

Efficacy of Nano Priming

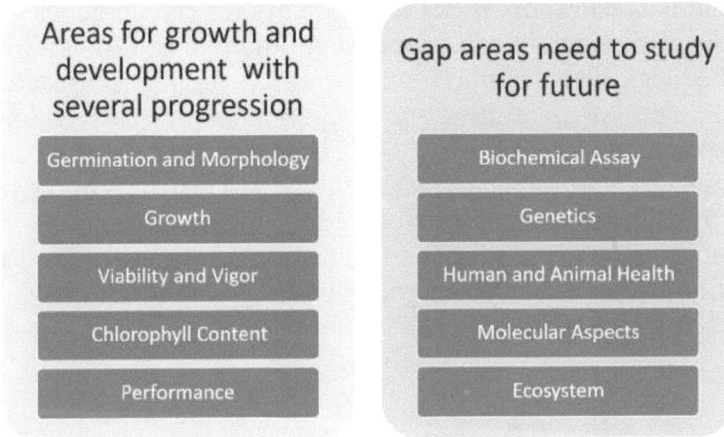

Fig. (3). Application of seed nano-priming.

TO ENHANCE SOIL QUALITY

Nanotechnology can improve the quality of soil by improving nutrient delivery systems to enhance soil fertility, improving soil structure, leading to better water retention, aeration, and root development, removing toxic substances from contaminated soil, creating water-retaining polymers to improve soil moisture levels, increasing drought resistance, creating targeted delivery systems for pesticides, thus reducing their overall use and potential harm to the environment [16].

ANTIMICROBIAL POTENTIAL OF NANOPARTICLES

Nanoparticles are being explored as a potential tool in the management of plant diseases due to their unique physical and chemical properties. The small size of nanoparticles and their ability to penetrate plant tissues make them attractive for use in plant disease management. The use of nanoparticles as bactericides, fungicides, and nano fertilizers in disease control techniques is growing [17]. For their antibacterial action and capacity to change host defence, NPs of Ag, Cu, and Zn have been the subject of most studies so far. Nanoparticles may significantly contribute to the control of infection in both the greenhouse and the field [18]. When compared to traditional metallic fungicides, the use of nanoparticles in disease control results in a significant decrease in the amount of active metals released into the environment. Nanomaterials may act differently in various plant/disease systems, necessitating the potential need for an independent study of each disease system [19]. When tested with harmful bacterial strains *Staphylococcus epidermidis, S. aureus, B. subtilis, Pseudomonas aeruginosa, etc.*

nanoparticles such as [Ag, Cu, Mn, Chitosan, Zn, Cao] frequently exhibit substantial bactericidal activity. Depending on the size of particles, capping approach, and concentration, metallic nanoparticles can be either bactericidal or bacteriostatic. The thin cell wall provides optimum penetration into the bacterial cell, and the negatively charged surface offers high electrostatic interaction between cells and NPs, resulting in the production of reactive oxygen species (ROS) and oxidative stress [20] Fig. (**4**). These characteristics are directly related to the antibacterial activity displayed by the NPs. This leads to the inhibition and destruction of bacterial cells, improved permeability to substances, and increased penetration of NPs by negatively charged superoxide radical anions and peroxide ions, which ensure cell death and cause oxidative stress via ROS and harm to microbial DNA and protein.

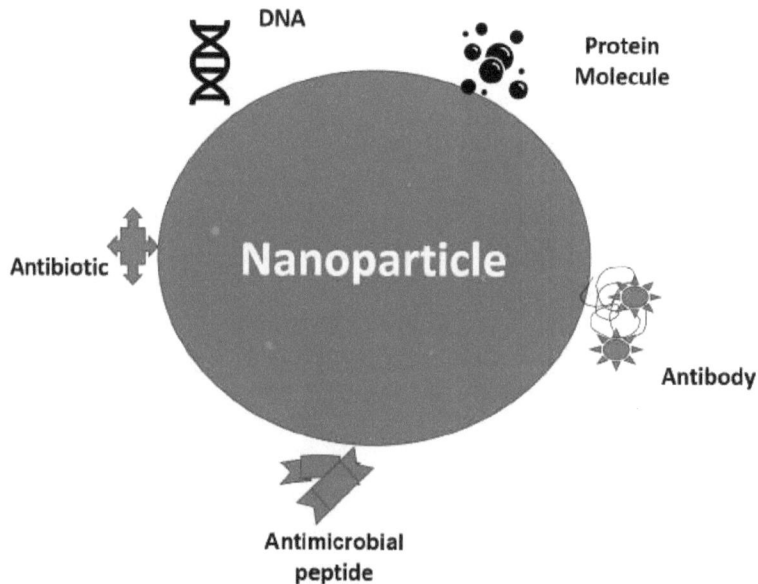

Fig. (4). Application of nanoparticles as antimicrobial agent.

The majority of crop damage—more than 70%—is caused by fungi, which are the most prevalent type of disease. Conventional fungicides can reduce these losses, but they can harm biodiversity since they target a wide variety of living things. As a result, other strategies are required to handle fungal diseases with more accuracy. The production of NMs as a microbial-fighting strategy is one of the greatest methods. The efficiency of AgNPs and Ag ions against the plant-pathogenic fungus *Bipolaris sorokiniana* and *Magnaporthe grisea* was tested in an experiment [21]. When administered for 3 hours, the ionic and nanoparticulate formulations greatly reduced illness severity and were potent against fungus. In an

MWCNT-g-PCA hybrid material, two popular pesticides, zineb and mancozeb, have been shown to be more efficient fungicides against *A. alternaria*. Both CuNPs and AgNPs can inhibit the growth of the fungi *Alternaria alternata* and *Botrytis cinerea*. *Alternaria alternative, Rhizopus stolonifer, Fusarium oxysporum,* and *Mucor plumbeus* have all been demonstrated to be resistant to the antifungal effects of ZnO and MgO nanoparticles. As a result, fungicides may be delivered via nanoparticles directly to affected plant tissues, which can improve treatment effectiveness and lower the quantity of fungicide required. Thus, the potential of nanoparticles in plant disease management is promising, more research is needed to fully understand their efficacy and safety.

NPS MITIGATE ABIOTIC STRESS RESPONSE

Respiration, seedling vigor, and shoot and root growth are significant aspects of the plant's growth and maturation that are all influenced by the application of NPs. Due to their widespread presence, plants must withstand extreme weather conditions such as drought, salt stress, hot temperatures, and UV radiation. Plants react to adverse conditions in a variety of ways, including alterations in molecular pathways, the expression of stress-responsive proteins, and production of enzymatic antioxidants [22]. Plants produce a variety of organic osmolytes, including polyols and trehalose to protect themselves against osmotic stress. It was discovered that plants had defensive mechanisms against abiotic stress. Similar results were discovered for how TiO_2 (anatase) modifies photo-reduction activity and inhibits linolenic acid in the electron transport chain. ROS production is carried out by cell organelles under stressful circumstances [23]. By turning on genes, storing osmolytes, and by transporting amino acids, NMs aid in reducing such stress. NPs offer an effective solution, or, to put it another way, help plants in reducing this defence mechanism. NPs not only promote plant growth but also shield it from abiotic stress. Due to the NP's huge surface area and tiny size, toxic metal binds to it, decreasing its availability. Drought, salinity, alkalinity, temperature swings, and metal and mineral toxicity are examples of abiotic stress. As nano-enzymes that can scavenge from oxidative stress, NPs can mimic the action of antioxidant enzymes.

NANOPESTICIDE

To increase and improve agricultural productivity and efficiency, pesticides are used to shield plants from harmful elements like insects and plant diseases. The use of pesticides is hazardous for the environment and deadly. As a result, numerous pesticides are outlawed by regional, governmental, or worldwide bodies. The widespread use of pesticides at high levels, which disrupt the environment, encourage bioaccumulation, render the soil infertile, and alter its microbiota, is one of the major challenges raised by this [24]. In order to improve agronomic productivity and efficiency and lessen adverse environmental effects, a number of things must be taken into consideration. Nanotechnology offers a new and more effective way to synthesize efficient and secure insecticides, which is both difficult and expensive Fig. (**5**), which is encouraging given the most recent advancements in the food and agriculture industries. Nanotechnology is not only used to defend plants from pests; it is also used to reduce waste, track plant growth, ensure better food quality, and assure growing levels of global food supply.

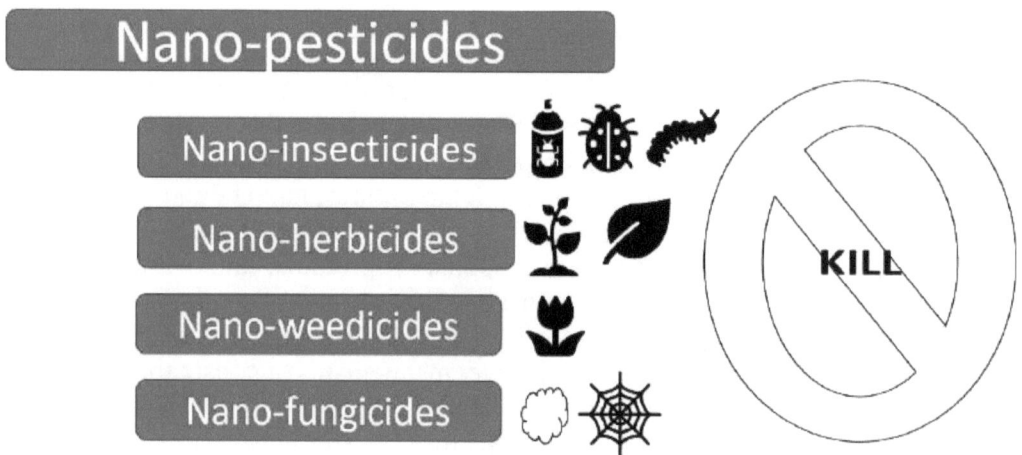

Fig. (5). Nanopesticide.

Several kinds of nano pesticides include nano-fungicides, nano-herbicides, and nano-insecticides. Nanostructured alumina (NSA) is a material designed to use its system to kill insects. Triboelectrification produces the electric charges that insect display. When pesticide molecules are mixed with metal-based NPs, the formulation's potency is increased and the required application dosage for pest management is decreased [25]. These NPs can be made simply from chemical components or in conjunction with biological things. Insecticidal qualities have been demonstrated by NPs made of silver, nanostructured alumina, aluminium oxide, zinc oxide, and titanium oxide Fig. (**6**). When exposed to an ethanol-based nanosilver colloid, *Tinea pellionella* case-bearing cloth moth larvae displayed 100% mortality. When exposed to the wettable dust formulation of nanostructured alumina, *S. oryzae* and *R. dominica* adults infesting wheat showed more than 95% mortality. Nevertheless, the shape, size, surface area of the particle, target species, exposure period, and dose, as well as biotic and abiotic variables, all affect the efficiency of the metal-based NPs (relative humidity). Target-specific nanopesticides are provided by nanotechnology for the creation of innovative eco-friendly pest control formulations. Target-specific nanopesticides often increase the efficacy of pesticides, minimize pollution, and eliminate unwanted pesticide residues in tea. As nanopesticides are manufactured utilizing high polymer materials that are enzyme-, temperature-, humidity-, and light-sensitive, they have a sluggish release and protective performance. These formulations enhance droplet adherence to plant surfaces, resulting in improved dispersion and bioactivity of the target pesticide molecules [26].

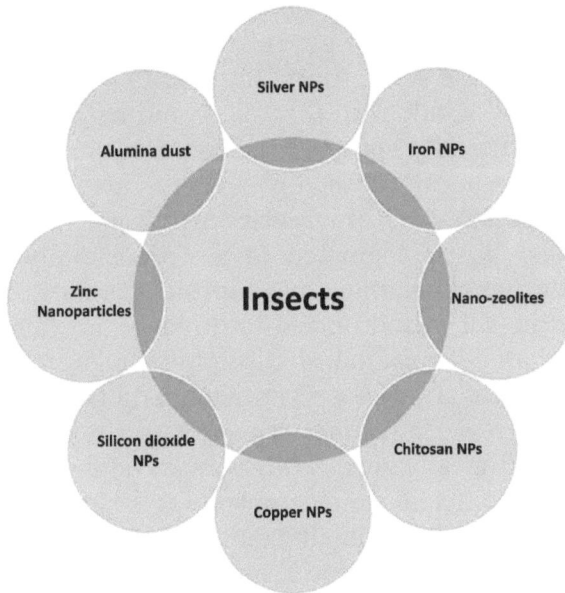

Fig. (6). Types of nanoparticles used to kill insects.

For these reasons, nanopesticides exhibit greater efficacy for the management of tea pests as compared to traditional pesticides. Due to their tiny sizes, effectiveness when sprayed outdoors, enhancement of droplet adherence on plant surfaces, wettability, and quick target absorption, nanopestides provide effective and environmentally friendly benefits. They improve plant protection, decrease hazardous residues, and use fewer chemicals overall. Depending on the physiochemical characteristics or dose of the nanomaterials, nanopesticides can disrupt the photosynthetic pathway in the biochemical or photochemical phases, which in turn affects the production of crops. The mechanical impact of nanoparticles on plants is associated with their size and shape, which limits the distribution of nanomaterials to particular locations on organ surfaces or in plants. The impacts, however, may be felt throughout the vegetative body, which can either impede or promote plant development. Nanopesticides interact with chloroplasts in the cytoplasm of the cell, which affects the photosynthetic processes at certain places in plants by binding to the photosynthetic equipment and impairing its functionality.

NANOHERBICIDES

Since weeds take up nutrients that would otherwise be available to plants, they are thought to pose the greatest risk for causing more extensive crop damage. The traditional approach to getting rid of weeds, which includes pulling them out by hand, takes a long time and a lot of work. There are many different pesticides available today that could perhaps get rid of weeds in a field but could harm different crops and decrease soil fertility. In order to effectively eradicate weeds without leaving harmful residues in the soil, Nano herbicides may be a better, more environmentally friendly option [27]. The weeds may develop resistance if the same herbicide is used repeatedly over a long period of time. The conventional herbicide has been transported using poly (epsilon-caprolactone) (PCL) Nanocapsules. When compared to a commercially produced atrazine formulation, the atrazine-containing nano capsules demonstrated more effective post-emergence herbicidal action on mustard plants. For regulated distribution dependent on glutathione levels, diuron disulphide linkages were used to cross-link chitosan NPs. Formulations based on nanoparticles or nanostructures may have more solubility, greater effectiveness, and lower toxicity than conventional herbicides. Early weed management with nanoparticle-based herbicide release systems could reduce the likelihood of herbicide resistance, maintain the potency of the active component, and prolong the duration of the herbicide's release [28]. A specific receptor located at the weed's root is the objective of the development of a herbicide chemical contained in a nanoparticle. The created nanoparticle penetrates the weed's root system and is transported there to carry out its effect, which in turn prevents the root system of the plant from undergoing glycolysis. The targeted action causes the plant to starve, which kills it. In addition to herbicides, adjuvants that are often employed to increase herbicidal efficacy now purport to include nanomaterials. A crop that was resistant to glyphosate was reportedly rendered sensitive to it when a nanotechnology-based surfactant was added to a soybean micelle [29]. Herbicides can be mixed with nanoparticles to make nanoformulations, which can act as efficient transporters. The rise of plants with herbicide resistance, which is the main issue facing the herbicide industry, is helped by these nano formulations. The majority of the polymeric materials in the nanoparticle systems for herbicide administration are biodegradable and include non-toxic by-products.

NANOTECHNOLOGY IN FOOD PACKAGING

For several years, silver has been utilized as an antibacterial agent to safeguard food and drinks. AgNPs may be incorporated into plastic polymers for packaging using a variety of techniques. For instance, porous zeolite may be used to deposit or trap silver ions, which can subsequently be applied to polymers. According to certain reports, orange juice may be preserved and have its shelf life extended by AgNPs and ZnO-NPs that include low-density PE (LDPE). Although ZnO-NPs are less expensive and harmful to both people and animals than AgNPs, it is especially appealing for packaging applications [30]. In addition, ZnO may generate a significant quantity of hydrogen peroxide when exposed to UV light, which can make bacterial cells more susceptible to oxidative stress. In order to evaluate pathogenic bacteria, the impact of UV radiation on the mechanical and antimicrobial capabilities of PLA/ZnONPs films was studied. Numerous studies show that zinc ions are more effective at preventing bacteria from growing—a process known as bacteriostasis—than at actually eradicating them—a process known as bactericidal Fig. (**7**). Nano-silica is generally utilized in waterproof coatings, especially for self-cleaning materials. Food can become a freely flowing substance within jars or containers by applying a non-adhesive covering. Beer, wine, and powdered soups are a few products that profit from this technique

Advantages

- Increase the physicochemical and microbial quality of food
- Helps in food packaging as barrier agent

Disadvantages

- Increase toxicity of food due to mitigation in low pH

Fig. (7). Application of nanotechnology in food industry.

TiO_2 is frequently used as a coating and pigment ingredient in food packaging. Visible light is efficiently scattered by the white TiO_2 particles, providing the coated item its brightness, brilliance, and power [31]. TiO_2NPs are typically added to PET and co-extruded HDPE bottles for milk product packaging to lessen the detrimental impact of light on milk quality. TiO_2 is absolutely inert, non-flammable, and a totally insoluble food ingredients.

Thus, increased food production with cost-effective returns, innovative products with customizable qualities to deliver smarter and more nutritious options, and clever packaging techniques with increased storage features for better food safety will benefit all nations in the world. With a better understanding of nanoparticles and the realization of their potential in the food sector, the introduction of nanotech meals will bring long-term economic advantages as well as answers to lingering food-related issues. The use of nanomaterials in food will considerably boost sustainability and have favourable benefits on both human health and the environment if it is properly managed. However, the development of innovative nanomaterials for application in the food sector has made it increasingly difficult to evaluate the safety of nanofoods and nano packaging. Therefore, more cooperation is required to make sure that when new goods are launched, social and environmental interests are not compromised. For a safer future, the rate of food technology introduction has to be sufficiently slowed down to allow for the identification and evaluation of possible dangers [32]. This basically means that in order for innovation to be balanced by regulatory guidelines, reliable and robust risk-assessment tools, which do not currently exist for nanofoods, must be made available. If nanofoods are to be implemented successively in our food cycle, the benefits of such foods must be accompanied by increased public transparency of the risks associated with them in order to foster consumer acceptance. This will enable the use of nanomaterials for improved food preservation and packaging.

NANOSENSOR

Agriculture has always been the primary means of providing food, money, and work for people all around the world. Due to the current era's increasing urbanization and erratic environment, precision farming has drawn a lot of interest from all around the world. This method of farming can increase soil quality, produce more crops per acre, and use less agrochemicals overall in an agricultural system (such as fertilizers, herbicides, pesticides, *etc.*) [33]. With careful observation of environmental factors and the use of directed action, precision farming is made feasible. Computers, global satellite positioning systems, sensors, and remote sensing techniques are also used in this kind of agricultural system. As a result, it is simple to monitor severely constrained environmental settings. In the end, it also makes use of smart sensors to deliver precise data that enhances production by assisting farmers in making informed recovery decisions. Smart nanosensors, which are among the most sensitive and carefully used sensors, have already begun to show promise as a crucial instrument for promoting agricultural sustainability in the future [34]. It has been discovered that the use of nanosensors and/or biosensors increases agricultural productivity. These real-time sensors can actually monitor temperature, soil nutrient status, moisture content, soil microbiology, and microclimate. It is noteworthy that these sensors can also detect residual pesticides, track plant illnesses, identify heavy metals, measure fertilizers, and identify toxins. These nanosensors provide instantaneous, precise, historical data that even aid in the forecast and mitigation of crop loss in agroecosystems. The use of nanotechnology-based biosensors is another way to advance the concept of sustainable agriculture. It has been noted that for nanosensors and/or biosensors to be employed as plant diagnostic tools, they need to be enhanced in terms of their sensitivity and specificity. Additionally, it is necessary to use multiplexed screening that is quick, precise, inexpensive, and able to find a range of plant-based bioproducts. Additionally, broad-spectrum nanosensor development that can recognize a variety of objects would boost mobile technology [35]. It has been suggested that developing ultra-efficient "new nanomaterials" that are available in the near future will enable biosensor effectiveness to be significantly improved. The convergence of agriculture sciences, rhizosphere engineering, and general plant engineering may, in the coming years, pave the way for the accomplishment of all Sustainable Development Goals by 2030 without having a negative impact on human health, the economy, or the environment.

ECOTOXICOLOGICAL IMPLICATIONS OF THE NANOPARTICLES

Each year, hundreds of research papers and studies on the use of nanoparticles in agriculture are released. For the identification of heavy metals, herbicides, plant pathogens, and other pollutants in an agroecosystem, however, only a few nanoparticles have yet been marketed. due to the improper conversion/transmission of these academic results to commercial or other regulatory platforms [36]. These nanoparticles are hampered from proof-o--concept to fully marketed devices by several scientific and non-scientific issues. The invasion of nanotech will give long-term economic advantages as well as answers to lingering agriculture-related issues, thanks to enhanced perspectives on nanomaterials and the realization of their potential in the agricultural business [37]. Nations all over the world will profit from improved agricultural output with cost-effective returns, new goods with adjustable qualities to offer smarter and healthier meals, and equally clever applicable solutions with improved properties for better agricultural safety [38]. Thus, if properly controlled, the presence of nanomaterials in agricultural will have a significant positive influence on sustainability, as well as positive effects on human health and the environment. Yet, the development of innovative nanomaterials for application in the agricultural sector has made it increasingly difficult to evaluate the safety of nanopesticides, nanofertilizers, nanofoods and nanopackaging [41 - 43]. However scale-up and real-world application (technical), validation as well as regulatory compliance (regulatory), management priorities and choices (political), standardization (legal), expense, request, and IPR protection (economic), security and safety (environmental health and safety), as well as a number of ethical concerns should be validated [44].

FUTURE PERSPECTIVES

The future perspectives of nanotechnology in agriculture hold immense promise for transforming the way we approach and enhance various aspects of food production. By leveraging the unique properties of nanoparticles, such as their high surface area and reactivity, nanotechnology can contribute to more efficient nutrient delivery systems, precision farming, and pest management. Nanoencapsulation of fertilizers and pesticides allows for controlled and targeted release, optimizing resource utilization and reducing environmental impact. Additionally, nanosensors can be employed for real-time monitoring of soil conditions, crop health, and environmental factors, enabling farmers to make data-driven decisions [45]. The integration of nanotechnology in agriculture not only offers the potential for increased crop yields and quality but also addresses challenges such as water scarcity and sustainable farming practices. As research in this field continues to advance, the future application of nanotechnology is poised to revolutionize agricultural practices, fostering a more resilient and sustainable global food system.

CONCLUSION

Nanotechnology in agriculture refers to the application of nanoscale science, engineering, and technology to improve agricultural production and address challenges in food security. It involves the use of nanomaterials, nanodevices, and nano-systems for various purposes, such as improving plant growth and resistance to pests and diseases, enhancing food quality and safety, increasing nutrient delivery, and conserving water and soil. The various functions that NPs perform in the agricultural sector are highlighted in this chapter. The application of nanotechnology in contemporary agriculture supports and develops the global economy in a variety of ways. Thus, compared to traditional resources, the effectiveness and agronomic efficiency of NPs have greatly increased. Pesticides can be applied more effectively when they are contained in various nano-formulations, and controlled release safeguards the environment. Thus, the benefits of nanotechnology within agricultural areas are numerous, but it is significant to consider the potential risks as well, including environmental and health impacts, and to assess these risks through ongoing research and development. To find and quantify NPs via the plant system, which provides an understanding of their transformation and safety issues in a complex system, new methodologies should be developed. In order to create such nanosensors for agroecosystem, product validation, intellectual protection, and their societal understanding and implementation, it is required to assist motivated researchers and institutions for research and development. Strategic consideration of these variables will aid in the development of nanosensor products and their integration into agroecosystems. For the purpose of monitoring chemical changes in water, soil, and the environment in real-time, the US-based firm Razzberry created portable chemical nanosensors. A similar nanosensor based on metal oxides was developed by the Italian firm Nasys to identify air pollution. Nanosensor technologies are being used by other firms, such as nGageIT and Tracense, to find biological and hazardous chemicals in agricultural settings.

ACKNOWLEDGEMENT

The authors are highly thankful to the Chancellor of Integral University for his support and encouragement.

REFERENCES

[1]   M. Gavrilescu, K. Demnerová, J. Aamand, S. Agathos, and F. Fava, "Emerging pollutants in the environment: present and future challenges in biomonitoring, ecological risks, and bioremediation", *New biotechnology,* vol. 32, pp. 147-156, 2015.

[2]   E. Fouilleux, N. Bricas, and A. Alpha, "'Feeding 9 billion people': global food security debates and the productionist trap", *J. Eur. Public Policy,* vol. 24, no. 11, pp. 1658-1677, 2017. [http://dx.doi.org/10.1080/13501763.2017.1334084]

[3]   M. Usman, M. Farooq, A. Wakeel, A. Nawaz, S.A. Cheema, H. Rehman, I. Ashraf, and M. Sanaullah, "Nanotechnology in agriculture: Current status, challenges and future opportunities", *Sci. Total Environ.,* vol. 721, p. 137778, 2020. [http://dx.doi.org/10.1016/j.scitotenv.2020.137778] [PMID: 32179352]

[4]   A.H. Gondal, and L. Tayyiba, *Prospects of using nanotechnology in agricultural growth, environment, and industrial food products.* vol. 10. Reviews in Agricultural Science, 2022, pp. 68-81.

[5]   Ahmed Thabet Mohamed, "Emerging Nanotechnology Applications in Electrical Engineering", IGI Global, Publisher of Timely Knowledge, Pages 318, June 2021. ISBN13: 9781799885368, ISBN10: 1799885364, EISBN13: 9781799885382, ISBN13 Softcover: 9781799885375. [http://dx.doi.org/10.4018/978-1-7998-8536-8]

[6]   M. Bala, S.K. Bansal, and F. Fatima, "Nanotechnology: A boon for agriculture", *Mater Today: Proc.,* vol. 73, no. 2, pp. 267-270, 2022. In press

[7]   F. Fatima, A. Hashim, and S. Anees, "Efficacy of nanoparticles as nanofertilizer production: a review", *Environ. Sci. Pollut. Res. Int.,* vol. 28, no. 2, pp. 1292-1303, 2021. [http://dx.doi.org/10.1007/s11356-020-11218-9] [PMID: 33070292]

[8]   A. Shahzad, S. Ullah, A.A. Dar, M.F. Sardar, T. Mehmood, M.A. Tufail, and M. Haris, "Nexus on climate change: Agriculture and possible solution to cope future climate change stresses", *Environmental Science and Pollution Research,* vol. 28, pp. 14211-14232, 2021.

[9]   J.W. Erisman, N.V. Eekeren, J.D. Wit, C. Koopmans, W. Cuijpers, N. Oerlemans, and B.J. Koks, "Agriculture, and biodiversity: a better balance benefit both", *AIMS Agriculture and Food,* vol. 1, pp. 157-174, 2016. [http://dx.doi.org/10.3934/agrfood.2016.2.157]

[10] A. Servin, W. Elmer, A. Mukherjee, R. De la Torre-Roche, H. Hamdi, J.C. White, P. Bindraban, and C. Dimkpa, "A review of the use of engineered nanomaterials to suppress plant disease and enhance crop

yield", *J. Nanopart. Res.,* vol. 17, no. 2, p. 92, 2015.
[http://dx.doi.org/10.1007/s11051-015-2907-7]

[11]   B.Z. Butt, and I. Naseer, "Nanofertilizers", *Nanoagronomy,* vol. 2015, pp. 125-152, 2020.
[http://dx.doi.org/10.1007/978-3-030-41275-3_8]

[12]   M.T. El-Saadony, "Vital roles of sustainable nano-fertilizers in improving plant quality and quantity-an updated review", *Saudi journal of biological sciences,* vol. 28, pp. 7349-7359, 2021.

[13]   K.K. Verma, X.P. Song, A. Joshi, D.D. Tian, V.D. Rajput, M. Singh, J. Arora, T. Minkina, and Y.R. Li, "Recent trends in nano-fertilizers for sustainable agriculture under climate change for global food security", *Nanomaterials (Basel),* vol. 12, no. 1, p. 173, 2022.
[http://dx.doi.org/10.3390/nano12010173] [PMID: 35010126]

[14]   D. Wang, N.B. Saleh, A. Byro, R. Zepp, E. Sahle-Demessie, T.P. Luxton, and C. Su, "Nano-enabled pesticides for sustainable agriculture and global food security", *Nature nanotechnology,* vol. 17, pp. 347-360, 2022.

[15]   S.H. Nile, M. Thiruvengadam, Y. Wang, R. Samynathan, M.A. Shariati, M. Rebezov, A. Nile, M. Sun, B. Venkidasamy, J. Xiao, and G. Kai, "Nano-priming as emerging seed priming technology for sustainable agriculture—recent developments and future perspectives", *J. Nanobiotechnology,* vol. 20, no. 1, p. 254, 2022.
[http://dx.doi.org/10.1186/s12951-022-01423-8] [PMID: 35659295]

[16]   M. Tondey, A. Kalia, A. Singh, G.S. Dheri, M.S. Taggar, E. Nepovimova, O. Krejcar, and K. Kuca, "Seed priming and coating by nano-scale zinc oxide particles improved vegetative growth, yield, and quality of fodder maize (Zea mays)", *Agronomy (Basel),* vol. 11, no. 4, p. 729, 2021.
[http://dx.doi.org/10.3390/agronomy11040729]

[17]   R.K. Ibrahim, M. Hayyan, M.A. AlSaadi, A. Hayyan, and S. Ibrahim, *Environmental application of nanotechnology: air, soil, and water.* vol. 23. Environmental Science and Pollution Research, 2016, pp. 13754-13788.

[18]   F. Fatima, P. Bajpai, N. Pathak, S. Singh, S. Priya, and S.R. Verma, "Antimicrobial and immunomodulatory efficacy of extracellularly synthesized silver and gold nanoparticles by a novel phosphate solubilizing fungus Bipolaris tetramera", *BMC Microbiol.,* vol. 15, no. 1, p. 52, 2015.
[http://dx.doi.org/10.1186/s12866-015-0391-y] [PMID: 25881309]

[19]   F. Fatima, S. Siddiqui, and W.A. Khan, "Nanoparticles as Novel Emerging Therapeutic Antibacterial Agents in the Antibiotics Resistant Era", *Biol. Trace Elem. Res.,* vol. 199, no. 7, pp. 2552-2564, 2021.
[http://dx.doi.org/10.1007/s12011-020-02394-3] [PMID: 33030657]

[20]   F. Fatima, N. Pathak, S.R. Verma, and P. Bajpai, "Toxicity and immunomodulatory efficacy of biosynthesized silver myconanosomes on pathogenic microbes and macrophage cells", *Artif. Cells Nanomed. Biotechnol.,* vol. 46, no. 8, pp. 1637-1645, 2018.
[PMID: 29022370]

[21]   F. Fatima, S.R. Verma, N. Pathak, and P. Bajpai, "Extracellular mycosynthesis of silver nanoparticles and their microbicidal activity", *J. Glob. Antimicrob. Resist.,* vol. 7, pp. 88-92, 2016.
[http://dx.doi.org/10.1016/j.jgar.2016.07.013] [PMID: 27689341]

[22]   O.M. Darwesh, and I.E. Elshahawy, "Silver nanoparticles inactivate sclerotial formation in controlling white rot disease in onion and garlic caused by the soil borne fungus Stromatinia cepivora", *Eur. J. Plant Pathol.,* vol. 160, no. 4, pp. 917-934, 2021.
[http://dx.doi.org/10.1007/s10658-021-02296-7]

[23]   Z. Khan, and H. Upadhyaya, *Impact of nanoparticles on abiotic stress responses in plants: an overview.* Nanomaterials in plants, algae, and microorganisms, 2019, pp. 305-322.
[http://dx.doi.org/10.1016/B978-0-12-811488-9.00015-9]

[24]   A. Salam, M. S. Afridi, M. A. Javed, A. Saleem, A. Hafeez, A. R. Khan, and Y. Gan, "Nano-priming against abiotic stress: A way forward towards sustainable agriculture", *Sustainability,* vol. 14, p. 14880,

2022.

[25] S. Yan, N. Gu, M. Peng, Q. Jiang, E. Liu, Z. Li, and M. Dong, "A preparation method of nano-pesticide improves the selective toxicity toward natural enemies", *Nanomaterials,* vol. 12, p. 2419, 2022.

[26] M. Kah, R.S. Kookana, A. Gogos, and T.D. Bucheli, "A critical evaluation of nanopesticides and nanofertilizers against their conventional analogues", *Nature nanotechnology,* vol. 13, pp. 677-684, 2018.

[27] M. Chaud, E.B. Souto, A. Zielinska, P. Severino, F. Batain, J. Oliveira-Junior, and T. Alves, "Nanopesticides in agriculture: Benefits and challenge in agricultural productivity, toxicological risks to human health and environment", *Toxics,* vol. 9, no. 6, p. 131, 2021.
[http://dx.doi.org/10.3390/toxics9060131] [PMID: 34199739]

[28] J. Dong, X. Liu, Y. Chen, W. Yang, and X. Du, "User-safe and efficient chitosan-gated porous carbon nanopesticides and nanoherbicides", *J. Colloid Interface Sci.,* vol. 594, pp. 20-34, 2021.
[http://dx.doi.org/10.1016/j.jcis.2021.03.001] [PMID: 33744730]

[29] J. Sangeetha, R. Hospet, D. Thangadurai, C.O. Adetunji, S. Islam, N. Pujari, and A.R.M.S. Al--, *Nanopesticides, nanoherbicides, and nanofertilizers: the greener aspects of agrochemical synthesis using nanotools and nanoprocesses toward sustainable agriculture.* Handbook of Nanomaterials and Nanocomposites for Energy and Environmental Applications, 2021, pp. 1663-1677.

[30] H. Chhipa, "Nanofertilizers and nanopesticides for agriculture", *Environmental chemistry letters,* vol. 15, pp. 15-22, 2017.

[31] P. Chaudhary, F. Fatima, and A. Kumar, "Relevance of Nanomaterials in Food Packaging and its Advanced Future Prospects", *J. Inorg. Organomet. Polym. Mater.,* vol. 30, no. 12, pp. 5180-5192, 2020.
[http://dx.doi.org/10.1007/s10904-020-01674-8] [PMID: 32837459]

[32] Ş. Sungur, *Titanium dioxide nanoparticles.* Handbook of nanomaterials and nanocomposites for energy and environmental applications, 2020, pp. 1-18.
[http://dx.doi.org/10.1007/978-3-030-11155-7_9-1]

[33] M. Sahoo, S. Vishwakarma, C. Panigrahi, and J. Kumar, "Nanotechnology: Current applications and future scope in food", *Food Front.,* vol. 2, no. 1, pp. 3-22, 2021.
[http://dx.doi.org/10.1002/fft2.58]

[34] M. Javaid, A. Haleem, R.P. Singh, S. Rab, and R. Suman, "Exploring the potential of nanosensors: A brief overview", *Sensors International,* vol. 2, p. 100130, 2021.
[http://dx.doi.org/10.1016/j.sintl.2021.100130]

[35] C.R. Yonzon, D.A. Stuart, X. Zhang, A.D. McFarland, C.L. Haynes, and R.P. Van Duyne, "Towards advanced chemical and biological nanosensors—An overview", *Talanta,* vol. 67, pp. 438-448, 2005.

[36] R. Bogue, "Nanosensors: a review of recent progress", *Sensor Review,* vol. 28, pp. 12-17, 2008.

[37] R.D. Handy, F. Von der Kammer, J.R. Lead, M. Hassellöv, R. Owen, and M. Crane, "The ecotoxicology and chemistry of manufactured nanoparticles", *Ecotoxicology,* vol. 17, pp. 287-314, 2008.

[38] G.D. Avila-Quezada, A.P. Ingle, P. Golińska, and M. Rai, "Strategic applications of nano-fertilizers for sustainable agriculture: Benefits and bottlenecks", *Nanotechnol. Rev.,* vol. 11, no. 1, pp. 2123--, 2022.
[http://dx.doi.org/10.1515/ntrev-2022-0126]

[39] V.I. Slaveykova, M. Li, I.A. Worms, and W. Liu, "When environmental chemistry meets ecotoxicology: Bioavailability of inorganic nanoparticles to phytoplankton", *Chimia (Aarau),* vol. 74, no. 3, pp. 115-121, 2020.
[http://dx.doi.org/10.2533/chimia.2020.115] [PMID: 32197668]

[40] J. Sangeetha, A. Mundaragi, D. Thangadurai, S.S. Maxim, R.M. Pandhari, and J.M. Alabhai, "Nanobiotechnology for agricultural productivity, food security and environmental sustainability", In: *Nanotechnology for agriculture: Crop Production & Protection.*, D. Panpatte, Y. Jhala, Eds., Springer: Singapore, 2019.
[http://dx.doi.org/10.1007/978-981-32-9374-8_1]

[41] K. Singh, V. Khanna, S. Sonu, S. Singh, S.A. Bansal, V. Chaudhary, and A. Khosla, "Paradigm of state-of-the-art CNT reinforced copper metal matrix composites: processing, characterizations, and applications", *J. Mater. Res. Technol.*, vol. 24, pp. 8572-8605, 2023.,
[http://dx.doi.org/10.1016/j.jmrt.2023.05.083]

[42] V. Khanna, V. Kumar, S.A. Bansal, C. Prakash, M. Ubaidullah, S.F. Shaikh, A. Pramanik, A. Basak, and S. Shankar, "Fabrication of efficient aluminium/graphene nanosheets (Al-GNP) composite by powder metallurgy for strength applications", *J. Mater. Res. Technol.*, vol. 22, pp. 3402-3412, 2023.,
[http://dx.doi.org/10.1016/j.jmrt.2022.12.161]

[43] M. Dahiya, V. Khanna, and S. Anil Bansal, "Effect of graphene size variation on mechanical properties of aluminium graphene nanocomposites: A modeling analysis", *Mater. Today Proc.*, vol. 73, pp. 249-254, 2022.,
[http://dx.doi.org/10.1016/j.matpr.2022.07.259]

[44] P. Gupta, N. Ahamad, D. Kumar, N. Gupta, V. Chaudhary, S. Gupta, V. Khanna, and V. Chaudhary, "Synergetic Effect of CeO 2 Doping on Structural and Tribological Behavior of Fe-Al 2 O 3 Metal Matrix Nanocomposites", *ECS J. Solid State Sci. Technol.*, vol. 11, no. 11, p. 117001, 2022.,
[http://dx.doi.org/10.1149/2162-8777/ac9c92]

[45] K. Singh, S.A. Bansal, V. Khanna, and S. Singh, "Effects of Performance Measures of Non-conventional Joining Processes on Mechanical Properties of Metal Matrix Composites", *Metal Matrix Composites*, No. Aug, pp. 135-165, 2022.,
[http://dx.doi.org/10.1201/9781003194897-7]

CHAPTER 6

The Effect of Economic Natural Dyes on the Performance and Efficiency of TiO$_2$ Nano-Structure Solar Cells

Nada M. O. Sid Ahmed[*, 1], **Nodar. O. Khalifa**[2], **Manahil E. Mofdal**[3] and **Nada H. Talib**[4]

[1] *Computer Engineering Department Computer Science and Engineering College, University of Hail, Hail, KSA*

[2] *Department of Physics, Sudan University of Science and Technology, Khartoum, Sudan*

[3] *Qassim University, Faculty of Science, Department of Physics, Buraydah, KSA*

[4] *Solar Energy Department, National Energy Research Center, Khartoum, Sudan*

Abstract: The aim of this research can be divided into two stages. The first stage is to synthesize and find a simple and less expensive method to produce titanium dioxide nanostructures with optimum properties that can be used in the construction of low-cost, nanoparticle-based solar cells as a replacement for custom silicon solar cells. The second stage is to determine the effect of natural dyes on the performance and efficiency of TiO$_2$ nano-structure dye synthesized solar cells (TiO$_2$ DSSC) *via* spin coating. In order to improve and enhance the performance and efficiency of dye solar cells, thin film TiO$_2$ nanostructure was synthesized using the sol-gel process, which is simple and inexpensive. Afterward, different natural dies were introduced in the fabrication process over the TiO$_2$ layer also *via* spin coating. The function of the dye is to confine a sufficient amount of light, for optimum performance and power conversion efficiency. In the last fabrication step, graphite contacts were evaporated on the top dye layer. The I-V characteristics of the different dyes were studied and the structural properties of the TiO$_2$ nanostructures were investigated through an X-Ray Diffraction (XRD) pattern. The TiO$_2$ nanoparticles' morphology and particle size were determined by a scanning electron microscope (SEM), while the optical band gap energy was found by employing UV-VIS-NIR diffuse absorption spectroscopy. Three types of natural dye were used which were Roselle, curcumin, and black tea and their conversion efficiencies were 8.46, 6.94, and 6.33 respectively, which is considered acceptable compared to the results obtained by other researchers.

* **Corresponding author Nada M. O. Sid Ahmed:** Computer Engineering Department Computer Science and Engineering College Hail University, Hail, KSA;
E-mail: n.sidahmed@liveuohedu.onmicrosoft.com

Virat Khanna, Suneev Anil Bansal, Vishal Chaudhary and Reddicherla Umapathi (Eds.)

Keywords: Band gap, Curcumin, Diffraction, Diffuse reflectance spectroscopy, Dye-Sensitized Solar Cells, Electrical properties, Fill factor, FTO, Hibiscus, Morphology, Nanostructures, Natural dyes.

INTRODUCTION

Photovoltaic (PV) is the process of converting direct sunlight into electrical energy. Solar cells have been proven to be very efficient PV devices in the process of converting solar energy into electric energy without emitting any toxic waste to the environment or producing noise. It is a clean energy source and can produce electrical energy from direct sunlight *via* the photovoltaic effect, which is the phenomenon of producing electrical energy from visible light. Solar cells have come a long way from the first-generation silicon-wafer solar cells to all types of modern high-efficiency solar cells. The implementation of dyes is one of many methods employed to enhance PV conversion efficiency (PCE) [1].

Recently, silicon solar cells were found to be the most effective and efficient solar energy source, with a PCE of about 25%. Although they are produced from a very cheap raw material, which is sand, the drawback is the high production cost and complicated fabrication process due to the requirement of innovative clean room technology, which makes it a relatively expensive energy source. The nanotechnology revolution led to great developments in solar cell technology. Dye-Sensitized Solar Cells (DSSCs) technology is considered one of the effective trends in solar cell technology. DSSCs are produced by a low-cost and simple process [2]. Due to their relatively low fabrication costs, metal oxide nanoparticles like TiO_2 have been found to be excellent replacements for silicon. TiO_2 nanoparticles are highly important in solar cell fabrication due to their n-type semiconductor properties, such as high energy conversion efficiency, physical and chemical stability, non-toxic and easy synthesis [3, 4]. Recently, TiO_2 has proved to be very efficient in photovoltaic cells, and in dye-sensitized solar cells (DSSCs). However, the rapid rate of its electron-hole recombination results in quite low efficiency when these solar cells are fabricated without dyes [5 - 8].

TiO_2 nanoparticles can enhance light scattering, and they also can increase electron transport. Their charge carrier recombination is quite low, but their surface area is quite large, which boosts the adsorption of large amounts of dye molecules, and these characteristics result in adequate electron injection into the conduction band of the TiO_2 layer [9, 10].

DSSCs in general and TiO_2-based DSSCs are considered to be essential solar energy sources for next-generation, due to their low production cost, low environmental impact, and relatively high power conversion efficiency [11].

DYE-SENSITIZED SOLAR CELLS (DSSCS)

DSSCs are considered third-generation solar cells. The determination of the efficiency of DSSCs is complicated, because of the effect of different parameters and layers included. Two main key layers that need more investigation and enhancement are the semiconductor photo-nodes layer and the dye layer. Both of these layers are effective in absorbing ultraviolet and visible light sensitizers [12].

In the DSSCs type of solar cells, the light absorption and separation of the electrical charges happen in the different processes involving four basic steps: light absorption, electron injection, transportation of carrier, and collection of current. Dye molecules perform the light absorption process, and the Nanocrystal inorganic semiconductor that has a wide band gap [13, 14] enables the separation of the electrical charge. An electric charge can be produced by the electron transfer procedure; when the DSSC is exposed to visible sunlight, the highest occupied molecular orbital (HOMO) electrons stimulate the lowest unoccupied molecular orbital (LUMO), and as a consequence, the electrons are infused into the conduction band. Then the electrons advance to the external circuit *via* the anode and cathode [15, 16]. DSSCs are inexpensive, thin-film solar cells. Their efficiency is relatively high in comparison to other types of solar cells, but the traditional silicon-based solar cells are an exception. In the construction of DSSCs, two layers of conductive transparent materials are used to enhance the substrate to be current collectors, and they can convert photons from sunlight into electrical energy [11, 17].

The advantages of DSSCs over regular silicon PV are the relatively uncomplicated production process, low production costs, and the ability to sensitize an extensive array of band gaps [5, 6, 9]. In order to achieve optimum conversion efficiencies, it is important for the DSSCs to collect the photo-generated charge carriers before recombination, and hence convert them into electric current. Furthermore, DSSCs can operate at very low luminous powers and subsequently produce high PCEs [18].

Researchers have been experimenting with natural dyes because they are environment-friendly, nontoxic, and much cheaper to fabricate, but they found their efficiency quite low. W. Ghana *et al* [6], managed to extract natural dyes from pomegranate and berry fruits pigments and they employed them in the fabrication process of the natural dye-sensitized solar cells (NDSSC), and the maximum PCE they achieved was 2%. Cari *et al* [19] prepared a green dye in a broccoli mixture, but the maximum PCE they managed to obtain was 0.072%. Ahmed M. Ammar *et al.* [20], obtained PCEs of 0.17%, 0.0647%, and 0.060% successively from natural dyes made out of spinach, onions, and red cabbage extracts. Sujan Kumar Das *et al.* [21] managed to achieve a maximum PCE of 9.23%, which is quite high compared to the results obtained by other researchers. Huizhi Zhou *et al.* managed to introduce twenty natural pigment extracts from natural sources in their local surrounding. Furthermore, Mangosteen pericarp extract has achieved the highest conversion efficiency of 1.17% and anthocyanin [22].

Experimentally, the dye-sensitized solar cell, which is a part of an electric circuit, is illuminated *via* a fluorescent lamp. Then the open-circuit voltage V_{oc} and short circuit current ($\mathbf{I_{sc}}$) were measured *via* an Avo-meter. The overall efficiency of solar energy was found from the ratio of the maximum output power received at the output to the sunlight power. The fill factor and overall conversion efficiency are the essential performance measurement parameters for the dye-sensitized solar cells. The fill factor (FF) is very important in the calculation of the cell. The maximum output power (P_{max}) can be obtained at a potential at a point between an open circuit and a short circuit. The highest output power is calculated from the product of the maximum output voltage (V_{max}) and the maximum output current (I_{max}).

The conversion efficiency (η), is also a parameter implemented to determine the overall performance of the cell. It is found from the ratio of maximum power obtained by the cell (P_{max}) to the power of the incident radiation on the illustrative area of the cell (Pin) [24].

The equations below demonstrate the maximum power, fill factor, efficiency, and energy of the dye's synthesis of TiO_2 nano-structure solar cells.

$$P_{max} = I_{max}V_{max} \tag{1}$$

$$FF = \frac{V_{max}I_{max}}{V_{oc}I_{sc}} \tag{2}$$

$$\eta = \frac{FFV_{oc}I_{sc}}{P_{in}} \tag{3}$$

$$E(eV) = \frac{hC}{\lambda} = \frac{hC}{\lambda} \tag{4}$$

Where: $\mathbf{V_{max}}$= maximum value of voltage, I_{max} = maximum value of current, V_{oc}= open circuit voltage, $\mathbf{I_{sc}}$= short circuit current, FF = fill factor, η = the power conversion efficiency of the solar cell, P_{in}= incident power, λ = wavelength of light, and C = Speed of light.

Working Principle of Dye Sensitized Cell (DSSC)

The adsorbed dye molecules on the TiO_2 film absorb photons that are stimulated from the highest occupied molecular orbitals (HOMO) to the lowest unoccupied molecular orbitals (LUMO) state upon illumination. The dye species that have been photo-excited injects an electron into the TiO_2 electrode's conduction band and gets oxidized. After accepting an electron from the electrolyte, the oxidized dye species returns to its ground state. The electron, which is introduced into the mesoporous TiO_2 film, percolates through it to the FTO layer and then passes through an external circuit to a load where the work is converted into electrical energy. The electrolyte system is refilled when the electron from the external load diffuses to the cathode and transfers to the electrolyte [23].

Titanium dioxide nanoparticles were prepared using different methods depending on the application. In this research, the so–gel and hydrothermal methods were used to prepare a Titanium dioxide thin film to fabricate TiO_2-Nano dye solar cells. In this chapter, three different types of economical natural dyes (Roselle, Curcumin, and black tea) have been used.

NATURAL DYES

There are three different natural dyes used in the fabrication of TiO_2 nano-structure solar cells in this research. The preparation procedures of the natural dyes as dye sensitizers are mentioned in section 7.4.

Hibiscus Sabdariffa (Roselle)

It is well known as the Roselle flower (Fig. **1**), which is implemented as a dye sensitizer for (DSSCs), belongs to the Malvaceae family, and is a wild tropical plant. Roselle is predicted to be a possible nominee for dye-sensitized solar cells since it is rich in anthocyanins, and the brilliant red colorant for many foods can be produced from it. The delphinidin and cyanidin complexes were obtained from Rosella [24].

Hibiscus sabdariffa (Roselle) is a red-stemmed, serrated, annual, or perennial shrub with crimson calyces. It has a variety of names, such as Florida cranberry, Jamaica sorrel, *etc.*, depending on where it grows. Its history of medicinal and dietary uses extends from South America to Egypt, Sudan, Trinidad and Tobago, Mexico, China, Thailand, Malaysia, and Indonesia. The calyces, or flower pots, are used to make tea, wine, cocktails, sauces, jams, marmalade, candies, pickles, curry condiments, *etc.* Its color is bright red, and it has an acidic taste. Its scent is distinct. All over the world, the drinks are consumed during customary rituals. In Nigeria, dried calyces are extracted with water to create the non-alcoholic beverage "soborodo." Constipation, heart disease, high blood pressure, urinary tract infections, cancer, diabetes, and nerve diseases are all treated with it as folk medicine. Numerous studies have been performed recently to confirm its scientific validity. Antioxidant, hypocholesterolemia, anti-obesity, hypotensive, antidiabetic, immunomodulatory, anticancer, hepatoprotective, antimicrobial, Renoprotective, diuretic, and anti-urolithiasis characteristics have received enough experimental support thus far. With the help of these discoveries, Western health stores have begun to sell it, boosting its global popularity [25].

Fig. (1). The roselle flower (Hibiscus Sabdariffa).

Turmeric (Curcumin)

Commonly known as 'haldi' in India (Fig. **2**), it is a spice derived from the roots of the curcumin longa plant. Curcumin is classified as a phytopolylphenol pigment, which is separated from the plant Curcuma longa, known as turmeric, and it has various pharmacologic properties. There are various names for turmeric curcumin in different languages, and it is native to tropical South Asia [26]. Curcumin is a yellow pigment. The shoot of this plant in powder form is called turmeric. It is applied in curries, different food-seasoning, and food coloring. Curcumin also has medical applications such as treatments of inflammations, skin wounds, and cough. Furthermore, its anticancer properties have generated great interest and results in many publications [12]. Extraction and optical characterization of Curcumin dye sensitizer are shown in Fig. (**2**).

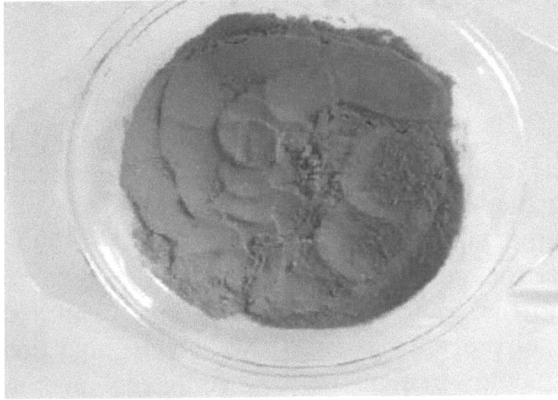

Fig. (2). The Turmeric- Curcumin powder.

Black Tea Leaves (red tea)

Black tea can be classified based on grade one to four scales of quality (Fig. **3**). Whole-leaf teas are the highest quality. Even though the consumption of tea was primarily for its central nerve-stimulating and calming effects, it has been associated with health-improving effects for several years. It has been found that the health benefits associated with tea drinking include: anti-oxidant, anti-inflammation, cancer prevention, decreased occurrence of health diseases, and so on. Polyphenols in black tea consist of amino acids, alkaloids, theanine, catechins, isomers of the avins, caffeine, and theanine. Tea is composed of polyphenols, polyphenol conjugates, and polymerized phenolic structures. In addition, black, white, and green tea contain an assorted combination of conjugated flavonoids [27]. Tea is not commonly used as a dye, but in this research, it has been implemented as a dye sensitizer.

Fig. (3). The black tea leaves.

EXPERIMENTAL TECHNIQUES

Synthesis

Sol-Gel Method

Sol-gel is a synthetic technique that produces homogenous component distribution, due to good mixing, and homogeneity of materials at the nanoscale. The process is carried out at temperatures ranging from 1,000 to 13500 °C, as opposed to 1,500 to 1,7000 °C for more conventional techniques such as reactive powder mixes [28]. Due to the controlled shape and size exhibited by the produced products, the Sol-gel process has been extensively investigated for producing metal oxide nanostructures in many fields including engineering, material science, and technological applications. Since Ebelman first created silica gel in 1846, this method has been gradually improved, and the synthesized materials have been used in numerous applications because of their superior optical, magnetic, electrical, thermal, and mechanical properties. The sol-gel method can be used to produce a variety of materials at a reasonable price, including thin films, nanoparticles, and glass. The outstanding conditions of this method are principally low-temperature chemistry, reproducibility, and the high surface-to-volume ratio of the synthesized materials. Moreover, the sol-gel process has opened some new avenues in bioengineering fields, including drug delivery, organ implantation, pharmaceuticals, and biomaterial synthesis due to the purity and quality of the yields from this process. These advantages have attracted many researchers and industrialists groups to utilize this method more widely in the past few decades [29]. Metal oxides are a class of functional materials that can be created using a sol-gel and have many uses. Compared to solid-state reactions, the sol-gel synthesis of metal oxide can be carried out at relatively low temperatures. In general, the sol-gel process entails the creation of sol from a uniformly mixed solution, polycondensation of the sol to create a gel and final heat treatment of the product to the desired material. The final heat treatment step determines the formation of crystalline materials like nanoparticles, thin films, and non-crystalline materials (ceramics, aerosols, and glasses) [29].

Hydrothermal Method

The hydrothermal method was described by Morey and Niggli in 1913 as a method in which the components are subjected to the action of water at high temperatures, usually above the critical temperature of water (370°C) in closed bombs, and under the corresponding high pressures developed by solutions. In addition, according to Yoshimura (1974), hydrothermal reactions occur under conditions of low temperatures and high pressures (>100°C, > 1 atm) in aqueous solutions in a closed system. Although most scientists think that hydrothermal synthesis is carried out above 100°C and 1 atm pressure, there is no definite lower limit for temperature and pressure conditions [30]. The hydrothermal synthesis method is used in many fields in materials science and solid-state chemistry, mainly in crystal growth and the synthesis of new materials with perfect properties. Also, it is attractive to scientific groups because using high pressure provides an additional parameter for obtaining fundamental information on the structures, behavior, and properties of solids. Hydrothermal synthesis is a chemical reaction in the presence of aqueous solvents above 100°C and at pressures greater than 1 atm in a closed system in order to dissolve and recrystallize materials that are relatively insoluble under normal conditions. [30].

The hydrothermal technique is a promising alternative synthetic method, due to low process temperature and very easy control of the particle size. There are many advantages to the hydrothermal process such as the use of simple equipment, catalyst-free growth, low cost, large area uniform production, environmental friendliness, and less hazardous. This method is attractive for microelectronics and plastic electronics due to low reaction temperature, and has also been successfully employed to prepare TiO_2 Nanoscale and other luminescent materials. The adjustment of the reaction temperature, time, and concentration of precursors, controls the morphology and particle size obtained *via* the hydrothermal process [31 - 35].

Preparation of the Natural Dyes

1- The flowers of Hibiscus sabdariffa were dried *via* the air-drying method. Then the dried flowers were soaked in water (Aq.) to extract the dye. Distilled water was used as a solvent for aqueous extraction. 5g of the sample (Hibiscus Sabdariffa) flower was dipped in 50 ml of water for 30 minutes, after which solid residues were filtered out to obtain clear dye solutions. Then 2.5g ground dried turmeric root was dissolved in 50 ml hot distilled water for 30 minutes until converted to a clear dye solution.

3-Black tea extract solution was used as a dye, by boiling 5g of local black tea powder in 50 ml of distilled water at 70°C for about 30 minutes until the color changed from watery to light in the aqueous solution. Later the mixture was maintained and cooled to room temperature. The extract was then left for 20 minutes and filtered. 4–50 ml of $TiCl_4$ solution was slowly added to 200 ml of distilled water in an ice bath. After the addition was completed, the mixture was stirred for 30 minutes at room temperature. The solution was heated in a water bath for 90 minutes under reflux. Then, it was filtered using a vacuum pump and calcined at 600°C in the muffle furnace for 2 hours. The dried solids were crushed to a fine powder in an agate mortar and calcined at 4000 °C for 6 hours in a muffle furnace. The paste of TiO_2 was prepared from 2g of TiO_2 powder, dissolved in 10 ml of acetone. Then the (FTO) substrate was cleaned with distilled water and acetone.

Preparation of the DSSC

Small quantities of the previously prepared TiO_2 paste were uniformly spread on three different FTO substrates *via* a glass rod and smoothed entirely. Next, the coated substrates were baked for one hour in an oven at 100 °C. Afterward, the three different dyes were prepared by dissolving each dye in acetone and filtering it. The cathodes were constructed by injecting the substrates into the dissolved dyes for 15 minutes. Three new FTO substrates were cleaned and coated with graphite on the conductive surface. These substrates were placed on top of the FTO substrates coated with TiO_2 and dyes, and two binder clips were used in order to attach each two substrates. Then an iodine electrolyte solution was inserted between each of the two substrate slides. Each of the fabricated solar cells was inserted into an electrical circuit containing a voltmeter and ammeter, a light source, *i.e.* a neon lamp. The I-V characteristics of the solar cells were measured and recorded due to the illumination of a neon lamp 0.55 (mW/cm^2).

RESULT AND DISCUSSION

The TiO_2 nanoparticles were characterized using XRD, SEM, UV, and FTIR techniques. By implementing structural characterization methods, the uniformly distributed single-phase nanoparticles were discovered to be impurity-free. The optical property exhibits good light sensitivity.

Characterization of Synthesized TiO_2 Nanoparticles

XRD, SEM with FTIR and absorption measurements were carried out. The crystal structure of tetragonal (anatase) crystalline TiO_2 nanoparticles is given in Fig. (**4**). The Lattice parameters were found to be: a=b= 3.73 A° with α=β=γ= 90° and the coordination numbers of Ti and O were (0.000, 0.000, 0.200 and 0.000, 0.000, 0.000), respectively.

X-Ray Diffraction (XRD)

X-ray diffraction with Cu K_α radiation is a useful technique for analyzing the crystal structure and crystal size of samples. The crystal structure of the prepared samples was analyzed. All of the characteristic peaks observed for the TiO_2NPs were in good agreement with those obtained from the joint committee on powder diffraction standards (JCPDS-211272) card, as shown in Fig. (**5**) and the XRD pattern, which depicts the relationship between intensity and 2 thetas. TiO_2

typically crystallizes in three different shapes: tetragonal (anatase), tetragonal (rutile), and orthorhombic. It is observed that the tetragonal (anatase) crystalline TiO_2 nanoparticles show characteristic peaks at (101), (004), (200), (211), (204), (220), and (215), respectively. TiO_2 nanoparticles' average particle size is estimated to be (t) at the peak (101) plane of greatest intensity. The Scherrer formula is used to calculate the estimated average particle size (t) of TiO_2 nanoparticles at the most intense peak (101) plane, where X-ray wavelength was 1.54 A°, with full-width half maxima, and the Bragg's diffraction angle. The estimated average particle size (t) of TiO_2 nanoparticles was found to be 7.4 nm.

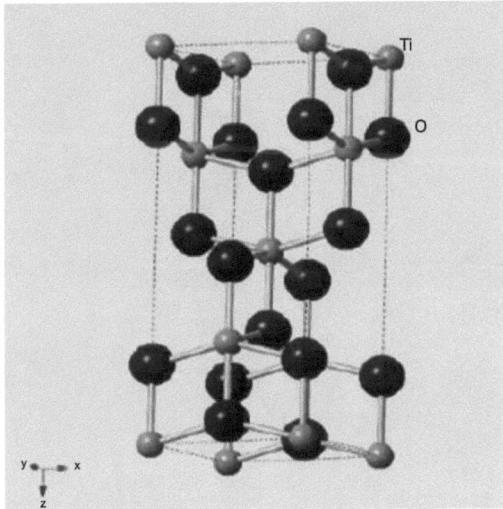

Fig. (4). The crystal structure of tetragonal crystalline TiO_2 nanoparticles.

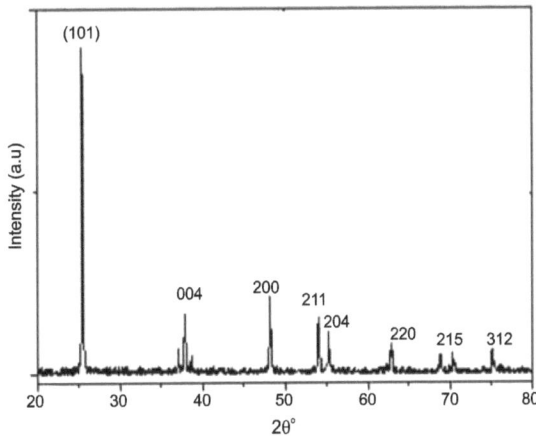

Fig. (5). The XRD pattern of TiO_2 Nano-crystalline powder.

Morphology – Scanning Electron Microscope (SEM)

The morphology obtained using SEM showed that the nanoparticles were aggregated and irregularly shaped as illustrated in Fig. (**6**). The particlesize was estimated using SEM micrographs, which was in the range of 7~50 nm (Fig. **7**).

Fig. (6). SEM image TiO$_2$ Nanoparticles.

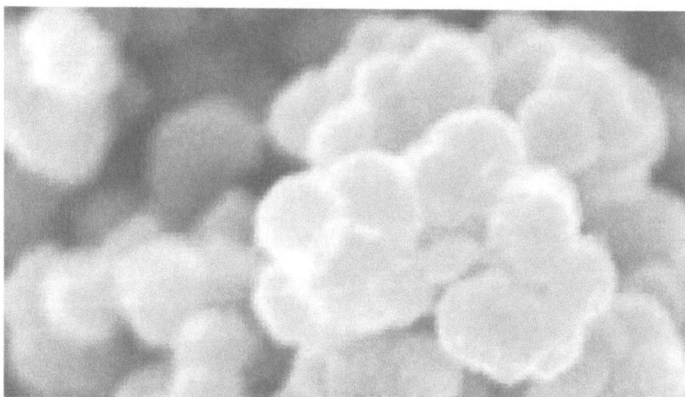

Fig. (7). shows the SEM micrographs of TiO$_2$ Nanoparticles.

Fourier Transform Infrared (FTIR)

The FTIR spectra of the TiO$_2$ nanoparticles are displayed in Fig. (**8**) below. In the TiO$_2$ nanoparticles' FTIR spectra, Ti-O bending mode, and deformities, the vibration of Ti-OH stretching mode can be seen at 525 cm^{-1} and 1530 cm^{-1}, respectively. Hydroxyl group (-OH) stretching vibrations can be seen at 3495 cm^{-1} both symmetrically and asymmetrically. Water adsorbed on the surface of TiO$_2$ may be the cause of the band at 1530 cm^{-1}. These results are consistent with earlier publications.

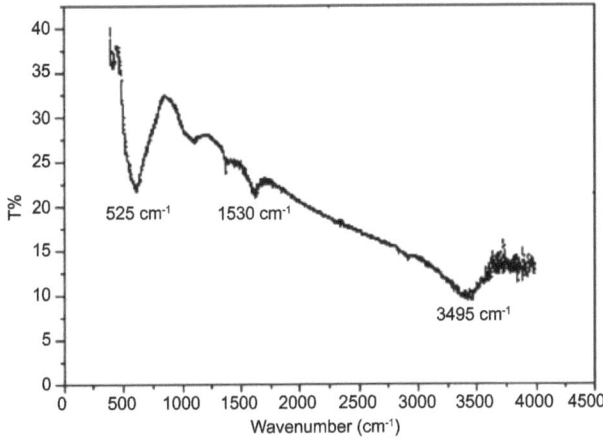

Fig. (8). The FTIR spectrum of TiO$_2$ nanoparticles.

Optical Properties

Absorption Spectra of TiO$_2$ Nanoparticles

Using UV-VIS-NIR diffuse absorption spectroscopy, the absorption spectra of the TiO$_2$ nanostructures synthesized by the sol-gel method were recorded in the wavelength scan region (100–500) nm. The optical band gap of TiO$_2$ nanoparticles was calculated from equation 7.4 and was found to be 3.86 and 3.32, respectively. The absorption spectra of TiO$_2$ nanoparticles exhibit two distinct peaks, the development of a well-defined shoulder located at 321 nm, and the other at 374 nm, as shown in Fig. (**9**). These results were found to be in good agreement with the theoretical values presented in [23].

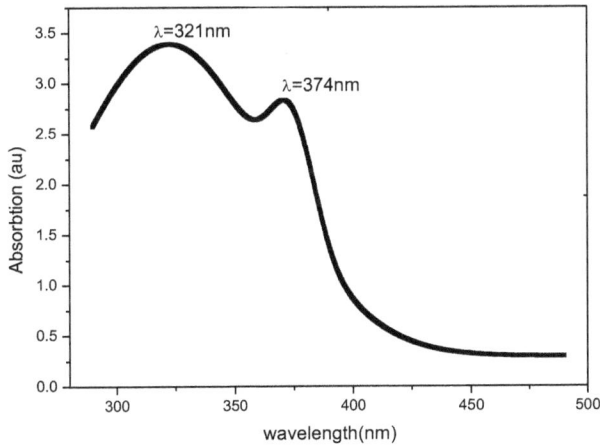

Fig. (9). The absorbance spectrum of TiO$_2$ Nanoparticles.

Absorption Spectra of the Natural Dyes (Roselle, Curcumin, and Black Tea)

The UV-VIS-NIR absorption spectra of the Roselle natural dye sample were determined using diffuse absorption spectroscopy (Fig. **10**). The FTO/TiO$_2$/Roselle solar cell's absorption showed a strong broadband absorption in the visible area from 430 nm to 670 nm, with the highest peak at 550 nm and an optical band gap of 2.25 eV (see equation 7.4). The presence of Roselle's anthocyanin pigment was demonstrated by this absorption spectrum, which ranged from 430 nm to 670 nm. The high anthocyanin content of Roselle makes it a possible good source of a good colorant.

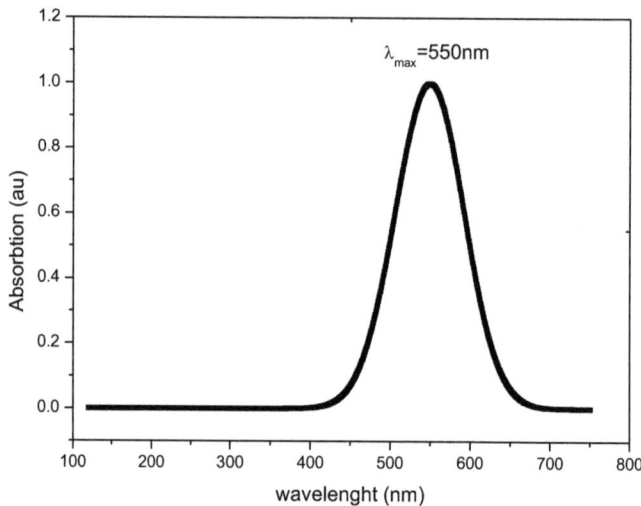

Fig. (10). The absorbance spectrum of Roselle.

The absorption spectra of the curcumin natural dye sample were measured using UV-VIS-NIR diffuse absorption spectroscopy, as shown in Fig. (**11**). The maximum absorption of FTO/TiO$_2$/curcumin was found to be at 430 nm, with absorption ranging from 110 nm to 750 nm. The maximum wavelength at 430 nm corresponds to an optical band gap equal to 2.88 eV (see equation 7.4).

Fig. (**12**) illustrates the measurement of the black tea natural dye sample's absorption spectra using UV-VIS-NIR diffuse absorption spectroscopy. The wavelength of 333 nm, which absorbs wavelengths from 110 nm to 750 nm, is determined to have the highest peak. At 333 nm, the maximum wavelength corresponds to an optical band gap of 3.72 eV (see equation 7.4).

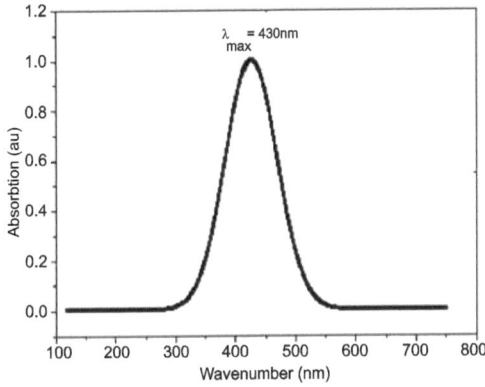

Fig. (11). The absorbance spectrum of curcumin natural dyes to synthesize Nanostructured solar cells.

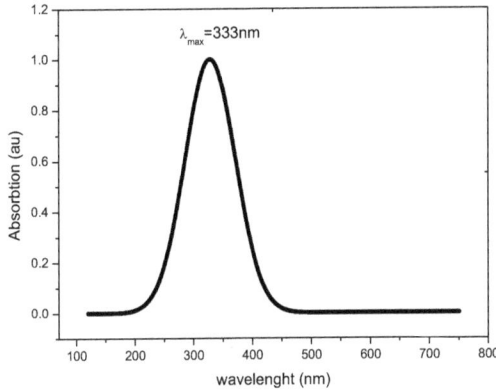

Fig. (12). The absorbance spectrum of black tea natural dyes to synthesize Nano-structured solar cells.

The spectra of different dyes adsorbed on TiO_2-coated FTO glasses are depicted in Fig. (**13**). Different dyes showed very similar spectra with different absorption maxima.

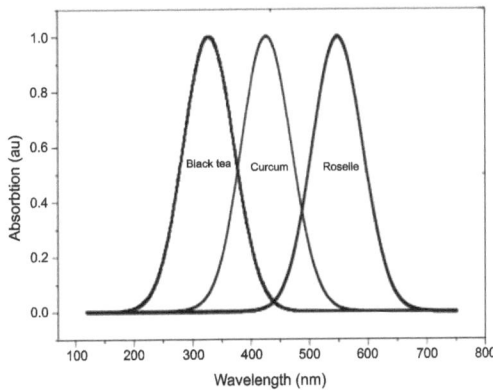

Fig. (13). Shows the comparison of wavelengths of the three dyes used.

The capability of the dye sensitizer to absorb light and the diffusion of the ejected electron through the mesoporous TiO_2 film determine how well a DSSC performs overall. Equation 7.4 was used to determine the optical band gap of the three dyes employed to fabricate the TiO_2 nanostructured solar cell in the FTO/TiO_2/Dye device. As shown in Table **1**, the FTO/TiO_2/Roselle light absorption spectra showed higher absorption and a smaller outside band gap than those of curcumin and black tea.

Table 1. Shows the wavelengths and calculated optical band gaps of the three natural dyes synthesized solar cells.

Type	Wavelength λ(nm)	Energy gap (E_g) eV
TiO_2	321 (peak 1)	3.86
	374 (peak 2)	3.32
Roselle	550	2.25
Curcuma	430	2.88
Black Tea	333	3.72

The I-V Characteristic of Dye Solar Cells (DSSCs)

The photovoltaic response of Natural Dyes Synthesizes Solar Cells FTO-TiO_2-DSSC, was estimated based on current-voltage (I-V) characteristics. According to equations (7.1–7.4) and Figures (7.9–7.11), the photovoltaic parameters were calculated, including the appropriate open circuit voltage (V_{oc}), short current (I_{sc}), fill factor (FF), and power conversion efficiency (PCE), and their values were then reported in Table **2**. Curcumin and black tea have similar open current voltage values; however, Roselle has significantly increased. In contrast, the short circuit current has distinct values, with Roselle having the highest value. The ratio of the photovoltaic-generated electric output of the cells to the luminous power falling on it was used to determine the efficiency of dye-synthesized solar cells, and the values for Roselle, Curcumin, and Black Tea were found to be 8.46, 6.94, and 6.33, respectively. We noticed that Roselle's efficiency is higher than all other dyes used because of its high absorption and small optical bandgap (Figs. **14**, **15** and **16**). Table **3** shows the comparison between efficiencies for different natural dyes obtained by different researchers.

Table 2. Shows the data obtained from the I-V characteristic of the three different dyes used in the TiO_2 Nano-structure solar cells.

Type of Dye	I_{max}/mA	V_{max}/v	V_{oc}/v	I_{SC}/mA	P_{max}/watt	FF	η%
Roselle	7.11	0.71	0.80	8.46	0.004899	83	8.46

(Table 2) cont.....

Type of Dye	I_{max}/mA	V_{max}/v	V_{oc}/v	I_{SC}/mA	P_{max}/watt	FF	η%
Curcumin	6.17	0.75	0.77	4.93	0.004639	80	6.94
Black Tea	4.98	0.75	0.77	6.34	0.003703	75	6.33

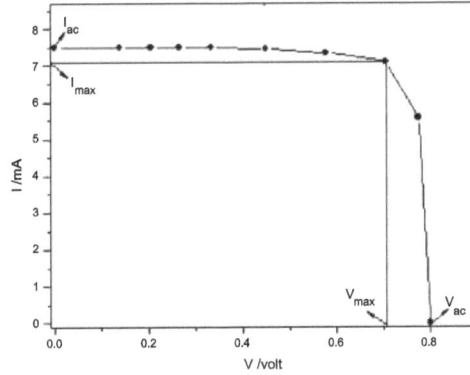

Fig. (14). Photovoltaic curve (I–V) of the DSSC based on Roselle natural dye used in the fabrication of TiO₂ Nano-structure solar cells.

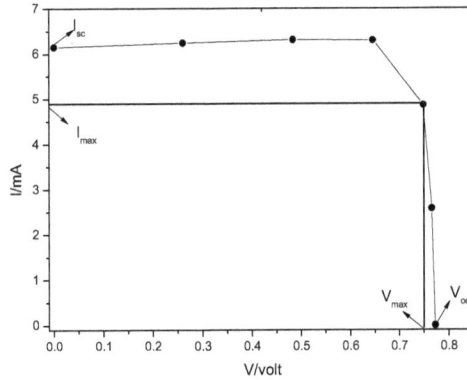

Fig. (15). Photovoltaic curve (I–V) of the DSSC based on curcumin natural dye used in the fabrication of TiO₂ Nano-structure solar cells.

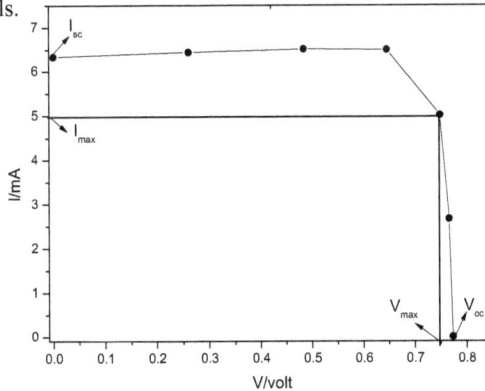

Fig. (16). Photovoltaic curve (I–V) of the DSSC based on black tea natural dye used in the fabrication of TiO₂ Nano-structure solar cells.

Table 3. The comparison between efficiencies for different natural dyes obtained by different researchers.

Natural dye	η%
Roselle	8.46
Curcumin	6.94
Black Tea	6.33
Pomegranate and berry fruits (W. Ghana *et al.*)	2.0
Broccoli (Cari et.)	0.072
Green Spinach (Ahmed M. Ammar *et al.*)	0.17
Onions (Ahmed M. Ammar *et al.*)	0.0647
Red cabbage (Ahmed M. Ammar *et al.*)	0.060
Malabar red spinach seeds (Sujan Kumar Das *et al.*)	9.23

Device Lay Out

The synthesized DSSC consisted of the following layers as shown in Fig. (**17**); a glass layer covered by FTO, then the TiO_2 seed layer, followed by the TiO_2 nanoparticle layer upon which the dye is applied. The last layer is graphite.

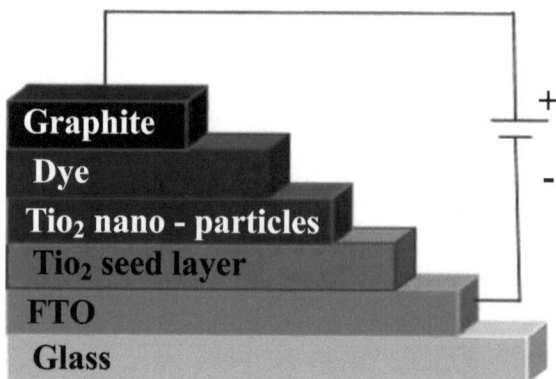

Fig. (17). structure of TiO_2 dye solar cell.

Natural dye-sensitized solar cells were successfully fabricated using the dyes of the three chosen plants (NDSSC). Natural dyes, including roselle, curcumin, and black tea dyes, were examined. The efficiency, voltage, and current of the solar cell's photovoltaic performance were evaluated. The absorption maxima on TiO_2 were bathochromically shifted in the majority of spectra. While the efficiencies attained by various dyes when applied in glass-based cells vary significantly. The data demonstrates that the efficiency of the Roselle sensitizer is significantly higher than that of the curcumin and black tea dyes used in the study. The most effective solar cell was the roselle dye-sensitized one (8.46%), followed by curcumin- and black tea-sensitized solar cells (6.94% and 6.33%, respectively). The Roselle-synthesized TiO_2 nanostructured solar cell's high efficiency, high absorption, and small band gap make it simple to convert photons into direct currents. However, they produce only modest efficiencies, making them most appropriate for use in large areas. Our opinion is that drinking tea, curcumin, and roselle are good additions to the process of making solar cells. The use of natural dye sensitizers in solar cell applications is still possible with ongoing, advanced research and studies.

REFERENCES

[1] M. Tawalbeh, A. Al-Othman, and F. Kafiah, "Environmental impacts of solar photovoltaic systems: A critical review of recent progress and future outlook", *Science of The Total Environment,* vol. 759, p. 143528, 2021.

[2] D. Yang, X. Zhang, Y. Hou, K. Wang, and T. Ye, "28.3%-efficiency perovskite/silicon tandem solar cell by optimal transparent electrode for highly efficient semitransparent top cell", *Nano Energy,* vol. 84, p. 105934, 2021.

[3] S. Ananthakumar, "Semiconductor nanoparticles sensitized TiO_2 nanotubes for high efficiency solar cell devices", *Renewable and sustainable energy,* vol. 57, pp. 1307-1321, 2016.

[4] P. Riente, and T. Noël, "Application of metal oxide semiconductors in light-driven organic transformations", *Catal. Sci. Technol.,* vol. 9, no. 19, pp. 5186-5232, 2019. [http://dx.doi.org/10.1039/C9CY01170F]

[5] Q. Liu, and J. Wang, "Dye-sensitized solar cells based on surficial TiO_2 modification", *Solar Energy,* vol. 184, pp. 454-465, 2019.

[6] Y. Yao, G. Li, S. Ciston, R.M. Lueptow, and K.A. Gray, "Photoreactive TiO_2/carbon nanotube composites: synthesis and reactivity. Environmental", *Environ. Sci. Technol.,* vol. 42, no. 13, pp. 4952-4957, 2008. [http://dx.doi.org/10.1021/es800191n]

[7] T.S. Wu, K.X. Wang, G.D. Li, S.Y. Sun, J. Sun, and J.S. Chen, "Montmorillonite-supported Ag/TiO(2) nanoparticles: an efficient visible-light bacteria photodegradation material", *ACS Appl. Mater. Interfaces,* vol. 2, no. 2, pp. 544-550, 2010. [http://dx.doi.org/10.1021/am900743d] [PMID: 20356203]

[8] K. Yim, *Computational discovery of p-type transparent oxide semiconductors using hydrogen descriptor.* Computational Materials, 2017, pp. 4-17.

[9] K. Gawlak, A. Knapik, G.D. Sulka, and L. Zaraska, "Improving the photoelectrochemical performance of porous anodic SnOx films by adjusting electrosynthesis conditions", *Int. J. Energy Res.,* vol. 46, no. 12, pp. 17465-17477, 2022. [http://dx.doi.org/10.1002/er.8414]

[10] W. Ghann, H. Kang, T. Sheikh, S. Yadav, T. Chavez-Gil, F. Nesbitt, and J. Uddin, "Fabrication, Optimization and Characterization of Natural Dye Sensitized Solar Cell", *Sci. Rep.,* vol. 7, no. 1, p. 41470, 2017. [http://dx.doi.org/10.1038/srep41470] [PMID: 28128369]

[11] U. Mehmood, S.H.A. Ahmad, A. Al-Ahmed, A.S. Hakeem, H. Dafalla, and A. Laref, "Synthesis and Characterization of Cerium Oxide Impregnated Titanium Oxide Photoanodes for Efficient Dye-

Sensitized Solar Cells", *IEEE J. Photovolt.,* vol. 10, no. 5, pp. 1365-1370, 2020.
[http://dx.doi.org/10.1109/JPHOTOV.2020.3010232]

[12]   B. Zhai, L. Yang, and Y.M. Huang, "Improving the Efficiency of Dye-Sensitized Solar Cells by Growing Longer ZnO Nanorods on TiO 2 Photoanodes", *J. Nanomater.,* vol. 2017, no. 9, pp. 1-8, 2017.
[http://dx.doi.org/10.1155/2017/1821837]

[13]   K. Sharma, V. Sharma, and S.S. Sharma, "Dye-Sensitized Solar Cells: Fundamentals and Current Status", *Nanoscale Res. Lett.,* vol. 13, no. 1, p. 381, 2018.
[http://dx.doi.org/10.1186/s11671-018-2760-6] [PMID: 30488132]

[14]   A.E. Shalan, A.M. Elseman, and M. Rashad, *"Controlling the Microstructure and Properties of Titanium Dioxide for Efficient Solar Cells in Titanium Dioxide", Material for a Sustainable Environment.* Intech: London, United Kingdom, 2017.
[http://dx.doi.org/10.5772/intechopen.72494]

[15]   K. Sharma, V. Sharma, and S.S. Sharma, "Dye-Sensitized Solar Cells: Fundamentals and Current Status", *Nanoscale Res. Lett.,* vol. 13, no. 1, p. 381, 2018.
[http://dx.doi.org/10.1186/s11671-018-2760-6] [PMID: 30488132]

[16]   R.T. Subramaniam, N.M. Saidi, F.S. Omar, and R. Kasi, "Effect of Nickel Oxide on the Conductivity of Polymer Blend Electrolyte Doped with Sodium Iodide and Its Application in Dye-Sensitized Solar Cell", *MJS,* vol. 38, no. 1, pp. 1-12, 2019.
[http://dx.doi.org/10.22452/mjs/vol38no1.1]

[17]   M. Shahid, "Recent advancements in natural dye applications: a review", *Journal of Cleaner Production,* vol. 53, pp. 310-331, 2013.

[18]   T. Jalali, P. Arkian, M. Golshan, M. Jalali, and S. Osfouri, "Performance evaluation of natural native dyes as photosensitizer in dye-sensitized solar cells", *Optical Materials,* vol. 110, p. 110441, 2020.

[19]   N. Amarachukwu, "Enhanced photovoltaic performance of dye-sensitized solar cells-based Carica papaya leaf and black cherry fruit co-sensitizers", *Chemical Physics Impact,* vol. 2, p. 100024, 2021.

[20]   C. Cari, D. Kurniawan, and A. Supriyanto, "The Effect of ZnO and TiO 2 with Natural Dye of Broccoli (Brassia oleracea var. italica) on Dye-Sensitized Solar Cell (DSSC)", *J. Phys. Conf. Ser.,* vol. 1842, no. 1, p. 012055, 2021.
[http://dx.doi.org/10.1088/1742-6596/1842/1/012055]

[21]   C. Bhargava, and P.K. Sharma, "Use of natural dyes for the fabrication of dye-sensitized solar cell: a review", *Polska Academia Nauk, Bulletin of the Polish Academy of Sciences: Technical Sciences,* vol. 69, Warsaw, no. 6, pp. 1-12, 2021.

[22]   S. Patel, "Hibiscus sabdariffa: An ideal yet under-exploited candidate for nutraceutical applications", *Biomedicine & Preventive Nutrition,* vol. 4, no. 1, pp. 23-27, 2014.
[http://dx.doi.org/10.1016/j.bionut.2013.10.004]

[23]   Basuki, R.L.L.G. Hidajat, Suyitno, B. Kristiawan, and R.A. Rachmanto, "Effect of sintering time on the performance of turmeric dye-sensitized solar cells", *AIP Conf. Proc.,* vol. 1788, no. 1, p. 030010, 2017.
[http://dx.doi.org/10.1063/1.4968263]

[24]   M.A. Brza, S.B. Aziz, H. Anuar, F. Ali, E.M.A. Dannoun, S.J. Mohammed, R.T. Abdulwahid, and S. Al-Zangana, "Tea from the drinking to the synthesis of metal complexes and fabrication of PVA based polymer composites with controlled optical band gap", *Sci. Rep.,* vol. 10, no. 1, p. 18108, 2020.
[http://dx.doi.org/10.1038/s41598-020-75138-x] [PMID: 33093604]

[25]   L.S. Cividanes, T.M.B. Campos, L.A. Rodrigues, D.D. Brunelli, and G.P. Thim, "Review of mullite synthesis routes by sol–gel method", *J. Sol-Gel Sci. Technol.,* vol. 55, no. 1, pp. 111-125, 2010.
[http://dx.doi.org/10.1007/s10971-010-2222-9]

[26]   S. Thiagarajan, A. Sanmugam, and D. Vikraman, "Facile Methodology of Sol-Gel Synthesis for Metal

Oxide Nanostructures", In: *Recent Applications in Sol-Gel Synthesis* InTech, 2017.
[http://dx.doi.org/10.5772/intechopen.68708]

[27]   B. Jalouli, A. Abbasi, and S.M. Musavi Khoei, "Conventional and Microwave Hydrothermal Synthesis and Application of Functional Materials: A Review", *Materials (Basel),* vol. 12, no. 21, p. 3631, 2019.
[http://dx.doi.org/10.3390/ma12213631] [PMID: 31694140]

[28]   K. Singh, V. Khanna, A. Rosenkranz, V. Chaudhary, Sonu, G. Singh, and S. Rustagi, "Panorama of physico-mechanical engineering of graphene-reinforced copper composites for sustainable applications," Materials Today Sustainability, vol. 24, p. 100560, Dec. 2023.
[http://dx.doi.org/10.1016/j.mtsust.2023.100560]

[29]   A. Thakur, V. Khanna, and Q. Murtaza, "Characterization, Strength and Wear Optimization of Aluminium Based Boron Carbide/Graphite Hybrid Composite by TOPSIS for Dry Condition," ECS Journal of Solid State Science and Technology, vol. 12, no. 12, p. 127005, Dec. 2023.
[http://dx.doi.org/10.1149/2162-8777/ad1209]

[30]   T. Nagarajan, N. Sridewi, W. P. Wong, R. Walvekar, V. Khanna, and M. Khalid, "Synergistic performance evaluation of MoS2–hBN hybrid nanoparticles as a tribological additive in diesel-based engine oil," Sci Rep, vol. 13, no. 1, p. 12559, Aug. 2023.
[http://dx.doi.org/10.1038/s41598-023-39216-0]

[31]   K. Singh, V. Khanna, S. Sonu, S. Singh, S.A. Bansal, V. Chaudhary, and A. Khosla, "Paradigm of state-of-the-art CNT reinforced copper metal matrix composites: processing, characterizations, and applications", *J. Mater. Res. Technol.,* vol. 24, pp. 8572-8605, 2023.
[http://dx.doi.org/10.1016/j.jmrt.2023.05.083]

[32]   V. Khanna, V. Kumar, S.A. Bansal, C. Prakash, M. Ubaidullah, S.F. Shaikh, A. Pramanik, A. Basak, and S. Shankar, "Fabrication of efficient aluminium/graphene nanosheets (Al-GNP) composite by powder metallurgy for strength applications", *J. Mater. Res. Technol.,* vol. 22, pp. 3402-3412, 2023.
[http://dx.doi.org/10.1016/j.jmrt.2022.12.161]

[33]   M. Dahiya, V. Khanna, and S. Anil Bansal, "Effect of graphene size variation on mechanical properties of aluminum graphene nanocomposites: A modeling analysis", *Mater. Today Proc.,* vol. 73, no. 2, pp. 249-254, 2022.
[http://dx.doi.org/10.1016/j.matpr.2022.07.259]

[34]   P. Gupta, N. Ahamad, D. Kumar, N. Gupta, V. Chaudhary, S. Gupta, V. Khanna, and V. Chaudhary, "Synergetic Effect of CeO 2 Doping on Structural and Tribological Behavior of Fe-Al 2 O 3 Metal Matrix Nanocomposites", *ECS J. Solid State Sci. Technol.,* vol. 11, no. 11, p. 117001, 2022.
[http://dx.doi.org/10.1149/2162-8777/ac9c92]

[35]   K. Singh, S.A. Bansal, V. Khanna, and S. Singh, "Effects of Performance Measures of Non-conventional Joining Processes on Mechanical Properties of Metal Matrix Composites", *Metal Matrix Composites,* no. Aug, pp. 135-165, 2022.
[http://dx.doi.org/10.1201/9781003194897-7]

CHAPTER 7

Investigation of the Effect of Annealing Conditions on Chemical Bath Deposited CdTe Thin- film from Non-Aqueous Bath

Sudeshna Surabhi[1,*], **Kumar Anurag**[2] and **S.R. Kumar**[3]

[1] *Brindavan College of Engineering, Yelahanka, Bengaluru, Karnataka-560063, India*

[2] *School of Energy Materials, Mahatma Gandhi University, Kottayam, Kerala-686560, India*

[3] *Thin Film Laboratory, National Institute of Advanced Manufacturing Technology(Formerly NIFFT), Ranchi-834003, India*

Abstract: This research investigates the consistency of chemical bath deposition (CBD) for CdTe thin films. Films were deposited using tellurium dioxide and cadmium acetate in a non-aqueous medium at 160°C. The impact of subsequent annealing on the optical, structural, and surface properties of these films was examined. XRD, FTIR, UV-Vis, SEM, and photoluminescence techniques were used to characterize the films. EDS analysis revealed a Cd:Te ratio of 1.27 before annealing, which improved to 1.06 (closer to the ideal 1:1 ratio) after annealing. The average crystallite size of annealed CdTe film was around 25nm. Photoluminescence peaks were observed at 566 nm and 615 nm.

Keywords: Electrodeposition, FTIR spectroscopy, Schockley-Queisser, Spray pyrolysis.

INTRODUCTION

For a long time, the semiconductor CdTe, which belongs to the II-VI family, has been researched. CdTe has primarily been investigated as a polycrystalline thin film over the past ten years. The only thin film technology currently included in the top 10 global producers is CdTe thin film technology, which was initially developed in the early 1970s [1], this is because CdTe is extremely durable and chemically stable. Additionally, it can be deposited using a wide range of techniques, making it perfect for large-scale production. According to the Schockley-Queisser limit, CdTe can achieve efficiencies of around 32% with an

---

* **Corresponding author Sudeshna Surabhi:** Brindavan College of Engineering, Yelahanka, Bengaluru, Karnataka-560063, India; E-mail: surabhisudeshna@gmail.com

Virat Khanna, Suneev Anil Bansal, Vishal Chaudhary and Reddicherla Umapathi (Eds.)

open circuit voltage of more than 1 V and a short circuit current density of more than 30 mA/cm^2 by having an ideal band gap close to 1.5 eV [2]. CdTe, a material for solar cells [1] and high-energy radiation sensors [3], and highly sensitive as well as selective fluorescent sensors [4] offers ideal electrical and optical characteristics. CdTe has the potential to be used as an absorber material in thin-film solar cells and light-emitting diodes [5], enabling the production of highly efficient solar cells using inexpensive methods [6]. With a CdTe absorption value of 10^4 cm^{-1}, 90% of the light may pass through thin sheets that are only 1 μm thick [1]. Another benefit is the potential for producing p- and n-type conductivity in the films, allowing for homojunction production. The material, however, has certain drawbacks when used in solar cells; for instance, it is highly resistant and does not enable superior carrier collection. Additionally, the semiconductor-metal junction is impacted by CdTe's high work function of 5.7 eV [1]. Finding a metal with a work function bigger than the CdTe is important to improve this connection. The CdTe homojunction's high surface recombination speed also prevents the production of homojunction-based devices. As a result, CdS/CdTe heterojunctions have been often used in the processing of CdTe solar cell systems. The n-type CdTe had to be changed since the CdTe homojunction failed. So that CdTe can absorb the most light, the material must be transparent for this application. The ideal solution is cadmium sulfide. Without phonon assistance, CdTe exhibits a band to band type transition and significant optical absorption. Cadmium telluride thin films have been developed using a variety of fabrication techniques, including close spaced vapor transport (CSVT) [4, 5], laser ablation [7 - 10], spray pyrolysis [11, 12], electrodeposition [13, 14], pulsed laser deposition [15, 16], chemical vapour deposition [17, 18], physical vapour deposition [14], hot wall epitaxy [19], thermal evaporation [20], and RF sputtering [21]. Depending on the type of use the film is intended for, each of these approaches has benefits and drawbacks. Re-sputtering and similar defects in the substrate surface and film growths are caused by the enormous kinetic energy of a small number of plume species during pulsed laser deposition. In physical vapour deposition, the rate of coating deposition is often quite slow. This process generates a lot of heat as well, necessitating the use of suitable cooling systems. Expensive power supplies and additional impedance-matching equipment are needed for RF sputtering. The expensive and technically complex equipment needed for the successive ionic adsorption and reaction process makes it uneconomical. For this process, high working pressure is also necessary. Due to the low vacuum, thermal evaporation has a decreased throughput, which results in inadequate step coverage. The operation of several CdTe film deposition methods is covered in the next section.

Deposition Techniques for CdTe Film

Close Spaced Vapor Transport (CSVT)

The CSVT is a physical method that involves sublimating the substance to move the gas and deposit it on the substrate. Films of a 500 nm thickness can be produced using this method [8]. Two blocks, either made of metal or graphite, constitute the deposition system. Each block has a temperature control since these blocks must be heated by a halogen light or electrical resistance. The block above is referred to as substrate, and the block below is known as the source. A graphite boat is set between the blocks. It is then filled with the CdTe. In this instance, several people employed CdTe pellets or powder before placing the substrate on the graphite boat. Everything is contained in a vacuum chamber. In the instance of CdTe, when the source temperature is higher and the block temperature is lower, the mechanics of development involve establishing a temperature gradient between the two. The CdTe is carried to the substrate, which has a lower temperature, during the sublimation process, causing the material to be deposited on the substrate. The thin-film deposition requires a high vacuum of around 10^{-6} Torr. The benefit of this technology is that a film thickness of 500 nm is feasible due to the controlled development rate; nevertheless, the drawback of this approach is the time costs associated with the vacuum that the process necessitates.

Laser Ablation

Laser ablation is a physical technique. Because the material is sublimated during the deposition process, it is comparable to CSVT. The method of sublimation differs because a laser beam is used in this instance to sublimate the substance. To achieve this, the energy of the laser beam must be low enough for the material to absorb light photons and be sublimated. The laser beam's pulse and intensity are used to regulate the deposit process. This method is significant in addition because it can be applied in industry. One benefit of this method is that flexible substrate can be used because the substrate maintains a low temperature, However, laser ablation includes burning and vaporizing material, which can produce a lot of fumes. As a result, the work area needs a good exhaust system to remove the fumes.

Spray Pryolysis

Spray pyrolysis is a chemical process in which the material in the solution is crushed by the pressure of a gas (argon, air, nitrogen, *etc.*). Controlling the solution's flow and the gas's pressure is crucial for this operation. To create the film, the pulverized solution is sprayed onto the hot substrate. This method's flaw is the possibility of a crack forming, which accounts for its lack of popularity.

Electrodeposition

In an electrolytic solution, an electric current is utilized in the chemical process known as electrodeposition with the goal of causing an ion migration toward the cathode. The substance is deposited after the ions reach the cathode. This technology offers a number of benefits, including the ability to use flexible substrates. The fact that the substrate is not heated is an additional benefit [14], and finally, the ability to deposit bigger areas makes this approach suitable for industrial use. There are a few drawbacks to this method, one of which is the difficulty in disposing of the waste products generated during electrodeposition. The environment is polluted by the waste conducting solution. Multiple coatings of the metal are required during the process, which takes time and has a significant setup cost.

Chemical Vapour Deposition

Thin films can be developed on a surface using chemical vapor deposition (CVD), which can deposit evaporated reactants. The basic idea behind it is to fill the reaction chamber with the vapor of a gaseous or liquid reactant that contains the film elements and other gases required for the reaction. The CVD process can create chemical reactions on the substrate surface and develop new solid materials to be deposited on the surface by raising the temperature, plasma action, laser radiation, or using other types of energy. The formation of dense and homogeneous films on the uneven substrate surface is better suited to CVD. Film quality is stable and CVD deposition speed is quick. This approach allows for large-area deposition and mass production, but it has the drawback of having a very high energy need additionally, CVD equipment is quite expensive, which hinders its industrialization.

Chemical Bath Deposition

The simplest and the most practical method of depositing thin films of binary, ternary metal chalcogenides is the chemical bath deposition (CBD). This is an extended process involving competing mechanisms of ion-to-ion condensation and cluster-to-cluster condensation leading to the controlled deposition of the desired molecule on the substrate in the presence of a suitable complexing agent. The thin-film deposition process is referred to as chemical bath deposition (CBD), which is also known as solution growth. The goal of chemical bath deposition is the controlled deposition of a component from solution onto a suitable substrate. The method is capable of coating a large area in a cost-effective and repeatable manner. The window layer for thin film solar cells is often produced using this method, which is suitable for creating thin films of many different materials. The basic idea behind CBD is that a molecule must have a higher ionic product than its solubility product in order to precipitate from a solution. When a cationic solution (metal) and an anionic solution (chalcogenide) come together, rapid precipitation occurs as the ions on the substrate and in the solution combine when the ion product is greater than the solubility product of the reactive species. However, if the same reaction occurs in the presence of a complexing agent such as NH_4OH, metal ions become complexed and their concentration decreases to the point where controlled precipitation of the desired chemical compound occurs, resulting in the deposition of a thin film on the substrate. The work related to CdTe is summarized in the above points. It is important to understand the advantages and disadvantages of these approaches. These methods often differ in two ways: price and results obtained.

SELECTION OF DEPOSITION METHOD

To get credible, high-quality films, the deposition procedure choice is crucial. The deposition technique chosen for a given material is determined by the material's characteristics and the purposes for which it is designed. Porosity, physical integrity, temperature stability, electrical characteristics, vast area deposition, controlling growth rate, and cost-effectiveness are a few examples of the factors that need to be considered.

The chemical bath deposition (CBD) technique is one of these technologies that is simple and least expensive to employ for developing the films. You can make nanocrystalline thin films in both aqueous and non-aqueous media. Because we can operate at higher temperatures in non-aqueous mediums, we have a larger assortment of substrates to choose from. Since hydrogen does not evolve as it does in an aqueous media, the film deposition is devoid of pinholes. This chapter has covered the development of cadmium telluride thin films on nickel substrates (Ni) as well as their identification. Various investigations on CdTe nanocrystalline thin films are described using the chemical bath deposition approach that we have selected here. By using XRD, SEM, and EDS, the structure and formation of the produced samples are verified. This chapter also discusses the optical characteristics of CdTe thin films using FTIR, PL, and UV-Visible spectroscopy.

EXPERIMENTAL PROCEDURE

Utilizing the CBD approach, CdTe films have been laid on Ni substrates.

Sample Preparation and Experimental Setup

Fig. (**1**) depicts the experimental configuration employed for the deposition of CdTe film. It comprises an electrolyte-filled reaction cell made of borosilicate glass. To agitate the electrolyte, a magnetic stirrer with a hot plate was employed. A stand was used to fix the substrate vertically.

Fig. (1). Experimental setup of CdTe thin film.

Preparation of Bath

In order to make the electrolyte, 40ml of ethylene glycol was dissolved in AR grade, with 0.1M cadmium acetate. An hour and a half of moderate stirring was performed with a magnetic stirrer. The bath was kept at a constant temperature of 160° C. Then a Ni substrate with dimensions of 2 cm × 1 cm was vertically submerged in the solution in a parallel position. Tellurium dioxide with an AR grade of 0.02M was added to the solution after dipping the Ni substrate in it. A 15-minute deposition was conducted. The films had a brownish gray color when they were being deposited. The film was rinsed with distilled water before being allowed to air dry. Organic contaminants and counter ions were eliminated by doing this. The as-deposited CdTe films on the nickel substrate demonstrated good adhesion and were physically stable. 400 C was used as the annealing temperature for the sample. To evaluate the solid-state properties, the annealed films' structure, composition, and optical properties were scrutinized.

RESULTS AND DISCUSSIONS

Structural Analysis

The XRD spectra of a film that has been deposited and annealed at 400°C are shown in Fig. (**2a** and **b**). Both figures have polycrystalline CdTe peaks that are visible. The favored peak of CdTe (111) can be seen at $2\theta = 23.43°$ in both films [19, 20]. The peak width at half maxima (FWHM) dropped and the peak height increased in annealed CdTe films. The reduction in FWHM made it obvious that the average grain size has increased and grain boundaries have shrunk. The peak of CdTe (200), CdTe (222), and CdTe (400) were also visible in the annealed films, as was abundantly obvious from Fig. (**2b**). The deposition condition can have an impact on the CdTe thin film's crystallite size, and XRD measurements provide extensive information about the thin film's structural characteristics. The TeO_2 peaks, however, may be found at 28.78, 30.18, 36.17, and 48.54°. It is anticipated that the film contains some free TeO_2 species and exhibits reflections. Table **1** lists preferred planes and observed d values for annealed and as-deposited cadmium telluride films. The crystallite size of the as-deposited and annealed CdTe thin films is determined using the FWHM and Debye-Scherrer equations [22, 23, 24].

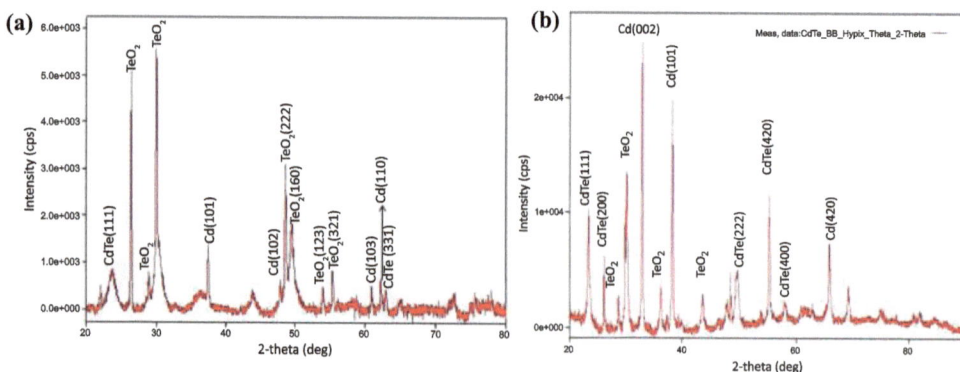

Fig. (2). XRD spectra of (**a**) as deposited and (**b**) annealed CdTe thin films.

$$\text{The crystallite size is given by D}=\frac{K\lambda}{\beta cos\theta} \tag{1}$$

Where, K is a constant (0.94), β=full width at half maximum of (002) peak, λ=X-ray wavelength (=1.54Å) and θ is a diffracted angle.

Table 2 displays the average crystallite size and micro strain of deposited and annealed CdTe thin films. It is shown that the CdTe crystallite size increases with annealing temperature, indicating an improvement in crystallinity [19]. Because the material's lattice becomes less stable and is more easily rearranged into larger crystallites as the annealing temperature rises, the crystallite size increases. Reduced surface energy accelerated diffusion rates, and an increase in the number of possible bonds between similar atoms contributes to an increase in crystallite size. The material has greater energy and can form larger, better-organized crystals more readily at higher temperatures. The diffraction peaks of Cd may also be seen in the XRD pattern. The "d" values, average crystallite size, micro strain, and "2θ" values of the as-deposited film and the annealed cadmium telluride film are provided in Tables **1** and **2**, respectively. The equation below is used to compute the micro strain [25]:

$$\text{Micro strain } (\varepsilon) = \frac{\beta Cos\theta}{4} \tag{2}$$

Table 1. 2θ, 'd' and miller-indices of as-deposited and annealed CdTe thin films.

Compound	Observed Value				Standard Value		
	As deposited		Annealed				
	2θ (degree)	d (Å)	2θ (degree)	d (Å)	2θ (degree)	d (Å)	Miller Indices
CdTe	23.59	3.769	23.429	3.79	24.072	3.700	(111)
			26.150	3.40	27.814	3.205	(200)
			49.721	1.83	49.201	1.850	(222)
			58.050	1.58	57.460	1.602	(400)
	63.046	1.473	-	-	63.177	1.470	(331)

Table 2. Average crystallite size and micro strain of as deposited and annealed CdTe thin films.

Compound	As deposited			Annealed		
	FWHM (degree)	Average Crystallite Size(nm)	Micro Strain(×10⁻³)	FWHM	Average Crystallite Size(nm)	Micro Strain
CdTe	1.28	6.60	5.40	0.345	24.6	1.46

The micro strain in as-deposited CdTe films is 5.4×10^{-3}, whereas in the annealed case, it is 1.46×10^{-3}. Thermal expansion and the recrystallization process in polycrystalline material [26] both contribute to this decrease. The film's surface temperature rises uniformly during annealing, and the thermal expansion of the film reduces micro strain and makes it more relaxed. As a result, the characteristics of the material and the homogeneity of the film thickness improve. The film's surface temperature rises uniformly during annealing, and the thermal expansion of the film reduces micro strain and makes it more relaxed. As a result, the characteristics of the material and the homogeneity of the film thickness improve.

Surface Morphology Studies

The surface morphology of the CdTe thin films as they were applied and annealed is shown in Fig. (**3a** and **b**). The visualization and characterization of surfaces are two tasks that are accomplished using scanning electron microscopy. SEM was used to examine the surface morphology at a 20kV accelerating voltage. For SEM analysis, 10,000X magnification is used. As deposited, CdTe exhibits a homogeneous, smooth, and crack-free texture with spherically organized grains dispersed throughout. We can observe that the grain size grows when the thin film is annealed at 400 °C for five minutes. Due to thermally stimulated recrystallization, the annealing procedure causes the grains on the thin film to enlarge. The metal atoms recombine and the lattice structure is altered during the annealing process, resulting in bigger grains and a smoother microstructure for the thin film. The annealing process raises the temperature, which increases the atoms' mobility and facilitates easier reorganization [27, 28]. The energy that the atoms gain from the higher temperature also contributes to the pace of recrystallization. Larger grains are produced on the thin film as a result of the annealing process's higher temperature and duration, which boosts the film's strength and improves its optical qualities. When CdTe thin films are heated to 400°C, they exhibit radically different morphologies that are uniform. In comparison to annealed CdTe films, the as-deposited CdTe film exhibits fewer dense surface morphologies all across the surface. We can infer from the figure that there were fewer pinholes, cavities, pits, *etc.* and that the grains or pieces were scattered on the film in more tightly packed, distinct cluster formations. This demonstrates how the fusing process, which occurs when the film is annealed at 400 °C improves the crystallinity of cadmium telluride thin films.

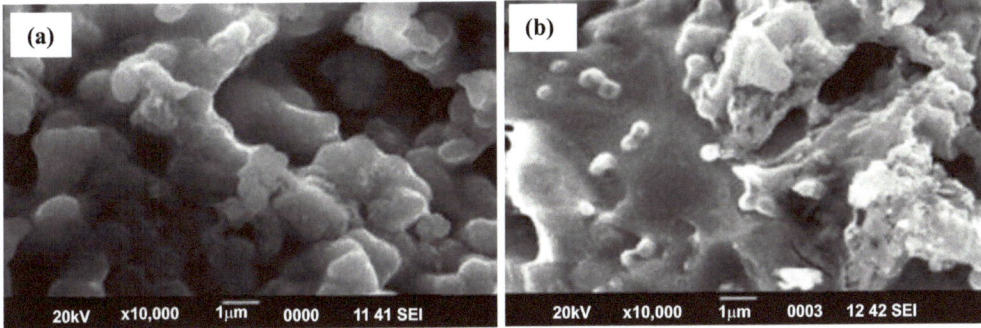

Fig. (3). SEM image of (**a**) as deposited and (**b**) annealed CdTe thin films.

Energy Dispersive Spectroscopy Analysis

With the use of energy dispersive spectroscopy (EDS), the elemental analysis of the CdTe thin film as it has been deposited and annealed is carried out, as shown in Fig. (**4a** and **b**). Energy dispersive spectroscopy (EDS), commonly referred to as energy dispersive X-ray spectroscopy (EDX or EDXS), is a method for conducting non-destructive chemical examinations of solids. It is based on how a sample reacts to an X-ray primary beam, which is typically produced by an X-ray tube. EDS can identify a wide variety of elements in a sample, and certain instruments can create elemental maps. EDS is most frequently used in conjunction with a scanning electron microscope (SEM) for elemental analysis since it is very quick and requires little sample preparation. Table **3** lists the atomic percentages of Cd and Te in CdTe films that have been deposited and annealed. The prepared CdTe films are made of Cd and Te, as shown by the spectra's appearance of the Cd and Te peaks. The proportion of the atomic percentage of Cd and Te of as-deposited film and annealed film of CdTe are 1.27 and 1.06, respectively. According to EDS analysis, the films become more stoichiometric after annealing, which is beneficial for device applications. By making the best use of the growing circumstances, the stoichiometry can be further reduced.

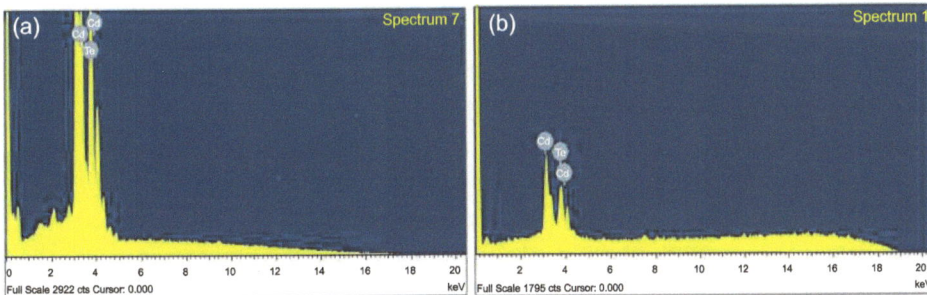

Fig. (4). EDS Spectra of (**a**) as-deposited and (**b**) annealed CdTe thin films.

Table 3. Atomic percentage of Cd and Te in as-deposited and annealed films.

Element	Atomic% As Deposited	Atomic% Annealed
Cd	56.07	51.52
Te	43.93	48.48
Total	100.00	100.00
Atomic% Ratio (Cd: Te)	1.27	1.06

FTIR Analysis

FTIR spectroscopy is a quantitative spectroscopic technique used to measure the absorption and transmission of infrared radiation of a material. It is typically used to recognize and examine a variety of substances, including both organic and inorganic chemicals. Infrared light and the fundamental vibrations of molecules have the same frequency. This light can be absorbed by molecules, who then transform it into molecular vibrations. We can track this absorption using FTIR spectrometers, and because each molecule vibrates somewhat differently, a distinct infrared spectrum can be obtained. A particular vibration is matched to the bands that can be seen in a spectrum. As a result, we may recognize particular chemical bonds by looking at the location (frequency) of these bands. We can determine how much light is absorbed from the bands' intensity, which is directly pertaining to the proportion of vibrating bands. In order to better understand reaction kinetics, reaction mechanism and routes, as well as catalytic cycles, academic and industrial laboratories use FTIR spectroscopy equipment. Chemical processes can be scaled up with the use of FTIR spectroscopy, which also helps to maximize the reaction yield and reduce contaminants. The FTIR transmission spectrum of deposited and annealed CdTe thin films are shown in Fig. (**5a** and **b**) in the 400–4000 cm^{-1} (mid-infrared range) spectral region. Due to C-O stretching, the sample exhibits the IR spectra of CdTe at 1010.9cm^{-1} [29]. The peak at 684 cm^{-1} is due to the vibration-symmetric stretching of Cd_2O_2 (Cd-O-Cd) [26]. The existence of C-O stretching caused by ethylene glycol is indicated by the peak at 1498 cm^{-1} [30]. The ester group (O=C-O-C) is responsible for the peak at 1625 cm^{-1}. The atmosphere is responsible for the peak at 3438.5 cm^{-1}, which is attributable to O-H [31]. As deposited, CdTe exhibits stretching with peaks at 1041 and 1084 cm^{-1}; however, after annealing, it exhibits a peak at 1010.9 cm^{-1}. This decrease in wavenumber is caused by an increase in particle size. The various functional groups present in deposited and annealed CdTe thin films are listed in Table **4**.

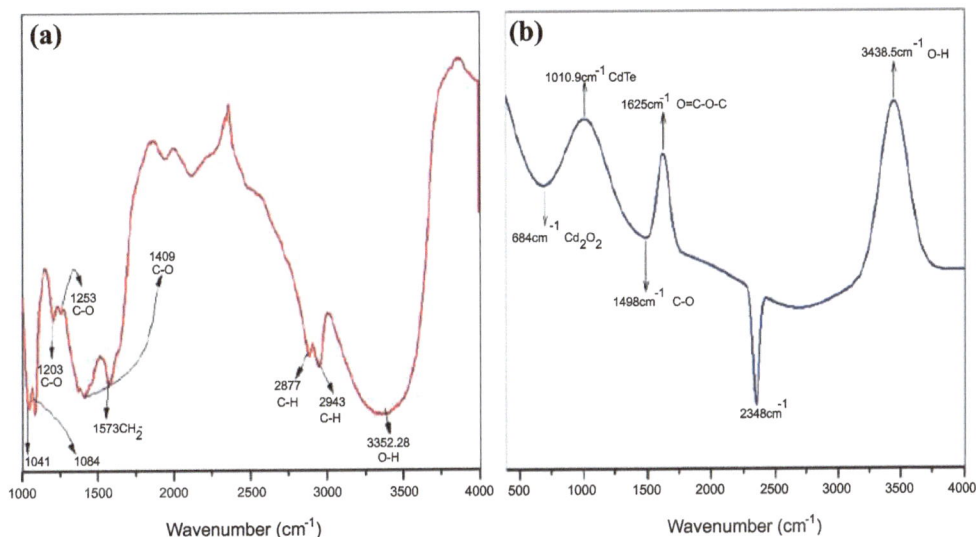

Fig. (5). FTIR spectra(**a**) as deposited and (**b**) annealed CdTe films.

Table 4. Different functional groups appeared in as-deposited and annealed CdTe thin films.

S.No.	Functional groups	Wavenumbers(cm^{-1}) As deposited	Wavenumbers(cm^{-1}) Annealed(400°C)
1.	CdTe	1041-1084	1010.9
.	Cd_2O_2	Not appeared	684
3.	O=C-O-C	Not appeared	1625
4.	C-O	1203-1409	1498
5.	O-H	3352.3	3438.5
6.	C-H	2877-2943	Not appeared

UV-visible Analysis of CdTe Thin Film

It is effective to use UV-visible absorption spectroscopy to keep track of the optical characteristics of quantum-sized particles. Plots of the $(\alpha h\upsilon)^2$ Vs hυ of the deposited and annealed CdTe thin film are shown in Fig. (**6a** and **b**). The Tauc formula [32, 33], which is as follows, $\alpha h\upsilon = A(h\upsilon-Eg)^{1/2}$ for h> Eg is used to calculate the film's band gap, where a, h, A, and Eg are the absorption coefficient, plank constant, vibration frequency, proportionality constant, and band gap, respectively. Light is used in UV-Visible spectroscopy to gauge a material's optical characteristics. The material absorbs light when it is exposed to light with an energy higher than its band gap, which causes electrons to be stimulated from the valence band to the conduction band. The band gap can be calculated from the energy difference between the two bands by measuring the absorption of light at various wavelengths. The computed energy gap is determined to be 1.43 eV for the as-deposited case and 1.37 eV for the as-annealed case, which agrees well with the reported values of CdTe thin layer cited by other authors [34]. As deposited, CdTe has an energy gap that is designed to be greater than in the annealed situation. Due to the thermal energy, the band gap of the film narrows as the annealing temperature rises. The atoms in the film vibrate more quickly and furiously as the temperature rises because it makes it possible for them to absorb more energy. As a result, the atoms' bonds are strengthened, which reduces the film's band gap. Thus, a reduction in the band gap of the film may result from higher annealing temperatures. The change in band gap is further influenced by annealing induced intrinsic CdTe film defects, quantum size effect, oxygen vacancies, high dislocation densities, *etc* [35]. Additionally, because the grain size increases with annealing temperature, the band gap is reduced, and the optical performance of the film is improved there.

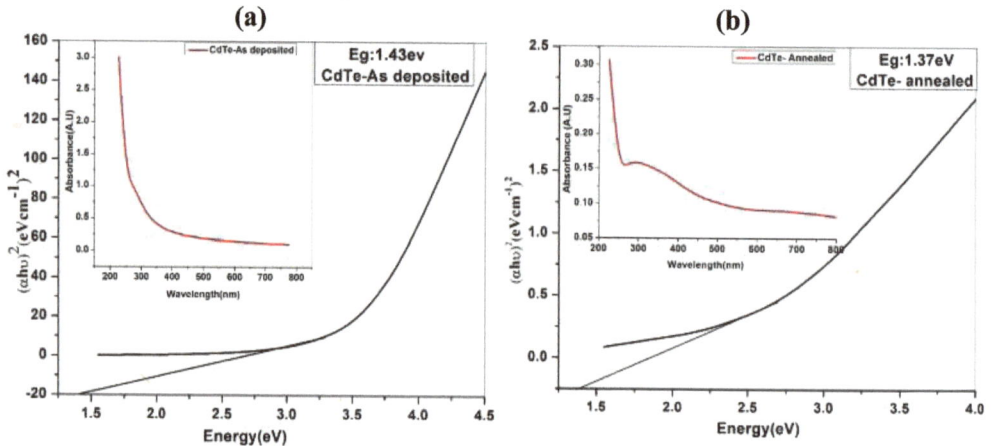

Fig. (6). Band gap of (**a**) as deposited and (**b**) annealed CdTe thin films.

Photoluminescence Analysis

A form of spectroscopy called photoluminescence spectroscopy is used to quantify how much light a sample emits when exposed to a light source with a predetermined wavelength. The electrical structure and optical characteristics of materials are studied using this methodology. It is possible to study a wide variety of optical properties, such as the spectra of excitation and emission, photoluminescence, and absorption. The method can be used with a variety of substances, including rare earth ions, single-molecule organic dyes, and nanomaterials including metal-organic frameworks and semiconductor nanocrystals. The investigation of samples' structure, composition, and optical characteristics can be carried out with the help of photoluminescence spectroscopy, which is also useful for sensing and material diagnosis. Studies on photoluminescence are conducted using a spectrophotometer with a 300-800nm wavelength range. The photoluminescence (PL) spectra of the chemical bath-deposited CdTe films on nickel substrate as well as those that have been annealed are shown in Fig. (7). The entire CdTe crystal is excited by the photon absorption's free carriers. The band gap edge of an excited CdTe crystal is represented by the excited PL spectra. For excitation in this study, a 250 nm Xenon lamp source is used. The strong and broad emission spectra, which are mostly found at around 471 nm in the case of the deposit, are depicted in this figure. Due to the size effect in a bulk CdTe system, the PL spectra of CdTe exhibit broad emission lines. The spectrum unequivocally demonstrates that the greater peak is situated on the high energy side after the blue shift [36]. While there is a sharp increase in emission intensity at 566 nm after annealing the film at 400 °C along with the emergence of additional peaks at 615 nm and 661 nm. The decrease in particle size may have an impact on the increase in thermoluminescent characteristics after annealing. Smaller particles include more accessible thermoluminescent carriers because they have higher surface-to-volume ratios and more surface states [35]. The lower wavelengths in the spectrum are attributed to CdTe quantum dots of a similar size, whereas the higher wavelengths are attributed to bulk CdTe crystals.

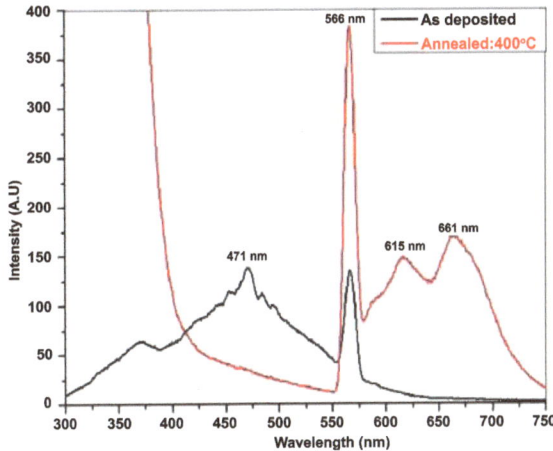

Fig. (7). PL spectra of as-deposited and annealed CdTe thin films.

By combining cadmium acetate and tellurium dioxide with ethylene glycol in a chemical bath, a polycrystalline CdTe thin film is formed. The as-deposited and annealed samples were characterized by various analyses like XRD, SEM, EDS, FTIR, and PL. The XRD results show that the cubic zinc blende (sphalerite) structure of the CdTe films is in the (111) plane. The average crystallite size is seen to rise after annealing, which suggests that crystallinity improves. CdTe is confirmed to be present at 1010cm^{-1} by the FTIR spectra. Additionally, surface morphology suggests an improvement in the grain structure after annealing at roughly 400 °C in air for 5 minutes. The fusion process is responsible for this improvement. After annealing, there is a red shift in the PL spectrum. EDS spectra indicate that the deposited layer is rich in Cd, yet the ratio of Cd to Te atomic proportions is found to be approximately equal to 1.27.

ACKNOWLEDGEMENTS

In order to conduct the trials, we acknowledge the director of NIAMT, Ranchi, for his assistance. We are grateful to CIF, BIT-Mesra for assistance in all characterizations.

REFERENCES

[1]   A. Romeo, and E. Artegiani, "CdTe-Based Thin Film Solar Cells: Past, Present and Future", *Energies,* vol. 14, no. 6, p. 1684, 2021.
[http://dx.doi.org/10.3390/en14061684]

[2]   S. Rühle, "Tabulated values of the Shockley–Queisser limit for single junction solar cells", *Sol. Energy,* vol. 130, pp. 139-147, 2016.
[http://dx.doi.org/10.1016/j.solener.2016.02.015]

[3]   "Nuclear Instruments and Methods in Physics Research Section A: Accelerators, Spectrometers, Detectors and Associated Equipment | Journal | ScienceDirect.com by Elsevier", Available from: https://www.sciencedirect.com/journal/nuclear-instruments-and-methods-in-physics-research--ection-a-accelerators-spectrometers-detectors-and-associated-equipment (Accessed on: Jan. 17, 2024). [Online]

[4]   "Highly sensitive selective sensing of nickel ions using repeatable fluorescence quenching-emerging of the CdTe quantum dots - ScienceDirect", Available from: https://www.sciencedirect.com/science/article/abs/pii/S0025540817319979 (Accessed on: Jan. 17, 2024). [Online]

[5]   H. Zare, M. Marandi, S. Fardindoost, V.K. Sharma, A. Yeltik, O. Akhavan, H.V. Demir, and N. Taghavinia, "High-efficiency CdTe/CdS core/shell nanocrystals in water enabled by photo-induced colloidal hetero-epitaxy of CdS shelling at room temperature", *Nano Res.,* vol. 8, no. 7, pp. 2317-2328, 2015.
[http://dx.doi.org/10.1007/s12274-015-0742-x]

[6]   D.K. Shah, D. Kc, M. Muddassir, M.S. Akhtar, C.Y. Kim, and O.B. Yang, "A simulation approach for investigating the performances of cadmium telluride solar cells using doping concentrations, carrier lifetimes, thickness of layers, and band gaps", *Sol. Energy,* vol. 216, pp. 259-265, 2021.
[http://dx.doi.org/10.1016/j.solener.2020.12.070]

[7]   R. Mendoza-Pérez, "Photovoltaic modules processing of CdS/CdTe by CSVT in 40 cm2", *34th IEEE Photovoltaic Specialists Conference (PVSC),* 2009pp. 002090-002095 Philadelphia, PA, USA
[http://dx.doi.org/10.1109/PVSC.2009.5411442]

[8]   O. Vigil-Galán, L. Vaillant, R. Mendoza-Pérez, G. Contreras-Puente, J. Vidal-Larramendi, and A. Morales-Acevedo, "Influence of the growth conditions and postdeposition treatments upon the grain boundary barrier height of CdTe thin films deposited by close space vapor transport", *J. Appl. Phys.,* vol. 90, no. 7, pp. 3427-3431, 2001.
[http://dx.doi.org/10.1063/1.1400090]

[9]   N.G. Semaltianos, S. Logothetidis, W. Perrie, S. Romani, R.J. Potter, M. Sharp, G. Dearden, and K.G. Watkins, "CdTe nanoparticles synthesized by laser ablation", *Appl. Phys. Lett.,* vol. 95, no. 3, p. 033302, 2009.

[http://dx.doi.org/10.1063/1.3171941]

[10]   S. Keitoku, and H. Ezumi, "Preparation of CdS/CdTe solar cell by laser ablasion", *Sol. Energy Mater. Sol. Cells,* vol. 35, pp. 299-303, 1994.
[http://dx.doi.org/10.1016/0927-0248(94)90154-6]

[11]   A. Ashok, S.V. Nair, and M. Shanmugam, "Spray pyrolysis-coated nano-clustered CdTe on amorphous Si thin films for heterojunction solar cells", *Appl. Nanosci.,* vol. 9, no. 7, pp. 1479-1486, 2019.
[http://dx.doi.org/10.1007/s13204-019-01022-4]

[12]   A. Arce-Plazaa, J.A. Andrade-Arvizub, M. Courel, J.A. Alvaradoc, and M. Ortega-Lópezd, "Study and application of colloidal systems for obtaining CdTe+Te thin films by spray pyrolysis", *J. Anal. Appl. Pyrolysis,* vol. 124, pp. 285-289, 2017.
[http://dx.doi.org/10.1016/j.jaap.2017.01.022]

[13]   D.P. Sali, and N.B. Chaure, "Electrodeposition and Characterization of CdTe thin films for photovoltaic applications", *Mater. Today Proc.,* vol. 42, pp. 1647-1650, 2021.
[http://dx.doi.org/10.1016/j.matpr.2020.07.475]

[14]   X. Mathew, J.P. Enriquez, A. Romeo, and A.N. Tiwari, "CdTe/CdS solar cells on flexible substrates", *Sol. Energy,* vol. 77, no. 6, pp. 831-838, 2004.
[http://dx.doi.org/10.1016/j.solener.2004.06.020]

[15]   P. Bhattacharya, and D.N. Bose, "Pulsed laser deposition of CdTe thin films for heterojunctions on silicon", *Semicond. Sci. Technol.,* vol. 6, no. 5, pp. 384-387, 1991.
[http://dx.doi.org/10.1088/0268-1242/6/5/012]

[16]   B. Ghosh, S. Hussain, D. Ghosh, R. Bhar, and A.K. Pal, "Studies on CdTe films deposited by pulsed laser deposition technique", *Physica B,* vol. 407, no. 21, pp. 4214-4220, 2012.
[http://dx.doi.org/10.1016/j.physb.2012.07.006]

[17]   X. Yi, L. Wang, K. Mochizuki, and X. Zhao, "The growth and the characteristics of cadmium telluride thin films prepared by chemical vapour deposition", *J. Phys. D Appl. Phys.,* vol. 21, no. 12, pp. 1755-1760, 1988.
[http://dx.doi.org/10.1088/0022-3727/21/12/015]

[18]   G. Kartopu, L.J. Phillips, V. Barrioz, S.J.C. Irvine, S.D. Hodgson, E. Tejedor, D. Dupin, A.J. Clayton, S.L. Rugen-Hankey, and K. Durose, "Progression of metalorganic chemical vapour-deposited CdTe thin-film PV devices towards modules", *Prog. Photovolt. Res. Appl.,* vol. 24, no. 3, pp. 283-291, 2016.
[http://dx.doi.org/10.1002/pip.2668]

[19]   D. Mohanty, Z. Lu, X. Sun, Y. Xiang, L. Gao, J. Shi, L. Zhang, K. Kisslinger, M.A. Washington, G-C. Wang, T-M. Lu, and I.B. Bhat, "Growth of epitaxial CdTe thin films on amorphous substrates using single crystal graphene buffer", *Carbon,* vol. 144, pp. 519-524, 2019.
[http://dx.doi.org/10.1016/j.carbon.2018.12.094]

[20]   T. Ablekim, J.N. Duenow, X. Zheng, H. Moutinho, J. Moseley, C.L. Perkins, S.W. Johnston, P. O'Keefe, E. Colegrove, D.S. Albin, M.O. Reese, and W.K. Metzger, "Thin-Film Solar Cells with 19% Efficiency by Thermal Evaporation of CdSe and CdTe", *ACS Energy Lett.,* vol. 5, no. 3, pp. 892-896, 2020.
[http://dx.doi.org/10.1021/acsenergylett.9b02836]

[21]   C. Doroody, K.S. Rahman, H.N. Rosly, M.N. Harif, K. Sopian, S.F. Abdullah, and N. Amin, "A comprehensive comparative study of CdTe thin films grown on ultra-thin glass substrates by close-spaced sublimation and RF magnetron sputtering", *Mater. Lett.,* vol. 293, p. 129655, 2021.
[http://dx.doi.org/10.1016/j.matlet.2021.129655]

[22]   S. Surabhi, KumarAnurag, S. Rajpal, and S.R. Kumar, "A new route for preparing CdTe thin films by chemical bath deposition", *Mater. Today Proc.,* vol. 44, pp. 1463-1467, 2021.
[http://dx.doi.org/10.1016/j.matpr.2020.11.635]

[23]   S. Surabhi, K. Anurag, and S.R. Kumar, "Effect of annealing on the structural, compositional and optical properties of CdTe films", *Mater. Today Proc.,* vol. 45, pp. 4477-4482, 2021.
[http://dx.doi.org/10.1016/j.matpr.2020.12.988]

[24]   S. Surabhi, "Development and Characterization of As-Deposited CdTe Thin Films in a Non-Aqueous Medium: Science & Engineering Book Chapter | IGI Global", Available from: https://www.igi-global.com/chapter/development-and-characterization-of-as-deposited-cdte-thin-fil-s-in-a-non-aqueous-medium/321046 (Accessed on: Jan. 17, 2024). [Online]

[25]   M.N.M. Daud, A. Zakaria, A. Jafari, M.S.M. Ghazali, W.R.W. Abdullah, and Z. Zainal, "Characterization of CdTe films deposited at various bath temperatures and concentrations using electrophoretic deposition", *Int. J. Mol. Sci.,* vol. 13, no. 5, pp. 5706-5714, 2012.
[http://dx.doi.org/10.3390/ijms13055706] [PMID: 22754325]

[26]   P.J. Carroll, and J.S. Lannin, "Vibrational properties of crystalline group-VI solids: Te, Se, S", *Phys. Rev. B Condens. Matter,* vol. 27, no. 2, pp. 1028-1036, 1983.
[http://dx.doi.org/10.1103/PhysRevB.27.1028]

[27]   A. Thabet, and N. Salem, "Experimental Progress in Electrical Properties and Dielectric Strength of Polyvinyl Chloride Thin Films Under Thermal Conditions", *Transactions on Electrical and Electronic Materials,* vol. 21, no. 2, pp. 165-174, 2020.
[http://dx.doi.org/10.1007/s42341-019-00163-1]

[28]   A. Thabet, F.A. Al Mufadi, and A.A. Ebnalwaled, "Synthesis and Measurement of Optical Light Characterization for Modern Cost-fewer Polyvinyl Chloride Nanocomposites Thin Films", *Trans. Electr. Electron. Mater.,* vol. 25, pp. 98-109, 2023.
[http://dx.doi.org/10.1007/s42341-023-00489-x]

[29]   C. Chotia, V. Sharma, T. Patel, U. Deshpande, and G. Okram, "Stable Dispersibility Study of Cadmium Telluride Nanoparticles in Deionized Water", *SciFed Nanotech Research Letters,* vol. 1, no. 1, pp. 1-4, 2017.

[30]   S. Kumar, S. Rajpal, S. Sharma, D. Roy, and R. Kumar, "Effect of Zn concentration on the structural, morphological and optical properties of ternary ZnCdS nanocrystalline thin films", *Dig. J. Nanomater. Biostruct.,* vol. 12, pp. 339-347, 2017.

[31]   S.R. Kumar, S. Kumar, S.K. Sharma, and D. Roy, "Structure, Composition and Optical Properties of Non Aqueous Deposited ZnCdS Nanocrystalline Film", *Mater. Today Proc.,* vol. 2, no. 9, pp. 4563-4568, 2015.
[http://dx.doi.org/10.1016/j.matpr.2015.10.071]

[32]   M. Saadati, O. Akhavan, and H. Fazli, "Single-Layer MoS2-MoO3-x Heterojunction Nanosheets with Simultaneous Photoluminescence and Co-Photocatalytic Features", *Catalysts,* vol. 11, no. 12, p. 1445, 2021.
[http://dx.doi.org/10.3390/catal11121445]

[33]   S. Kawar, and B. Pawar, "Synthesis and characterization of CdS n-Type of semiconductor thin films having nanometer grain size", *Chalcogenide Lett.,* vol. 6, pp. 219-225, 2009.

[34]   K.M. Garadkar, S.J. Pawar, P.P. Hankare, and A.A. Patil, "Effect of annealing on chemically deposited polycrystalline CdTe thin films", *J. Alloys Compd.,* vol. 491, no. 1-2, pp. 77-80, 2010.
[http://dx.doi.org/10.1016/j.jallcom.2009.10.146]

[35]   S. Rajpal, and S.R. Kumar, "Thermoluminescent properties of nanocrystalline ZnTe thin films: Structural and morphological studies", *Physica B,* vol. 534, pp. 145-149, 2018.
[http://dx.doi.org/10.1016/j.physb.2018.01.046]

[36]   S. Rajpal, and S.R. Kumar, "Development of Electrodeposited ZnTe Thin Films in Non-Aqueous Medium", *Journal of Nanoelectronics and Optoelectronics,* vol. 13, no. 2, pp. 190-194, 2018.
[http://dx.doi.org/10.1166/jno.2018.2294]

CHAPTER 8

Applications, Biomedical Necessities, and Green Future of Metallic Nanoparticles

Jyoti Bhattacharjee[1] and **Subhasis Roy**[1,*]

¹ Department of Chemical Engineering, University of Calcutta, Kolkata 700009, India

Abstract: Metallic nanoparticles like gold nanoparticles (AuNPs), magnetic iron oxide nanoparticles (Fe_3O_4), and cysteine-capped silver nanoparticles (Cyanopes) are changing the face of green nanotechnology. Their photonic capabilities, ultrafine size (10-100 nanometers), biocompatibility, diamagnetic strength, antibacterial activity, and photochemical qualities make them extremely useful in medical applications, radiotherapies, drug delivery, cosmetics, and solar cell coatings. This chapter provides a comprehensive outlook on the applications, biomedical necessities, and green future of metallic nanoparticles. The current discussion revolves around graphene-based nanofillers, focusing on their ability to enhance the tribological properties of aluminum and its alloys within the realm of materials research. Thin metallic tin sulfide nanoparticles and titanium oxide nanorods, on the other hand, play an important role in photochemical water splitting. Modern nanotechnology is advancing biological processes by allowing for a thorough examination of metallic nanoparticle forms as highlighted in the chapter. A notable application incorporates a nanoscale metallic lattice that facilitates the transfer of cisplatin and siRNA, showing great promise in re-sensitizing ovarian tumors. This chapter provides an exhaustive analysis of the potentials, benefits, and challenges associated with metallic nanoparticles, emphasizing their extensive applications and crucial role in the advancement of various fields.

Keywords: Biomedical, Quantum dots, Green synthesis, Nanoparticles, Nanofillers.

INTRODUCTION

The realm of metallic nanoparticles presents a captivating intersection of diverse applications, critical biomedical necessities, and the promising potential for a sustainable, green future. This introduction sets the stage for a comprehensive exploration of the multifaceted role played by metallic nanoparticles in various domains [1]. At the forefront of our discussion are the applications that underscore the versatility of metallic nanoparticles. From enhancing the tribo-

----

* Corresponding author Subhasis Roy: Department of Chemical Engineering, University of Calcutta, Kolkata 700009, India; E-mail: subhasis1093@gmail.com/srchemengg@caluniv.ac.in

Virat Khanna, Suneev Anil Bansal, Vishal Chaudhary and Reddicherla Umapathi (Eds.)

logical properties of materials, as seen in graphene-based nanofillers optimizing aluminum and its alloys, to the crucial role played by thin metallic tin sulfide nanoparticles and titanium oxide nanorods in photochemical water splitting, the breadth of applications is vast and impactful. In the biomedical sphere, metallic nanoparticles are proving indispensable [2, 3]. Modern nanotechnology is revolutionizing biological processes, allowing for a meticulous examination of metallic nanoparticle forms. Notably, the incorporation of nanoscale metallic lattices is demonstrating great promise, particularly in the targeted delivery of therapeutic agents like cisplatin and siRNA, showing potential in the re-sensitization of ovarian tumors. Moreover, the chapter delves into the essential biomedical necessities met by metallic nanoparticles, addressing challenges and showcasing benefits across various applications. The potential for precise drug delivery, diagnostic advancements, and therapeutic breakthroughs underscores the pivotal role of metallic nanoparticles in shaping the future of healthcare [1, 2]. Understanding the behaviour of dielectric losses is crucial for optimizing the performance of power cables, ensuring they meet the demanding requirements of biomedical applications. By comprehensively studying the electric field distribution in nanocomposite insulation and reliability, it is well-suited for critical biomedical applications [2]. The utilization of advanced materials, such as gold nanoparticles, magnetic iron oxide nanoparticles, and others, not only improves the electrical properties of the insulation but also aligns with sustainable and environmentally friendly practices. Looking forward, the narrative extends to the ecological front, exploring the green aspects of metallic nanoparticles. As advancements unfold, understanding how metallic nanoparticles contribute to environmentally sustainable practices becomes imperative. This introduction sets the groundwork for an in-depth analysis of the applications, biomedical necessities, and the promising trajectory toward a green future for metallic nanoparticles.

The article on green synthesis of metallic nanoparticles by Iravani *et al.* (2011) concentrates on the eco-friendly and sustainable method of producing metallic nanoparticles. The significance of hybridizing these nanoparticles using plant extracts, microbes, and other natural resources is highlighted in this chapter. It also covers green nanospheres' potential biomedical and environmental uses [1 - 3].

NMs are particularly sought-after in biomedical applications due to their unique physical and chemical features. A hybrid substance in nanospheres typically consists of a biological monolayer on top of an inorganic core. Controlled plant-mediated synthesis is more reliable, eco-friendly, economical, and less toxic than the conventional physicochemical synthesis of nanoparticles. Due to its distinctive optical, thermodynamic, magnetic, and mechanical properties, NM research has increased during the past few decades. For prospective medicinal applications, the excellent biocompatibility of nanoparticles has been acknowledged. Nanoparticles (NPS) with a range of acceptable dimensions have reportedly been shown to enter the bloodstream and undergo endocytosis in cells (Fig. **1**).

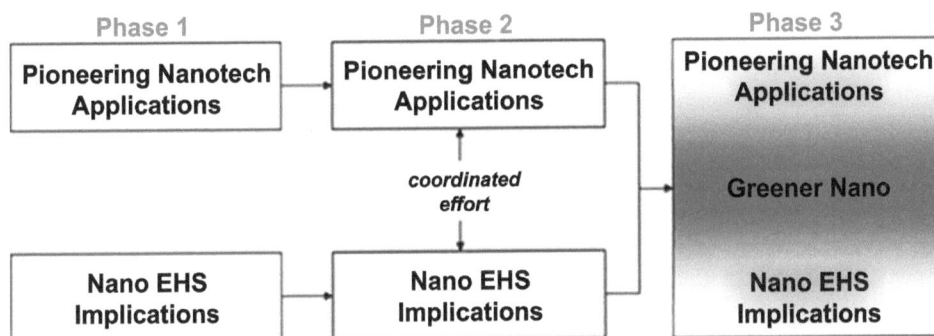

Fig. (1). Classification of green nanoscience using metallic nanoparticles [3]. (Reproduced with permission).

Nano fertilizers boost crop production efficiency and sustainability in agribusiness. Due to their nanosized characteristics, they have been found to boost productivity through targeted delivery or a gradual release of nutrients, lowering the chemical application rate. Through the use of nanosensors, plant nutrition, and plant protection, nanotechnology has applications in crop management and has the potential to be employed in precision farming. The main benefits of green nanotechnology include increased energy efficiency, decreased waste and greenhouse gas emissions, and less reliance on non-renewable raw materials. Nanoparticles offer complex antibacterial capabilities at low doses. For instance, zinc oxide nano-powders may eliminate Gram-negative and Gram-positive germs effectively. There has been an increase in research activities evaluating the technology for plant development and protection as evidence of nanoscale applications in agriculture has emerged. Nanostructured pesticide (nano pesticides) and nanostructured fertilizer (nano fertilizer) research are progressing, as shown in Fig. (**2**).

Once Michael Faraday discovered silver nanoparticles (AgNPs) (ruby-colored) in an aqueous solution in 1857 as a result of an Au salt reaction, he also discovered MNPs. These nanoparticles are useful in biomedicine due to their distinct physicochemical characteristics, such as their higher surface area-to-volume ratio. Surveys on early cancer detection are vital because the yearly prevalence of cancer is rising. Additionally, the possibility of biosensor-based cancer treatment monitoring offers promise for a customized course of treatment. This is the reason there is still a need for more practical, sensitive, and cost-effective approaches that can reveal even more details about a particular condition [4, 5].

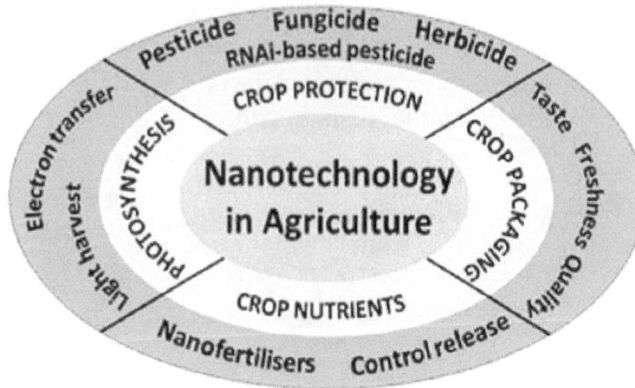

Fig. (2). Metallic nanoparticles' usage in forestry and agricultural sectors [4]. (Reproduced with permission).

Recently, there has been a lot of interest in using natural nanomaterials as alternatives to manmade products (Fig. **3**). One is nano-cellulose, composed of cellulose fibers or crystals produced by bacteria or plants at the nanoscale, for instance, bacterial-synthesized cellulose nanofibers which affect the growth times and culture mediums on the structural characteristics, and the pressure-engineered electrophoretic deposition for gentamicin loading within osteoblast-specific cellulose nanofiber scaffolds [6]. High strength, transparency, lightweight, and exceptional biocompatibility are all characteristics of these materials. Nano-cellulose aerogel has also been used as a sacrifice material for metal oxide nanotubes [6, 7].

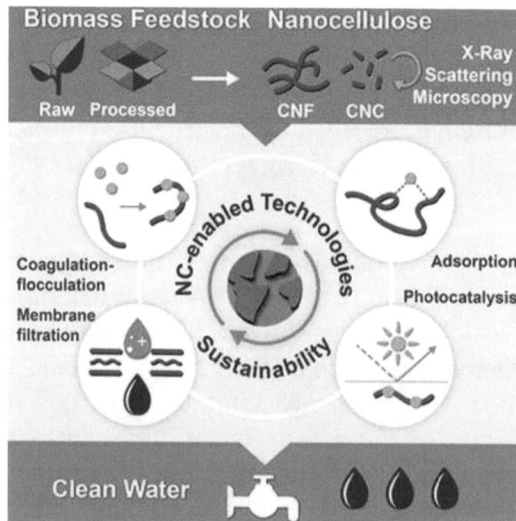

Fig. (3). Presentation of effective water purification using nano-cellulose [7]. (Reproduced with permission).

GRAPHENE-ALUMINUM NANOPARTICLES

The global demand for graphene and its environmentally friendly production processes faces everyday obstacles. One part involves synthesizing graphene from natural and industrial carbonaceous wastes, while another focuses on generating graphene from unusual sources such as food, insects, and waste materials. These problems have an impact on the overall availability and sustainability of graphene in a variety of industries. In recent years, various cutting-edge methodologies, including powder metallurgy, melting and solidification, electrodeposition, and some unique deposition techniques, have been developed to produce novel aluminum-graphene composites. Even though graphene gives outstanding electrical, thermal, and mechanical properties, all these Graphene contributions can be made if single-layer Graphene is created and present in the composite. This chapter also focuses on many noteworthy characteristics and influencing factors for Al/graphene composite.

A very prevalent type with a broad range of uses is aluminum. Due to its weak mechanical and limited electrical properties, its prospective usage was constrained. By successfully incorporating the supermaterial Graphene in the aluminum matrix, these aluminum drawbacks are eliminated. However, this composite has not yet been used commercially. Once such a composite is created, many fields will profit. Al/graphene composites can be utilized in the automotive sector, aerospace industries, transportation materials, military components, energy carrier materials, and material for consumer industries due to their exceptional strength, low density, surface-enhanced Raman scattering, and great electrical conductivity. The application of silane-grafted fumed silica nanoparticles for the development of stable transformer oil-based nanofluids holds promise for both biomedical and environmentally sustainable applications, contributing to the green future of metallic nanoparticles [8].

In addition to the roof, which is already being utilized, engineers are working to create new, lightweight, and flexible solar panels that might cover a structure's exterior. Graphene has sparked their curiosity. Reduced graphene oxide (rGO) generates reactive oxygen species (ROS) in the presence of visible light via a singlet oxygen-superoxide anion radical pathway, which quickly kills Enterobacter sp. GO is not hazardous, but as it ages, it acquires a surface coating of rGO and becomes toxic. This can be used to battle hospital bacteria [9]. This, combined with the electrical conductivity of Graphene, results in transparent, thin, flexible, and reasonably priced electrical conductors that are exceedingly effective. Fig. (**4a**) gives the Presentation of the wet etching of the core Ni-NPs and its magnification, SEM pictures of graphene balls with folded and flat structures were captured due to the collapse of a few graphene layers and Fig. (**4b**) shows A thin graphene layer has low mechanical stability, which causes it to collapse after wet etching instead of maintaining its original spherical shape.

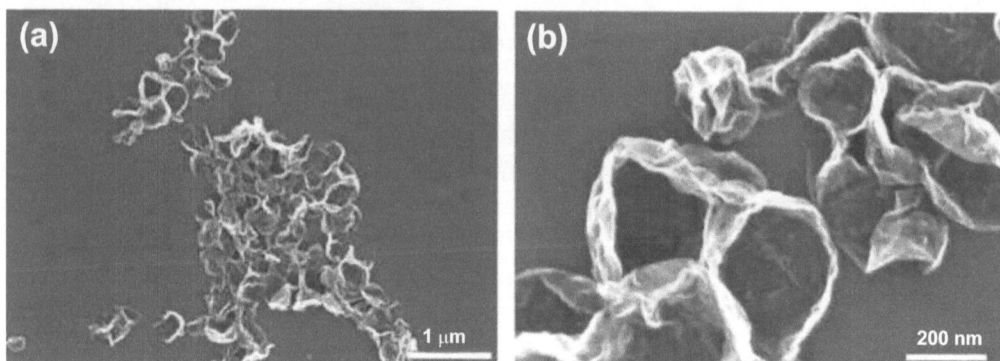

Fig. (4). (**a**) Presentation of the wet etching of the core Ni-NPs and its magnification, SEM pictures of graphene balls with folded and flat structures were captured due to the collapse of a few graphene layers [10]. (**b**) A thin graphene layer has low mechanical stability, which causes it to collapse after wet etching instead of maintaining its original spherical shape [10]. (Reproduced with permission).

GOLD NANOPARTICLES

When a reducing agent is available, gold salts can be reduced to produce gold nanoparticles. During the catalytic reaction, a stabilizing agent is typically added to aid in minimizing aggregation. Due to their stability, gold nanoparticles (AuNPs) have been fabricated in various shapes and configurations, including nanospheres, nanorods, nano-cubes, nano-fibers, nano-bipyramids, nanoflakes, nano-shells, nanotubes, and nanocages. A recent revolutionary study revealed the intravenous delivery of gold nanoparticles packed with tumor necrosis factor (TNF) to colon cancer. These radioactive gold nanoparticles have the potential to cure cancer of the prostate successfully. Using beta-emitting gold nanoparticles ensures a higher therapeutic dosage to treat breast and lung malignancies. By considering their biological applications as both diagnostic and therapeutic agents, the review emphasizes the biological characteristics of AuNPs in preclinical and clinical trials; their potential health risks, and the methods they employed to overcome their constraints, for instance, gold nanoparticles for antimicrobial purposes against multi-drug resistance bacteria. Their potential in medicine is addressed as a final point as shown in Fig. (**5**) [11].

Utilizing the Aqueous Extract of Garcinia Mangostana Fruit Peels for Green Synthesis of AuNPs

Mangosteen, scientifically known as Garcinia mangostana (G. mangostana), is a member of the Guttiferae family in Asian countries. At room temperature and under static circumstances, 20 mL of the peel extract was combined with ten mM of tetrachloroaurate in a conical flask. The color of the prepared solution quickly shifts from pale brownish to purple as [Au/G. mangostana] grows. The resulting Au-NPs nanoparticle emulsion was maintained at 4 °C. UV-visible spectra show that Au-NPs are present; tetrachloroaurate causes a solid peak to develop between 540 and 550 nanometers.

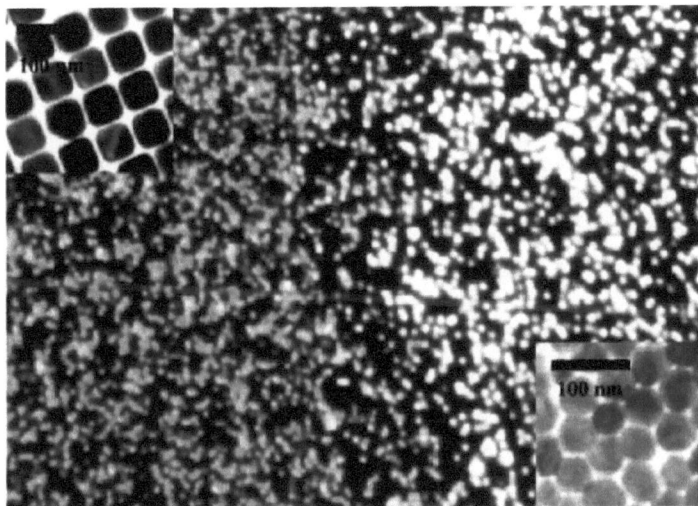

Fig. (5). The presentation of gold "nano-hexagons" and nanocubes was captured in a dark field microscope image on liquids dried down on a microscope slide. There is a 1 mm field of view. Nanoparticle transmission electron micrographs inset; 100 nm scale bars [11]. (Reproduced with permission).

The possible chemical equation for synthesizing the Au-NPs is-

$HauCl_4$(aqueous)+ $\longrightarrow$ G.magnostana

Au/G.mangostana.

Using plant peels maximizes the potential of undesirable waste materials and is inexpensive, effective, and secure. Synthetic Au-NPs can be applied to biomedical applications, tissue engineering, calorimetric detection of biomolecules, *etc.*

APPLICATION OF NANOPARTICLES TO REFRIGERATION SYSTEMS

From the literature review, it can be inferred that nanofluids and nano-refrigerants are desirable for our home refrigeration systems. Nanofluids are produced synthetically and, even at low concentrations, drastically change the characteristics of those fluids. Compared to traditional solid-liquid suspensions, nanofluids feature greater surface areas of nanoparticles, increased relative motion with improved dispersion stability, and better heat transmission between particles and fluids. There are fewer pumping forces and particle blockages compared to the base fluid. Nano lubricant is the result of mixing a lubricant (compressor oil) with nanoparticles. Scientists have discovered a way to blend pure refrigerant with tiny lubricants to produce nano-lubricant-refrigerant. Three different categories of nanofluids are lubricant, refrigerant, and lubricant-refrigerant. While nano-lubricant improves tribological characteristics, improving turbine performance in refrigeration systems, nano-refrigerant promotes thermo-physical properties, enhancing refrigerating action [12].

Metal oxides like zirconia, alumina, titania, and silica, metal carbides like silicon carbide, carbon in a variety of forms like fullerene, graphite, carbon nanotubes, and diamond, stable metals like copper and gold, metal nitrides like silicon nitride and oxide ceramics like copper oxide and aluminum oxide are among the most widely used nanoparticles (Fig. **6**).

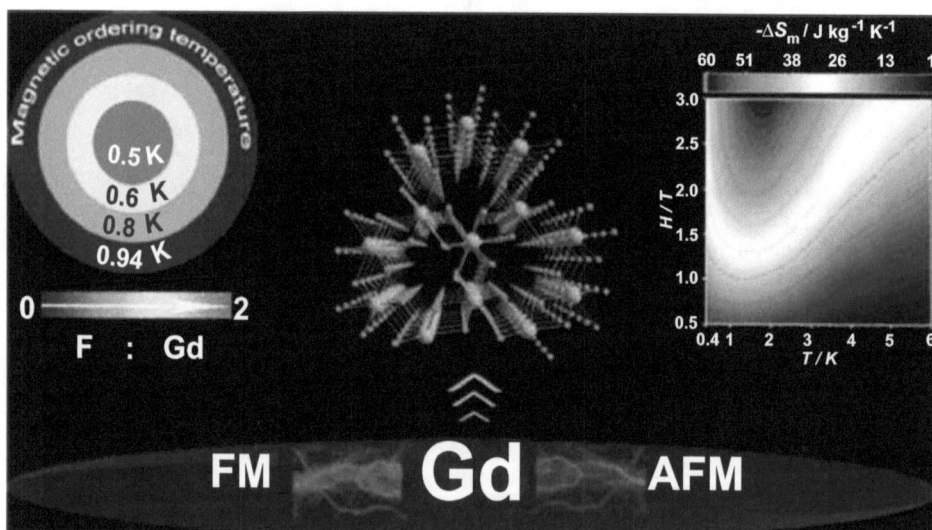

Fig. (**6**). Cryogenic magnetic refrigerant [12]. (Reproduced with permission).

SILVER NANOPARTICLES

In terms of inorganic materials that are friendly to the environment, AgNps are among the most appealing. Manufacturers use it, including photography, diagnostics, photocatalysts, biosensors, and antiviral drugs. Silver nanoparticles (AgNPs) are currently being investigated as a possible source of novel antimicrobial drugs due to their broad-spectrum activity and lesser potential to cause resistance, among other benefits. There have been studies of the extracellular production of Ag-NPs from silver nitrate solution using the fungus Trichoderma virus. This suggests that the Trichoderma virus is a critical biological element in the extracellular production of stable Ag-NPs. Due to their unique surface area-to-volume ratio and physicochemical characteristics, AgNPs are capable of various remarkable biological activities [13, 14]. In particular, nano-containing materials for bone implants and orthodontic equipment, as well as anti-bacterial agents, antivirals, antiprotozoal, anti-arthropods, and anti-cancerous agents, have found utility in the treatment of wounds and burns. Generally, AgNPs are synthesized in the liquid phase using chemical methods like classical reduction with citrate, gallic acid, photochemical, sonochemical methods, and pulse synthesis process (PSP) method (Fig. **7**).

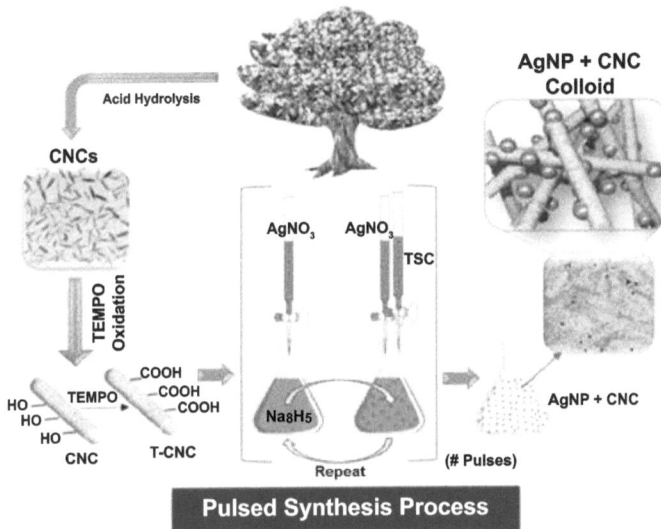

Fig. (7). Scheme showing pulsed synthesis of silver nanoparticles [13]. (Reproduced with permission).

Several extracts from plants, including those from Ocimum tenuiflorum (Black tulsi), Azadirachta indica (Neem), and Solanum Lycopersicum (Tomato), have been used to generate yellow-hued silver nanoparticle (Ag-NP)-based colloids [14]. These extracts act as excellent antioxidants and give metallic nanoparticles permanence.

According to many scientists, using consumer sprays based on nanotechnology and containing AgNPs can result in the emission of NPs and the production of nanosized aerosols proximal to the human breathing zone. In addition, regardless of whether they have undergone conventional or nanoscale Ag treatment, fabrics exposed to silver (Ag) can aid in forming AgNPs in washing solutions. Despite numerous studies evaluating lung exposure and toxicity, there is still no data on the long-term toxicity of silver nanoparticles, consumer exposure, or effects on human health. However, the proposed $0.19 \ g/m^3$ occupational exposure limit for AgNPs is based on subchronic rat inhalation poisoning research and kinetic analysis of the human equivalent dose. AgNPs have been shown to suppress a broad spectrum of bacteria in dressings for wounds and other applications to regulate the makeup of the skin's microbiome, including Pseudomonas aeruginosa, Escherichia coli, and Staphylococcus aureus [15]. As per numerous findings, AgNPs facilitate wound healing by suppressing an inflammatory reaction.

FACE MASKS AND NANOPARTICLES

The future of clothing technology will be nanoengineered functional textiles, in which nanoparticles are inserted into textiles to confer extra capabilities while keeping the comfort of the substrate. Due to their high surface-to-volume proportion and resemblance in size to virus particles, nanomaterials exhibit different biological, chemical, and physical properties than larger substances. Nanomaterials made of barium titanate ($BaTiO_3$) have been utilized to create inorganic ferroelectric materials for the electrospinning process that creates filter membranes in face masks [16].

Harmful health and environmental impacts that copper and silver nanoparticles may cause include immunotoxicity in mice's dams and offspring and toxic effects on marine animals. TiO_2 was once believed to be biologically safe, but researchers recently uncovered a growing body of data denoting TiO_2 toxicity to people and animals that are not its original target. A rare variety of multi-walled carbon nanotubes, known as an MWCNT, has also been connected to human carcinogenesis. Nanomaterials' aggregative state, both in the environment and the human body, significantly impacts their carcinogenicity. All nanotechnology products for direct contact with facial skin should undergo extensive dermatological laboratory tests. Although vaccine studies are ongoing, nanotechnology-based antiviral face masks may be a crucial ally in the fight against SARS-CoV-2 infections (Fig. **8**).

NANOSILICA

Nanosized silica, or SiO_2, has been produced thanks to advancements in nanotechnology and is frequently utilized as filler in industrial composites. Chemical vapor condensation (CVC) technologies are employed for manufacturing silica nanoparticles. Based on various fascinating optical phenomena caused by point defects developed from any defective SiO4 continuous network, including oxygen and silicon vacancies, silica nanoparticles have been the subject of extensive research. To monitor optical changes brought on by structural faults at the bulk and surface of nanoparticles, optical amplification and photoluminescence (PL) emerge as two valuable tools (Fig. **9**).

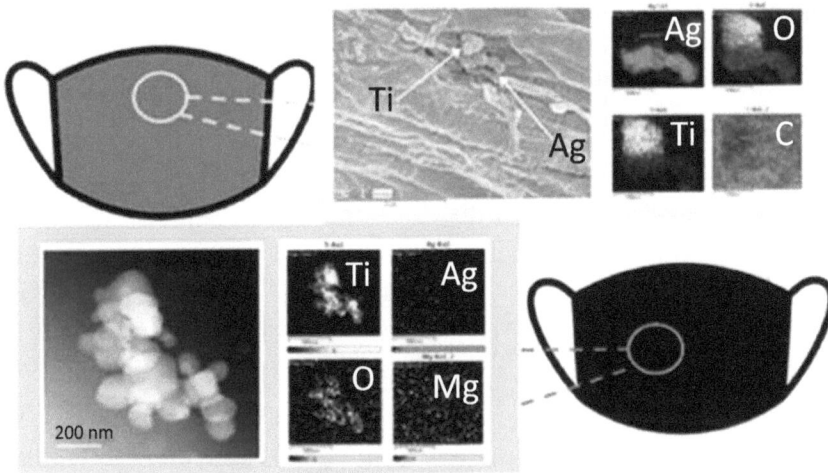

Fig. (8). Scheme showing the cloth face masks containing silver nanoparticles [16]. (Reproduced with permission).

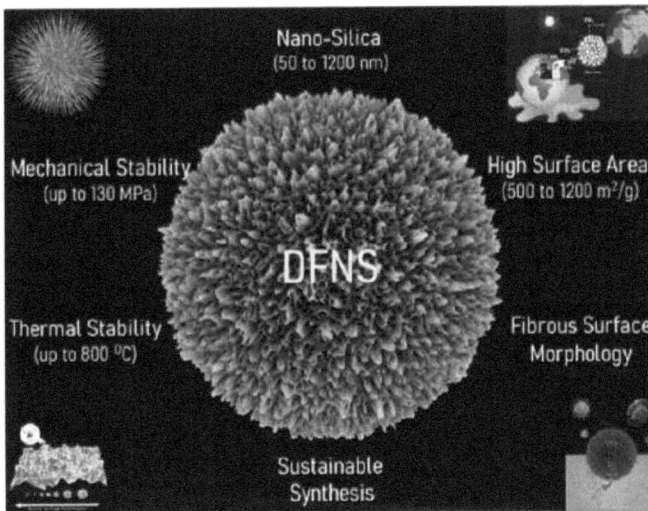

Fig. (9). Dendritic fibrous nano-silica (DFNS) [16]. (Reproduced with permission).

Spectra of Nanocomposites of Varying Diameters

Among the many benefits of nanoparticles are their remarkable mechanical strength, thermal stability, reduced loss of volume, thermal expansion, residual stress, and increased optical and electrical conductivity. In silica-polymer nanocomposites, silica nanoparticles are frequently used as fillers. The sol-gel method is the technique most frequently used to manufacture silica nanoparticles because, under ideal conditions, it can provide monodispersed nanoparticles with a constrained size distribution.

CHITOSAN NANOPARTICLES

Chitosan nanoparticles have received much attention in recent years because of their unique properties like biocompatibility, biodegradability, and low toxicity. Chitosan is a natural biopolymer derived from chitin, a polysaccharide found in crustaceans, insects, and fungal exoskeletons. Chitosan has been thoroughly studied for its prospective applications in fields such as medicine delivery, gene therapy, wound healing, and tissue engineering. Chitosan nanoparticles in gene therapy techniques have advanced in recent years [17]. This article reviews ionic gelation, which is a cutting-edge green methodology. This approach is simple, effective, and environmentally beneficial because no toxic chemicals or solvents are employed.

Characteristics of chitosan nanocubes can be readily altered by varying the concentration of chitosan, the crosslinking agent, and the pH of the medium. Chitosan nanoparticle size and morphology can be controlled by adjusting the stirring rate, temperature, and sonication duration throughout the synthesis process. One of the most promising biomedical applications of chitosan nanoparticles is drug delivery (Fig. **10**).

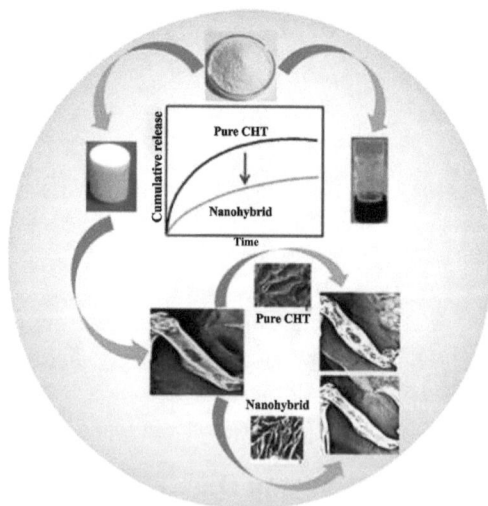

Fig. (10). Application of substratum and hydrogel based on chitosan for controlled drug release to bone regeneration [20]. (Reproduced with permission).

Chitosan nanoparticles can be engineered to release drugs under controlled conditions and encapsulate a wide range of drugs, such as chemotherapeutic agents, peptides, proteins, and nucleic acids. For instance, researchers have a chitosan nanoparticle-based system to improve the bioavailability of curcumin, a poorly water-soluble and poorly absorbed antioxidant with several health benefits. Tiny chitosan particles have also been researched for their potential use in gene therapy [18]. These can carry nucleic acids to target cells and tissues for gene regulation and manipulation, such as siRNA, plasmid DNA, and antagonistic oligonucleotides. Hence, it can be said that novel green manufacturing has a wide spectrum of applications [19].

QUANTUM DOTS

These nanometer-sized particles are extremely small and are composed of countless atoms. Germanium or silicon can be used to create these semiconductor materials. The Band theory in transistors can describe quantum dots. When discussing, HOMO and LUMO are typically the two molecular orbitals that interact [20 - 22]. Quantum dots have made it possible to study cell functions at the level of a single molecule, which may greatly improve the identification and treatment of illnesses. Due to its variable band gap, which enables harvesting solar light in both the infrared and visible wavelength areas, quantum dots (QDs) like cadmium sulfide, cadmium selenide, lead sulfide, *etc.* play a vital role in solar cell applications. Cadmium sulfide nanocrystals are used to prepare quantum dots solar cells without using any perovskite materials [21]. Quantum dots (QDs) are used in either inactive label testers, in which specific receptor atoms, such as antibodies, have been conjugated to the surface of the dots, or as active detector components in excellent quality cellular imaging, in which the quantum dots' fluorescence characteristics change following the analyte. As a result, graphene, the unwrapped planar form of a carbon nanotube, has emerged as one of the most exciting and intriguing substances for nanoscale electronics. Researchers have proven that a single graphene crystal may generate tiny transistors (graphene quantum dots) (Fig. **11**).

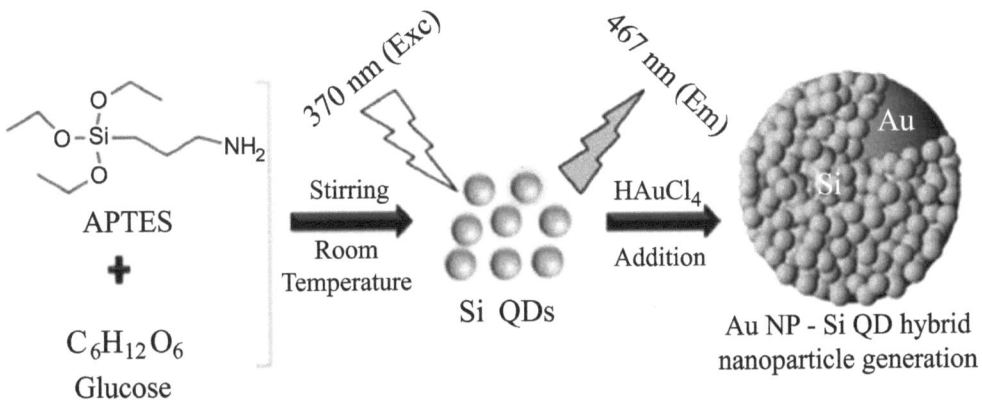

Fig. (11). Green synthesis of silicon quantum dots-silicon quantum dot nanocomposite [22]. (Reproduced with permission).

HYBRID METALLIC NANOPARTICLES

Hybrid nanoparticles (hybrid NPs) integrate different individual nanocomponents in a well-organized manner. The bio-nano interface and simultaneous growth of nanomaterials and molecular biology have led to the usage of hybrid NPs in various nanomedicine applications. The hybrid NPs exhibit the traits of the separate elements and their synergistic effects for specialized applications [23]. (Fig. **12**) portrays the new classes of hybrid nanoparticles (NPs) with the ability to deliver the maximum amount of drug loading to tumor cells while causing little or no harm to cells in good health.

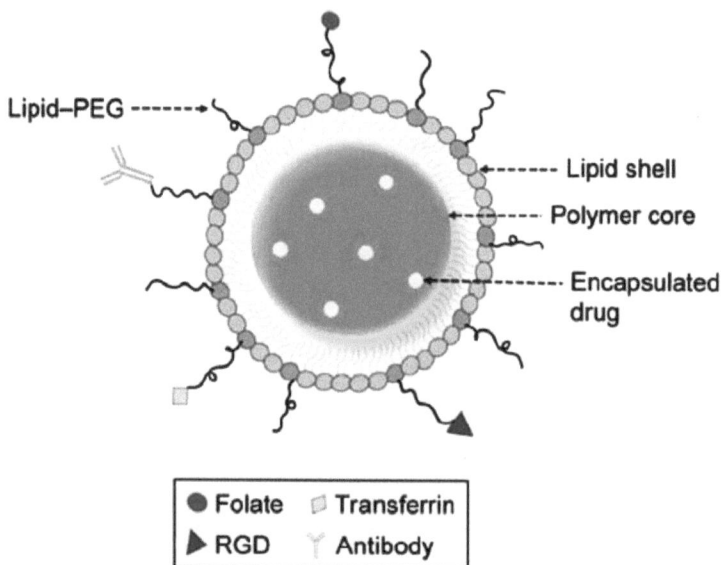

Fig. (12). A lipid-polymer hybrid nanoparticle (LPHNP) is the basic design of a drug-encapsulated core formed of polymer chains. The lipid component can be linked to various target molecules, including antibodies, folate, transferrin, and other short lipid-PEG chains. It is mostly found in the outermost layer [23]. (Reproduced with permission).

A hybrid composite is employed as a photocatalyst and solar cell made of titania-nano particles [24]. Rice husk/titania extraction yields pure crystalline nano-silica, and the hybrid composite is applied as a stain for photocatalyst and solar cell applications.

Due to the challenge of creating materials with more potential multifunctional uses, adding nanoscale components has increased the unique qualities of hybrid materials even more. Due to the physical and chemical attributes of the materials, which result in innovations with significant added value, multifunctional applications have piqued the curiosity of numerous researchers during the last 20 years. These functionalized graphene platforms for anticancer drug delivery substances are regarded as cutting-edge discoveries that can be used in various sectors, such as optics, electronics, sensors, ionics, energy storage and conversion, engineering, membranes, protective coatings, and catalysis [25]. An appropriate mix of two phases is used for magnetoelectric hybrid composites, such as piezomagnetic and piezoelectric or magnetostrictive and piezoelectric phases [26].

ADVANTAGES OF METALLIC NANOPARTICLES

Based on the nano concept, nanomedicine is developing new therapy methods and improving diagnostics by utilizing nanoparticles across several diagnostic types of equipment.

Optical and Electrical Functions

Applications based on the visual characteristics of metal nanoparticles include imaging sensors, displays, solar cells, photocatalysis, biomedical devices, optical detectors, and lasers. The key variables include form, size, surface area, doping, and environmental interactions. The optical characteristics of CdSe semiconductor nanoparticles might vary depending on their size. Adsorption Plasmon Au & Ag may alter color at the surface based on particle size, shape, and form, as well as condensation rate. An electric current runs through tiny tubes of carbon that are electrically conductive and have only one ion wave mode. Because of differences in length and orientation, electrically conducting carbon nanotubes interact with the outermost layer of the mercury at various moments. Because of their exceptional mechanical, electrical, and heat conductivity, carbon nanotubes, more commonly known as CNTs, are at the apex of multiple usage domains [27].

Fuel Catalysts and Medical Treatment

The fuel cell generates water, electricity, and heat. Because antibodies bind to the Au nanoparticle, healthy and malignant cells can be differentiated. Vegetable shelf life is also extended to a week; antibacterial zinc nano packaging can change the environment inside the packaging with controlled oxygen exchange.

Used as Paints and Sunscreen Lotions

Paint uses nano titanium dioxide to take advantage of two excellent qualities: UV protection and photocatalytic activity. Nano silicon dioxide can increase paints' abrasion, scratch, and macro and microhardness. Rutile titanium oxide nanocrystal analyses showed more photoactivity than large-sized standard rutile powder, both in photovoltaic and photocatalytic performance [28].

Nanomaterials are excellent in sunscreen protection because they effectively filter UV light for a long time. Skin burns are triggered by extended UV exposure. By using sunblock lotions with nano-TiO_2, one can enhance their sun protection factor (SPF) [29] (Fig. **13**).

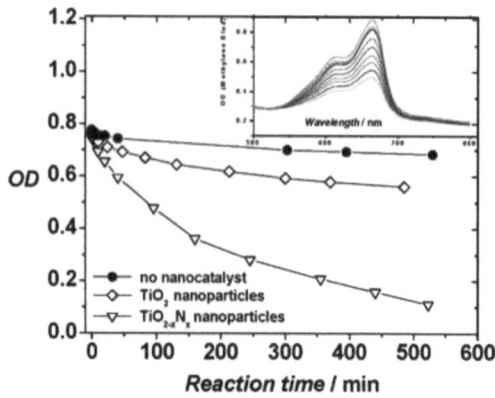

Fig. (13). TiO_2 Nanoparticles doped with nitrogen [30]. (Reproduced with permission).

The anti-angiogenic activity of green metallic nanoparticles is linked to their distinct physicochemical properties. Due to their small size, these nanoparticles can reach the cancer microenvironment and interact with endothelial cells, which are the building blocks of blood vessels. The surface chemistry of these nanoparticles can be modified to boost their affinity for specific target cells, such as endothelial cells. Numerous studies have demonstrated the usefulness of green metallic nanoparticles as anti-angiogenic agents in various tumor models. Endothelial cell apoptosis is mediated by the creation of silver nanoparticles from garlic extract. Similarly, gold nanoparticles derived from turmeric extract have been shown to inhibit melanoma cell proliferation by inhibiting VEGF expression. A mouse ear model was administered with an adenoviral vector conveying VEGF to assess the influence of gold nanoparticles on VEGF-mediated angiogenesis (Ad-VEGF- mimics the subsequent angiogenic response described in malignancies). The outcomes are shown in the (Fig. **14**).

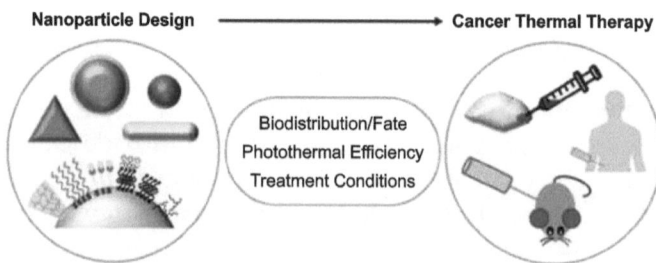

Fig. (14). Latest developments in gold nanoparticle thermotherapy for cancer [31]. (Reproduced with permission).

Nanoparticles are used in an emerging class of anti-arthritic medications that are generally less hazardous. Rheumatoid arthritis is an autoimmune condition that occurs when the immune system of an individual behaves inappropriately and attacks the joints. According to new research, gold particles can penetrate macrophages and prevent them from swelling the cells without killing them. According to research published in the Journal of Inorganic Biochemistry, making gold into smaller nanoparticles (50 nm) allowed for increased metal uptake by immune cells while minimizing toxicity.

CHALLENGES

Due to their outstanding chemical and physical properties, metal nanoparticles are perfect for use as self-modified therapeutic agents or as antibacterial drug carriers *in vitro* and *in vivo*. However, there are still a few limitations when working with metallic nanoparticles.

Each method of chemical and physical synthesis of nanoparticles has its setbacks, including using risky chemicals, producing poisonous by-products, and requiring high-energy processes. An alternate strategy of cheap, environmentally friendly biological synthesis mediated by plants or microbes has been chiefly embraced to address these shortcomings.

Existing medications still have several obstacles to getting where they desire to go. Treatment for most CNS and brain diseases is hampered by the blood-brain barrier (BBB), which prevents large or hydrophilic molecules from migrating into the cerebrospinal fluid. Several nanoparticle-based strategies are being developed to transfer medications across the BBB, including cationic albumin nanoparticles, nanospheres, and liposomes.

Metallic nanoparticles may require toxic and environmentally harmful methods to generate medicines and hydrophobic drugs. The absence of harsh chemicals gives nanomedicines a distinct advantage over their traditional ones. nab-paclitaxel, an albumin-bound nanostructures formulation of paclitaxel without cremophor, is the first nutrient nanotechnology-based chemotherapy to receive federal approval. Its average crystallite size is approximately 130 nm. Selecting a targeting moiety appropriate for treating the target disease may be crucial for improving a nanomedicine's effectiveness and minimizing side effects.

Nanoparticle, nitride, and oxide production can be worsened by a polluted environment while being manufactured. Due to their strong reactivity, nanoparticles also pose a high risk of contamination. Nanoparticles should be created as encapsulated particles in solution form. Therefore, eliminating contaminants in nanoparticles becomes challenging. Nanomaterials are reportedly harmful, cancer-causing, and irritable because they become transparent to the cell dermis (Fig. **15**).

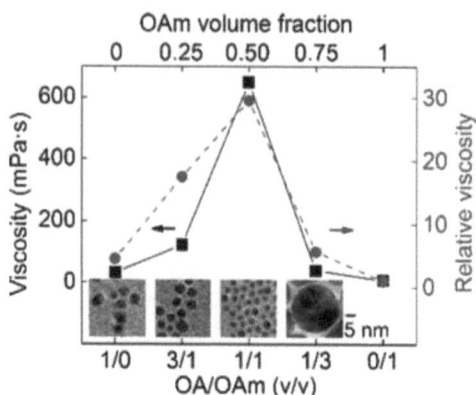

Fig. (15). Synergistic effects of oleic acid on sputtered metallic nanoparticles [32]. (Reproduced with permission).

FUTURE PROSPECTS

Presently metallic nanoparticles (MNPs) play an essential part in the growth of the commercial sector. Indeed, the development of nanotechnology may lead to numerous innovations as well as fresh economic opportunities. MNPs could have a broad range of applications in the future, including substantial use in many different sectors, particularly tumor therapy.

Due to their size, cytocompatibility, surface chemistry, relatively good stability, and controllable toxicity in biological systems, MNPs can be used for clinical diagnosis and target tissue therapy by Langmuir isotherm, which is a concept in physical chemistry that describes the relationship between the adsorption of molecules onto a surface and the concentration of those molecules in the surrounding solution [31, 32].

The bioconjugation of MNPs in tumor therapy is anticipated to significantly advance the state-of-the-art techniques for tumor-cell detection, tumor imaging, and tumor therapy.

MNPs are particularly helpful in the treatment of cancer. Additionally, they can enter cells more readily than non-metallic NPs of the same size, which is helpful for the treatment of cancer. Metals like silver, gold, and copper are examples of pure MNPs. Iron oxide, titanium dioxide, silicate, zinc oxide, and other metal oxide nanoparticles (NPs) also have several medical and therapeutic benefits and play an important role in green photonics and green chemistry principles.

Due to their small size, nanometals can interact with various cell surface components. Thus, they offer cancer patients receiving chemotherapy an efficient form of care (Fig. **16**).

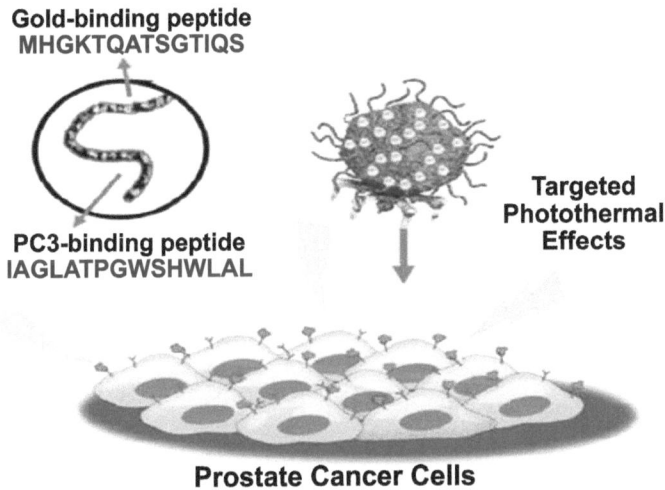

Fig. (16). Genetically modified clusters of nanoparticles for photothermal therapy of cancer Cells [33]. (Reproduced with permission).

Electrical Insulation: Nanocomposites, which are nanoparticles dispersed in a polymer matrix, can improve the electrical insulating qualities of a cable. This is critical for maintaining the integrity of power transmission, especially in biomedical devices where precision and reliability are critical. Nanocomposite-insulated power cables can help with the integration of smart grid technology, allowing for better monitoring, control, and optimization of power distribution. This leads to the creation of more efficient and responsive energy systems [34].

Oxygen-Rich Graphene/ZnO_2-Ag nanoframeworks: Zinc oxide (ZnO) is a semiconductor with antibacterial properties, and silver (Ag) nanoparticles are well known for their antimicrobial capabilities. The combination of ZnO and Ag in nano-frameworks shows a synergistic effect, perhaps increasing antibacterial activity. The nanomaterial system is proposed to manufacture oxygen nanobubbles. This process could entail the catalytic breakdown of hydrogen peroxide into oxygen and water. The nanobubbles may improve oxygen transport to specific bacterial locations, resulting in enhanced oxidative stress and bacterial inactivation. The combined impacts of oxygen-rich graphene, ZnO_2-Ag nano frameworks, and pH-switchable catalase/peroxidase activity aim to generate a hazardous environment for bacteria. The oxygen nanobubbles produced could potentially affect bacterial structures and metabolism, contributing to bacterial inactivation [35]. During the production process, graphene aerogel nanoparticles can be functionalized with binding sites or pores suited for trapping anticancer medicines.

In-situ loading entails integrating the drug molecules directly into the structure of the graphene aerogel during its production. pH-sensitive drug release guarantees that the loaded pharmaceuticals are released in a controlled manner in response to the unique pH environment of the target site (*e.g.*, acidic tumor microenvironment). Green synthesis approaches, such as environmentally friendly reducing agents and renewable resources, can be used to create graphene aerogel nanoparticles [36].

CONCLUSIVE REMARKS

In conclusion, the exploration of metallic nanoparticles reveals a captivating landscape rich with applications, biomedical advancements, and the promise of a sustainable future. The versatility exhibited by metallic nanoparticles, from enhancing material properties to playing a pivotal role in biological processes, highlights their multifaceted significance. The biomedical implications of metallic nanoparticles are particularly noteworthy, with nanotechnology offering unprecedented insights and solutions. The targeted delivery facilitated by nanoscale metallic lattices, as exemplified in the transfer of cisplatin and siRNA for ovarian tumor treatment, underscores their transformative potential in healthcare. As we navigate the challenges and harness the benefits of metallic nanoparticles, it becomes evident that these materials address critical biomedical necessities. From precise drug delivery to diagnostic innovations, metallic nanoparticles are instrumental in shaping the landscape of modern medicine. Numerous studies have been published on the ecologically friendly production of metallic nanoparticles (MNPs) and their potential biomedical applications. This chapter investigates natural and environmentally friendly methods of producing MNPs, such as employing plant extracts, microorganisms, and green solvents.

Due to its perfect chemical, electrical, optical, and magnetic properties, metallic nanoparticle creation is crucial. This chapter demonstrates that producing nanoparticles with a restricted size distribution using radiolytic and photolytic technologies has several benefits. In addition, electron pulse radiolysis and ultrafast lasers can create nanoparticles. Looking ahead, the incorporation of metallic nanoparticles into a green future is a promising avenue. Understanding and optimizing the environmental impact of these materials will be integral to ensuring that their continued use aligns with sustainability goals. As research and development progress, the responsible application of metallic nanoparticles can contribute to eco-friendly practices, further enhancing their appeal in diverse fields. In essence, this exploration illuminates the expansive role played by metallic nanoparticles in advancing science, medicine, and environmental consciousness. The journey from applications to biomedical necessities and towards a green future exemplifies the dynamism of metallic nanoparticles, positioning them as integral components in the ongoing pursuit of progress and sustainability.

ACKNOWLEDGMENT

Author (S. Roy) would like to acknowledge 'Scheme for Transformational and Advanced Research in Sciences (STARS)' (MoE-STARS/STARS-2/2023-0175) by the Ministry of Education, Govt. of India for promoting translational India-centric research in sciences implemented and managed by Indian Institute of Science (IISc), Bangalore, for their support.

REFERENCES

[1]   S. Maitra, S. Halder, T. Maitra, and S. Roy, "Superior light absorbing CdS/vanadium sulfide nanowalls@TiO2nanorod ternary heterojunction photoanodes for solar water splitting", *New Journal of Chemistry,* vol. 45, no. 16, pp. 7353-7367, 2021.
[http://dx.doi.org/10.1039/D0NJ06082H]

[2]   A. Thabet, and N. Salem, "Experimental investigation on dielectric losses and electric field distribution inside nanocomposites insulation of three-core belted power cables", *Advanced Industrial and Engineering Polymer Research,* vol. 4, no. 1, pp. 19-28, 2021.
[http://dx.doi.org/10.1016/j.aiepr.2020.11.002]

[3]   "Biosynthesized metallic nanoparticles as fertilizers: An emerging precision agriculture strategy", *Journal of Integrative Agriculture,* vol. 21, no. 5, pp. 1225-1242, 2022.
[http://dx.doi.org/10.1016/S2095-3119(21)63751-6]

[4]   Z. P. Xu, "Material Nanotechnology Is Sustaining Modern Agriculture", *ACS Agricultural Science & Technology,* vol. 2, no. 2, pp. 232-239, 2022.
[http://dx.doi.org/10.1021/acsagscitech.1c00204]

[5]   M. Khazaei, M. S. Hosseini, A. M. Haghighi, and M. Misaghi, "Nanosensors and their applications in early diagnosis of cancer", *Sensing and Bio-Sensing Research,* vol. 41, p. 100569, 2023.
[http://dx.doi.org/10.1016/j.sbsr.2023.100569]

[6]   R. Rahighi, M. Panahi, O. Akhavan, and M. Mansoorianfar, "Pressure-engineered electrophoretic deposition for gentamicin loading within osteoblast-specific cellulose nanofiber scaffolds", *Materials Chemistry and Physics,* vol. 272, p. 125018, 2021.
[http://dx.doi.org/10.1016/j.matchemphys.2021.125018]

[7]   A.T. Mohamed, and K.E.A. Ahmed, "o A. Thabet, and A.A. Ebnalwaled," Controlling on attraction forces of water droplets on surfaces of polypropylene nanocomposites coatings", *Transactions on Electrical and Electronic Materials,* vol. 19, no. 5, pp. 387-395, 2018.
[http://dx.doi.org/10.1007/s42341-018-0054-4]

[8]   M. I. Qureshi, and B. Qureshi, "Probing the Use of Silane-Grafted Fumed Silica Nanoparticles to Produce Stable Transformer Oil-Based Nanofluids", *Materials,* vol. 14, no. 24, p. 7649, 2021.
[http://dx.doi.org/10.3390/ma14247649]

[9]   S. Maity, "Microcarbon-based facial creams activate aerial oxygen under light to reactive oxygen species damaging cell", *Applied Nanoscience,* vol. 7, no. 8, pp. 607-616, 2017.
[http://dx.doi.org/10.1007/s13204-017-0604-9]

[10]   N. R. S. Sibuyi, "Multifunctional Gold Nanoparticles for Improved Diagnostic and Therapeutic Applications: A Review", *Nanoscale Research Letters,* vol. 16, no. 1, p. 174, 2021.
[http://dx.doi.org/10.1186/s11671-021-03632-w]

[11]   K. T. Nguyen, and Y. Zhao, "Engineered Hybrid Nanoparticles for On-Demand Diagnostics and Therapeutics", *Accounts of Chemical Research,* vol. 48, no. 12, pp. 3016-3025, 2015.
[http://dx.doi.org/10.1021/acs.accounts.5b00316]

[12]   S.-M. Yoon, "Synthesis of Multilayer Graphene Balls by Carbon Segregation from Nickel Nanoparticles", *ACS Nano,* vol. 6, no. 8, pp. 6803-6811, 2012.
[http://dx.doi.org/10.1021/nn301546z]

[13]   D. C. Ferreira Soares, S. C. Domingues, D. B. Viana, and M. L. Tebaldi, "Polymer-hybrid nanoparticles: Current advances in biomedical applications", *Biomedicine & Pharmacotherapy,* vol. 131, p. 110695, 2020.
[http://dx.doi.org/10.1016/j.biopha.2020.110695]

[14]   B. Fonseca-Santos, and M. Chorilli, "An overview of carboxymethyl derivatives of chitosan: Their use as biomaterials and drug delivery systems", *Materials Science and Engineering: C,* vol. 77, pp. 1349-1362, 2017.
[http://dx.doi.org/10.1016/j.msec.2017.03.198]

[15]   S. S. Nanda, D. K. Yi, and K. Kim, "Study of antibacterial mechanism of graphene oxide using Raman spectroscopy", *Scientific Reports,* vol. 6, no. 1, p. 28443, 2016.
[http://dx.doi.org/10.1038/srep28443]

[16]   S. Saha, "Unveiling the Electrocatalytic Activity of Crystal Facet-Tailored Cobalt Oxide-rGO Heterostructure Toward Selective Reduction of CO_2 to Ethanol", *ACS Applied Nano Materials,* vol. 5, no. 8, pp. 10369-10382, 2022.
[http://dx.doi.org/10.1021/acsanm.2c01703]

[17]   R. Jayakumar, K. P. Chennazhi, R. A. A. Muzzarelli, H. Tamura, S. V. Nair, and N. Selvamurugan, "Chitosan conjugated DNA nanoparticles in gene therapy", *Carbohydrate Polymers,* vol. 79, no. 1, pp. 1-8, 2010.
[http://dx.doi.org/10.1016/j.carbpol.2009.08.026]

[18]   H. Mousazadeh, Y. Pilehvar-Soltanahmadi, M. Dadashpour, and N. Zarghami, "Cyclodextrin based natural nanostructured carbohydrate polymers as effective non-viral siRNA delivery systems for cancer gene therapy", *Journal of Controlled Release,* vol. 330, pp. 1046-1070, 2021.
[http://dx.doi.org/10.1016/j.jconrel.2020.11.011]

[19]   Y. Ma, P. Liu, C. Si, and Z. Liu, "Chitosan Nanoparticles: Preparation and Application in Antibacterial Paper", *Journal of Macromolecular Science,* vol. 49, no. 5, pp. 994-1001, 2010.
[http://dx.doi.org/10.1080/00222341003609542]

[20]   N. Islam, and V. Ferro, "Recent advances in chitosan-based nanoparticulate pulmonary drug delivery", *Nanoscale,* vol. 8, no. 30, pp. 14341-14358, 2016.
[http://dx.doi.org/10.1039/C6NR03256G]

[21]   A. K. Mahanta, S. Senapati, P. Paliwal, S. Krishnamurthy, S. Hemalatha, and P. Maiti, "Nanoparticle-Induced Controlled Drug Delivery Using Chitosan-Based Hydrogel and Scaffold: Application to Bone Regeneration", *Molecular Pharmaceutics,* vol. 16, no. 1, pp. 327-338, 2018.
[http://dx.doi.org/10.1021/acs.molpharmaceut.8b00995]

[22]   G. Campisi, C. Paderni, R. Saccone, O. Fede, A. Wolff, and L. Giannola, "Human Buccal Mucosa as an Innovative Site of Drug Delivery", *Current Pharmaceutical Design,* vol. 16, no. 6, pp. 641-652, 2010.
[http://dx.doi.org/10.2174/138161210790883778]

[23]   B. Sharma, S. Tanwar, and T. Sen, "One Pot Green Synthesis of Si Quantum Dots and Catalytic Au Nanoparticle–Si Quantum Dot Nanocomposite", *ACS Sustainable Chemistry & Engineering,* vol. 7, no.

3, pp. 3309-3318, 2019.
[http://dx.doi.org/10.1021/acssuschemeng.8b05345]

[24]   A. I. Ribeiro, A. M. Dias, and A. Zille, "Synergistic Effects Between Metal Nanoparticles and Commercial Antimicrobial Agents: A Review", *ACS Applied Nano Materials,* vol. 5, no. 3, pp. 3030-3064, 2022.
[http://dx.doi.org/10.1021/acsanm.1c03891]

[25]   S. Sattari, M. Adeli, S. Beyranvand, and M. Nemati, "Functionalized Graphene Platforms for Anticancer Drug Delivery", *Int. J. Nanomedicine,* vol. 16, no. Aug, pp. 5955-5980, 2021.
[http://dx.doi.org/10.2147/IJN.S249712] [PMID: 34511900]

[26]   J. Bhattacharjee and S. Roy, "Utilizing a Variable Material Approach to Combat Climate Change," *Material Science Research India*, vol. 20, no. 3, pp. 141–145, Oriental Scientific Publishing Company, 2024.
[http://dx.doi.org/10.1021/nl034332o]

[27]   J. E. Hutchison, "Greener Nanoscience: A Proactive Approach to Advancing Applications and Reducing Implications of Nanotechnology", *ACS Nano,* vol. 2, no. 3, pp. 395-402, 2008.
[http://dx.doi.org/10.1021/nn800131j]

[28]   S. A. Moyez, and S. Roy, "Dual-step thermal engineering technique: A new approach for fabrication of efficient CH3NH3PbI3-based perovskite solar cell in open air condition", *Solar Energy Materials and Solar Cells,* vol. 185, pp. 145-152, 2018.
[http://dx.doi.org/10.1016/j.solmat.2018.05.027]

[29]   J. E. Hutchison, "The Road to Sustainable Nanotechnology: Challenges, Progress and Opportunities", *ACS Sustainable Chemistry & Engineering,* vol. 4, no. 11, pp. 5907-5914, 2016.
[http://dx.doi.org/10.1021/acssuschemeng.6b02121]

[30]   C. Burda, Y. Lou, X. Chen, A. C. S. Samia, J. Stout, and J. L. Gole, "Enhanced Nitrogen Doping in TiO2 Nanoparticles", *Nano Letters,* vol. 3, no. 8, pp. 1049-1051, 2003.
[http://dx.doi.org/10.1021/nl034332o]

[31]   K. R. Raghupathi, R. T. Koodali, and A. C. Manna, "Size-Dependent Bacterial Growth Inhibition and Mechanism of Antibacterial Activity of Zinc Oxide Nanoparticles", *Langmuir,* vol. 27, no. 7, pp. 4020-4028, 2011.
[http://dx.doi.org/10.1021/la104825u]

[32]   V. Farahani E, "pH-Sensitive Chitosan Hydrogel with Instant Gelation for Myocardial Regeneration", *Journal of Tissue Science & Engineering,* vol. 8, no. 3, pp. 1-10, 2017.
[http://dx.doi.org/10.4172/2157-7552.1000212]

[33]   S. Bera, D. Sengupta, S. Roy, and K. Mukherjee, "Research into dye-sensitized solar cells: a review highlighting progress in India", *Journal of Physics: Energy,* vol. 3, no. 3, p. 032013, 2021.
[http://dx.doi.org/10.1088/2515-7655/abff6c]

[34]   M. Fouad, A. Thabet, A.-M. Ahmed, and A. Elnodi, "High Performance of Power Cables Using Nanocomposites Insulation Materials", *Indonesian Journal of Electrical Engineering and Informatics,* vol. 9, no. 1, pp. 8-21, 2021.
[http://dx.doi.org/10.52549/ijeei.v9i1.1774]

[35]   M. Jannesari, O. Akhavan, H. R. Madaah Hosseini, and B. Bakhshi, "Oxygen-Rich Graphene/ZnO2-Ag nanoframeworks with pH-Switchable Catalase/Peroxidase activity as O2 Nanobubble-Self generator for bacterial inactivation", *Journal of Colloid and Interface Science,* vol. 637, pp. 237-250, 2023.
[http://dx.doi.org/10.1016/j.jcis.2023.01.079]

[36]   H. Ayazi, O. Akhavan, M. Raoufi, R. Varshochian, N. S. Hosseini Motlagh, and F. Atyabi, "Graphene aerogel nanoparticles for in-situ loading/pH sensitive releasing anticancer drugs", *Colloids and Surfaces B: Biointerfaces,* vol. 186, p. 110712, 2020.
[http://dx.doi.org/10.1016/j.colsurfb.2019.110712]

CHAPTER 9

Silver Nanoparticles with Enhanced Cytotoxicity and Biological Activity Produced from Green Methods

Celin. S. R.[1,*] and **R. Ajitha**[1]

[1] *Department of Chemistry, WCC (Affiliated to MS University, Abishekapatti, Tirunelveli-627012), Nagercoil, Tamilnadu, India*

Abstract: Research in the fields of physics, chemistry, and engineering is all facing more important challenges as a result of the rapid development of nanotechnology. The green synthesis of metallic nanoparticles opened the door for improvements and protections to be made to the environment by lowering the amount of harmful chemicals used and avoiding the biological dangers that were present in biomedical applications. Simple, fast, and environmentally friendly, plant-mediated production of metal nanoparticles is rising in popularity. We show an easy and environmentally friendly way to make silver nanoparticles using biomolecules found in an aqueous extract of the leaves of the plant *Kalanchoe gastonis-bonnieri.* No other chemical-reducing or stabilizing agent is needed in this way. The reaction is carried out in an aqueous solution in a process that is benign to the environment. This chapter examines the anti-oxidant, diabetic, inflammatory, cancer, and cytotoxic properties of silver nanoparticles that were generated utilizing the aqueous extract of the leaves of the plant *Kalanchoe gastonis-bonnieri.* The results of the investigation are presented and discussed in this chapter.

Keywords: AGS cells, Alpha-amylase, Green synthesis, Silver nanoparticles, Metal nanoparticles, Medicinal plants, Phytochemicals, Protein denaturation, SKMEL cells, Zeta potential.

INTRODUCTION

The term "green chemistry" refers to the process of producing chemical products and processes that minimise the use of hazardous compounds and the manufacture of those substances, decrease waste, and lessen the demand placed on resources that are in short supply [1]. In recent years, one emerging potential new area of research has been the environmentally friendly synthesis of metallic nano-

---

* **Corresponding author Celin S.R.:** Department of Chemistry, WCC (Affiliated to MS University, Abishekapatti, Tirunelveli-627012), Nagercoil, Tamilnadu, India; E-mail: celin.csr@gmail.com

Virat Khanna, Suneev Anil Bansal, Vishal Chaudhary and Reddicherla Umapathi (Eds.)

particles. A new way of thinking is used in the green synthesis method, which uses environmentally friendly solvents, less energy, and the removal of hazardous waste. Green synthesis has garnered a lot of attention as an alternative way to gain metal and metal oxide nanoparticles. Because conventional practices in chemical synthesis result in the production of potentially hazardous chemical species, which then adsorb onto the surfaces of nanoparticles, green synthesis is an environmentally friendly method of chemical synthesis. This is due to the fact that conventional approaches to chemical synthesis result in the creation of nanoparticles that include potentially harmful chemical species. Green synthesis does not result in the production of chemical species that are damaging to the environment [2].

Nanotechnology is the study and practice of designing, fabricating, and using nanostructured materials for a wide range of applications [3]. Richard Feynman, an American physicist who later won the Nobel Prize, is credited with having first proposed the concept of nanotechnology in 1959. At the annual conference of the American Physical Society, Richard Feynman delivered a speech with the title "There's Plenty of Space at the Bottom". During this talk, he addressed the topic, "Why can't we write the entire 24 volumes of the Encyclopedia Britannica on the head of a pin?" and presented a picture of the future wherein machines could be utilized to produce ever more miniature devices on the molecular level. As a result of this novel concept, which demonstrated that Feynman's ideas were accurate, Feynman is now widely recognized as the "father" of modern nanotechnology.

There has been a meteoric rise in the number of places around the globe where nanoscience and nanotechnology are being put to use, which has resulted in significant progress being made in the development of novel nanomaterials [4]. Scientific advancements and ground-breaking discoveries in this emerging sector have added to its reputation and attracted more funding for several labs in academic and business establishments [5]. Research institutes and businesses throughout the globe are now engaged in a mad dash to develop new goods and services that use nanotechnology. Many experts believe that this technology will shape the world's future as part of a new industrial revolution [6]. This boost in development is largely attributable to the unique features of these nanoscale materials. All of these recent advancements in the research and development of innovative uses for nanoparticles will have an immediate impact on business and society [7]. The majority of commercial items, which included anything from toys for children to products used for personal care, included nanoparticles made of metals and metal oxides.

The next industrial revolution will revolve around nanotechnology. Global governments have now come to terms with the fact that nanotechnology and its byproducts are here to stay; this cutting-edge technology is crucial since, unlike more antiquated methods, it does not break the bank and yields substantial financial benefits [8]. It is an amalgam of scientific knowledge and technological know-how with an eye towards practical scientific use. It all starts with atoms and molecules, the building blocks of matter. As a result, it influences pretty much every area of applied science and technology [9].

For plant genetic transformation, nanotechnology-based gene delivery technologies have recently been developed. This nano-approach demonstrates tremendous transformation performance, functionality, proper preservation of foreign nucleic acids, and the capacity to renew plant material [10]. Nanotechnologies have improved therapies for a number of illnesses, such as cancer [11], cardiovascular diseases [12], musculoskeletal disorders [13], mental and neurological diseases [14], bacterial [15], and viral infections [16], and diabetes [17]. They also hold great promise for the future of nanomedicine [18], including the development of imaging methods and diagnostic tools [19], drug [20] and gene delivery mechanisms [21], tissue-engineered constructions [22], implants [23], and pharmacological medicines [24]. Specific nanoparticle drug delivery systems can be made by attaching pharmaceuticals to radioactive antibodies that are on the same wavelength as antigens on cancer cells [25]. This method has shown promise in the past.

Because of their unique physicochemical properties, silver nanoparticles (AgNPs) are suited for a broad range of innovative profit-oriented and industrial applications. Some of these applications include antiseptic agents in the healthcare sector, beauty products, alimentary wrapping, genetic engineering, electrochemistry, and catalytic processes [26]. Even though AgNPs and NPs in general could be good for the economy, there are social concerns about their use. The most common and difficult issue to resolve prior to the general application of AgNPs synthesized by green synthesis techniques is the fine-tuning of the final properties [27].

It has been shown that a methodology that generates metal nanoparticles via the use of environmentally friendly synthesis processes is both dependable and cost-effective. Although microorganisms can also produce nanoparticles, their sluggish pace of synthesis and methodology only allows for a small range of sizes and shapes in comparison to plant-based techniques. Compared to other processes, plants create more stable nanoparticles, and scaling up is easy. There is also less worry about contamination [28].

In the present study, an aqueous leaf extract of the plant *Kalanchoe gastonis-bonnieri* (Fig. **1**) was used to synthesize AgNPs. This flowering plant species belongs to the *Crassulaceae* family. Because of the beneficial effects that *Kalanchoe* species have on inflammations, wounds, and abscesses, traditional medical practices in many parts of the globe, including India, Africa, China, and Brazil, make considerable use of *Kalanchoe* species [29].

Fig. (1). Photograph of *Kalanchoe gastonis-bonnieri.*

MATERIALS AND METHODS

Preparation of Extract

After gathering the young leaves of *Kalanchoe gastonis-bonnieri* (Fig. **1**), they were given a thorough cleaning in sterile distilled water and then washed again under running tap water. A knife was used to chop the leaves, and then 20 grams of leaf fragments and 200 mL of deionized water were placed in a conical flask with a capacity of 250 mL. The combination was heated on a heating mantle for 30 minutes in order to create the extract. After that, the extract was allowed to cool down before being filtered using Whatmann No. 1 filter paper.

Green Synthesis of Ag-NPs Using Leaf Extract

In a conical flask, 20 mL of the produced plant extract was mixed with 80 mL of a solution containing silver nitrate at a concentration of 1 mM. The beginning of the reaction was immediately identifiable by the cloudiness that appeared in the mixture after it had been stirred. The temperature of the reaction was not changed from room temperature. After 15 minutes, the color of the mixture turned brown, which showed the formation of Ag nanoparticles. Then the mixture was put aside for six hours. The Ag-NPs thus formed were collected by centrifugation.

Characterization

The size and form of the manufactured AgNPs were analyzed with the assistance of a high-resolution transmission electron microscope known as the Jeol/JEM 2100. The stability of the nanoparticles was investigated (using a Malvern Zeta Sizer Nano Series) to see how well they performed in water as a dispersion.

Phytochemical Screening

Phenolic Compounds

Together with a solution containing 1% lead acetate, 2 mL of plant extract was put into a test tube. The production of a white precipitate is indicative of the presence of phenolic chemicals in the sample.

Tannin

2 mL of plant extract should have a few drops of $aFeCl_3$ solution that is 0.1% added to it. The appearance of tannins may be recognized by the formation of a brownish-green color.

Flavonoids

10% lead acetate was applied to 2 mL of plant extract. The presence of flavonoids was revealed by the color's yellowish-green appearance.

Saponins

A test tube containing 2 mL of distilled water, 1 mL of each extract, and a few drops of olive oil was quickly shaken. The persistence of foam was seen as proof that saponins were present.

Terpenoid

Each extract from a plant sample accounted for 2 mL, and 2 mL of chloroform was added. 2 mL of concentrated sulfuric acid was added after allowing the liquid to evaporate, and then heated for 2 minutes. Terpenoids are present, as shown by the greyish tint.

Alkaloids

2 mL of Wagner's reagent was added to 2 mL of plant extract. We looked for the presence of a reddish-brown precipitate in test tubes.

Glycoside

It was decided to add a drop of 5% $FeCl_3$ and concentrated H_2SO_4 to a combination that already included 5 milliliters of plant extract, 2 milliliters of glacial acetic acid, and 5 milliliters of plant extract. When glycosides were present in a solution, we noticed that they formed a brown ring when we shook the container.

Anti-oxidant

The test sample's ability to scavenge free radicals from stable DPPH was assessed using a technique somewhat modified from that of Brand-William *et al.*, (1995). When an antioxidant chemical and DPPH interact, the antioxidant can contribute hydrogen and diminish DPPH. With a UV-visible spectrophotometer, the color shift (from deep violet to bright yellow) was observed at an optical density of 515 nm. Ascorbic acid served as the reference standard for the DPPH test. With distilled water, a stock solution of ascorbic acid was prepared. A newly prepared 200 µL of a 60 µM DPPH solution in methanol was combined with 50 µL of the test material at varied concentrations. The plates were left at room temperature in the dark for 15 minutes, and the reduction in absorbance was detected at 515 nm. The blank was made of 95% methanol [30].

For the purpose of determining the amount of radical scavenging activity, the following equation was used.

$$Percentage\ inhibition$$

$$= \left[\frac{Absorbance\ of\ control\ at\ 0\ min - Absorbance\ of\ test}{Absorbance\ of\ control\ at\ 15\ min} \right] \times 100$$

Anti-inflammatory

Along with 0.4 mL of bovine serum albumin, the reaction mixture (0.5 mL) included varying concentrations of the test sample. A total volume of 2.5 milliliters of phosphate-buffered saline with a pH of 6.3 was carefully added to each tube. The samples had previously been incubated at 37 °C. After that, the tubes were heated for ten minutes at a temperature of 80 °C. Using a spectrophotometer, the absorbance was determined to be 660 nm in wavelength

[31, 32]. For the purpose of determining the percentage of protein denaturation that was prevented, the following equation was utilized:

$$Percentage\ of\ inhibition = \left[\frac{Abs\ Control - Abs\ Sample}{Abs\ Control}\right] \times 100$$

Anti-diabetic

The method described in the Worthington Enzyme Manual was somewhat modified in order to conduct the analysis required to determine the anti-diabetic activity of the samples that were tested [33]. In a nutshell, the following was done: Various concentrations of the sample were pre-incubated at 37 °C for 10 min with 500 mL of 0.02 M Na_3PO_4 buffer, pH 6.9, with 0.006 M NaCl as enzyme inhibitors. After the first incubation, each tube was given 500 µL of a 1% starch solution that was suspended in a 0.02 M Na_3PO_4 buffer with a pH of 6.9. The tubes were then left to rest at room temperature for 5 minutes. The process was brought to a stop by using 1.0 mL of a reagent called dinitro salicylic acid. The temperature of the test tubes was brought down to room temperature after being heated to boiling for 5 minutes. After bringing the volume of the reaction mixture up to 10 mL with the addition of distilled water, an absorbance reading at 540 nm was taken using a spectrophotometer that measures UV-visible light.

Calculation:

$$Percentage\ inhibition = \frac{(B - A) \times 100}{(B - C)}$$

C- Absorbance of the Control with starch and without alpha-amylase

B- Absorbance of the Control with starch and alpha-amylase

A- Absorbance of the Test.

Anti-cancer and Cytotoxicity

The Human Gastrointestinal Epithelial (HGAEPC) cell line, the SKMEL cell line (human skin cancer), and the AGS-Human Gastric Adenocarcinoma cell line were all obtained from the National Centre for Cell Sciences in Pune, India. Through direct observation of the cells using an inverted phase contrast microscope, the vitality of the test sample-treated cells was assessed. By using the MTT test technique, the vitality of the treated cells was further evaluated.

RESULTS AND DISCUSSION

Phytochemical Screening

In the phytochemical screening conducted on the aqueous extract of the leaf of the *Kalanchoe gastonis-bonnieri* plant, the presence of the phytoconstituents namely, phenol, tannin, and saponins was identified. It is possible that these biomolecules are responsible for the decreased amount of nanoparticles as well as their stability. Table **1** displays the findings obtained from the preliminary examination.

Table 1. Phytochemical screening of *Kalanchoe gastonis-bonnieri* leaf extract

Constituents	Aqueous extract of *K.gastonis bonnieri* leaves
Phenol	+
Tannin	+
Flavonoid	−
Saponin	+
Terpenoid	−
Alkaloid	−
Glycoside	+

High-resolution Transmission Electron Microscopy

HR-TEM analysis provided conclusive evidence that nanoparticles were successfully synthesized. The HR-TEM picture that was acquired at a magnification of 20 nm (Fig. **2**) demonstrates that the produced nanoparticles are polydispersed and spherical in the suspension media, with just a small amount of agglomeration. Particle sizes ranging between 20 and 30 nm are observed. When the particles were zoomed in even further, to a magnification of 2 nm, Fig. (**3**) shows a single particle that has distinct lattice fringes and a d-spacing of 0.19 nm.

Zeta Potential

The zeta potential value was recorded for the AgNPs synthesized using *Kalanchoe gastonis-bonnieri* aqueous leaf extract. From the obtained result, we can see that the synthesized AgNPs are highly stable in a colloidal medium with a -40.9 mV zeta potential value. The peak obtained is shown in Fig. (**4**).

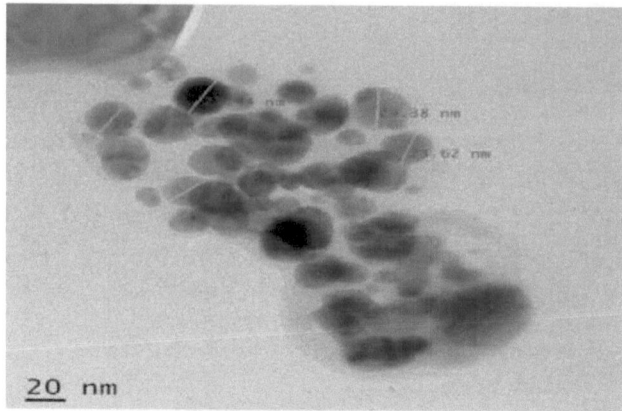

Fig. (2). HR-TEM image of AgNPs at magnification 20 nm.

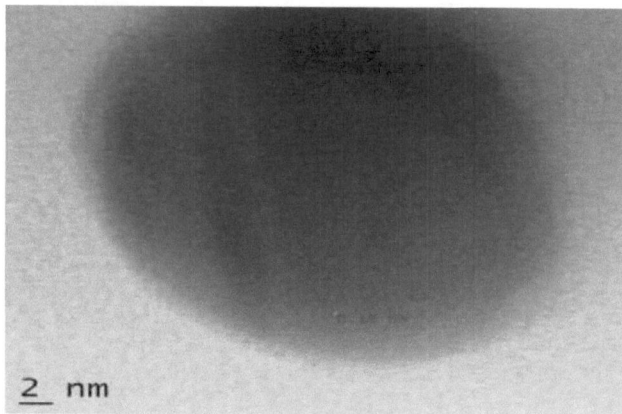

Fig. (3). HR-TEM image of AgNPs at magnification 2 nm.

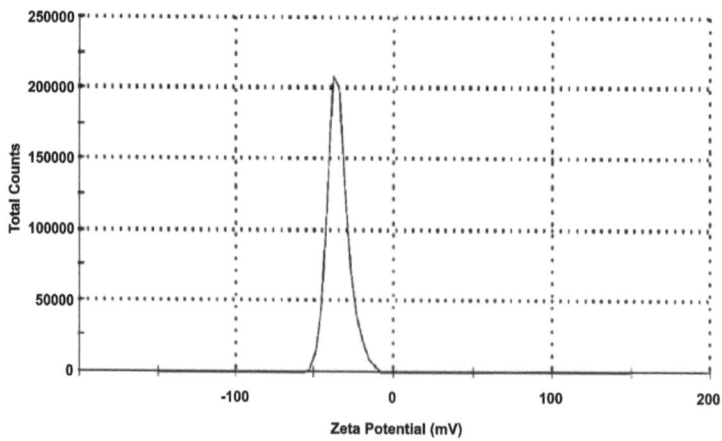

Fig. (4). Zeta potential of AgNPs synthesized using *K. gastonisbonnieri.*

Biological Applications

Anti-oxidant Activity

The activity of radical scavenging of AgNPs produced from the aqueous leaf extract of *Kalanchoe gastonis-bonnieri* was studied utilizing the DPPH test method. As the concentration is increased from 6.25 to 100 µg/mL, there is a slow but steady rise in the percentage of inhibition. The percentage of inhibition was found to be higher than 75% at a concentration of 50 µg/mL, which demonstrates that the produced AgNPs have strong anti-oxidant activity at low doses. This is essential for the AgNPs in order to function as medicines without producing toxicity in people. Fig. (5) displays a graphical depiction of the anti-oxidant activity that was measured using the DPPH test.

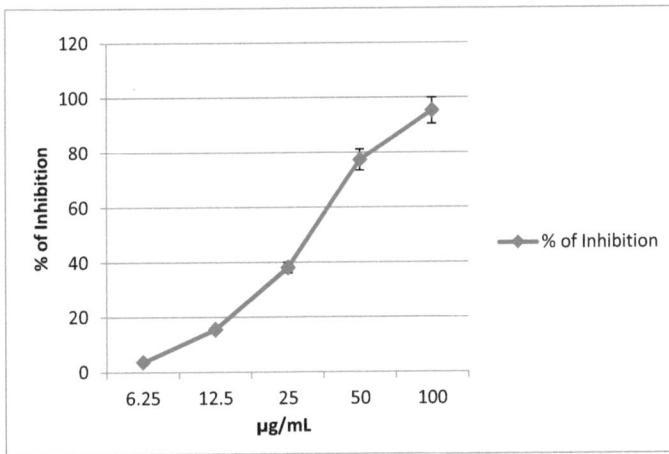

Fig. (5). Anti-oxidant activity of AgNPs synthesized using *K. gastonisbonnieri*.

Anti-inflammatory Activity

Denaturation of proteins is one of the primary processes that lead to inflammation. Therefore, in order to treat inflammatory disorders, it is possible to make use of medications that have the potential to stop the denaturation of proteins. Swelling, redness, pain, and heat are all results of the inflammation that occurs in the tissue. In the treatment of inflammation in traditional medicine, several components of plants are used. The protein denaturation test was used in this investigation to investigate the potential anti-inflammatory effects of AgNPs that had been produced using *Kalanchoe gastonis-bonnieri*. The percentage of inhibition offered by the AgNPs is 72.97% throughout a range of concentrations that extend from 6.25 to 100 µg/mL. Fig. (6) shows a graphical illustration of the anti-inflammatory activity of AgNPs that were produced utilizing *Kalanchoe gastonis-bonnieri*.

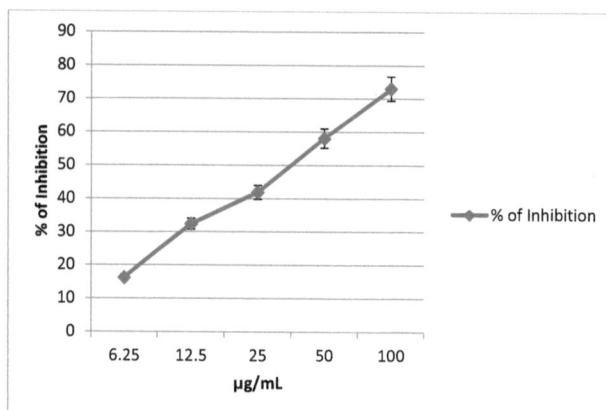

Fig. (6). Anti-inflammatory activity of AgNPs synthesized using *K.gastonisbonnieri.*

Anti-diabetic Activity

α-amylase and α-glucosidase in the pancreas and intestines break down oligos and disaccharides into monosaccharides [34]. Inhibiting these two digestive enzymes is an extremely useful therapeutic strategy for treating non-insulin-dependent diabetes [35]. Alpha-amylase is the primary enzyme responsible for the breakdown of polysaccharides and the release of sugar into the bloodstream [36]. Because of this, there is a spike in the amount of glucose in our bloodstream, which eventually culminates in diabetic complications. The anti-diabetic efficacy of the produced AgNPs was evaluated using an alpha-amylase inhibition test at a range of doses, including 10, 25, 50, 75, and 100 µg/mL and 71% of inhibition was seen at the highest concentration of 100 µg/mL. Fig. (7) shows a graphical illustration of the anti-diabetic activity of AgNPs that were synthesized using *Kalanchoe gastonis-bonnieri.*

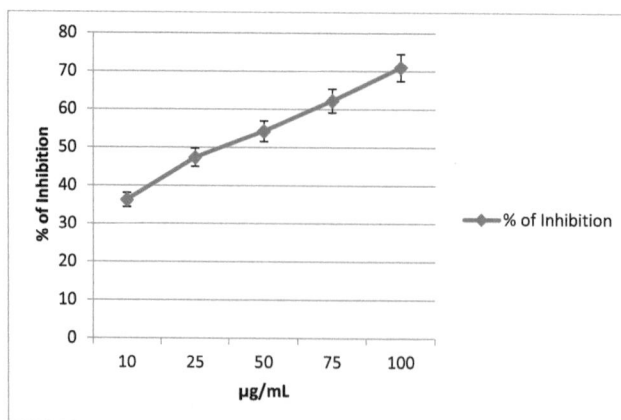

Fig. (7). Anti-diabetic activity of AgNPs synthesized using *K. gastonisbonnieri.*

Anti-cancer Activity

The MTT test is what is used to assess whether or not the substance inhibits the growth of cancer cells. The SKMEL (human skin cancer) cell line and the AGS (human gastric adenocarcinoma) cell line were both put through tests to see whether or not they have any anti-cancer capabilities. After 24 hours of incubation, the direct microscopic observations of AgNPs improved by *Kalanchoe gastonis-bonnieri* leaf extract-treated pictures of SKMEL and AGS cells were displayed in Fig. (**8** and **9**). Fig. (**10**) displays the proportion of viable cells that were produced by the AgNPs when tested against the SKMEL and AGS cell lines at doses of 10, 25, 50, 75, and 100 µg/mL. The fact that the IC_{50} value for the SKMEL (human skin cancer) cell line was discovered to be 90.11 µg/mL demonstrates that AgNPs are moderately cytotoxic toward the skin melanoma cancer cell line. The fact that the IC_{50} value for AGS cells was determined to be 50.26 µg/mL demonstrates that the produced AgNPs were successfully cytotoxic in their natural state on human stomach/gastric cancer cells.

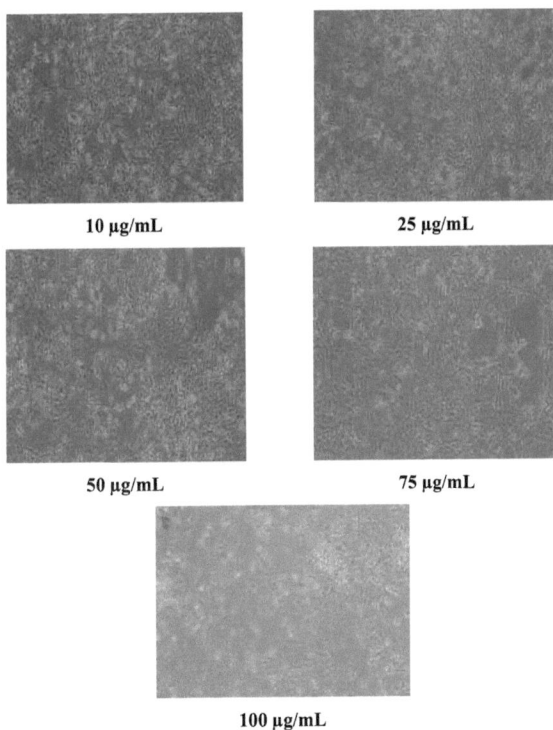

10 µg/mL

25 µg/mL

50 µg/mL

75 µg/mL

100 µg/mL

Fig. (8). Anti-cancer activity of AgNPs on SKMEL (Human Skin Cancer) cell line at five different concentrations.

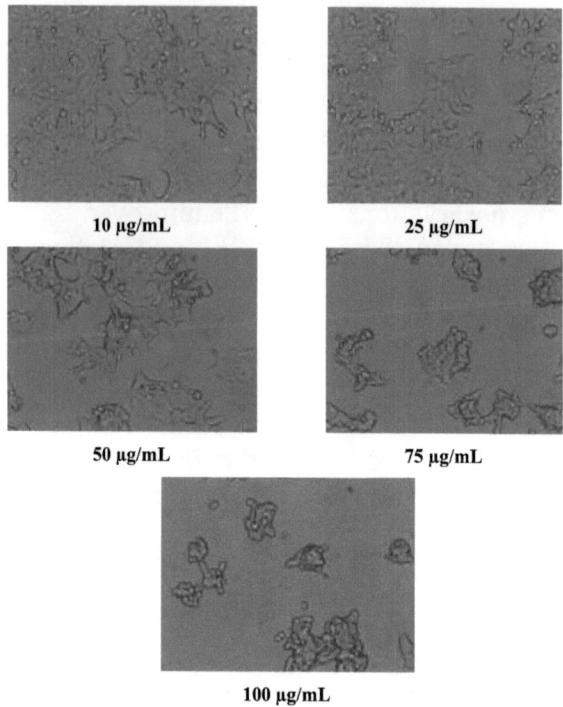

10 µg/mL   25 µg/mL

50 µg/mL   75 µg/mL

100 µg/mL

Fig. (9). Anti-cancer activity of AgNPs on AGS-Human Gastric adenocarcinoma cell line at five different concentrations.

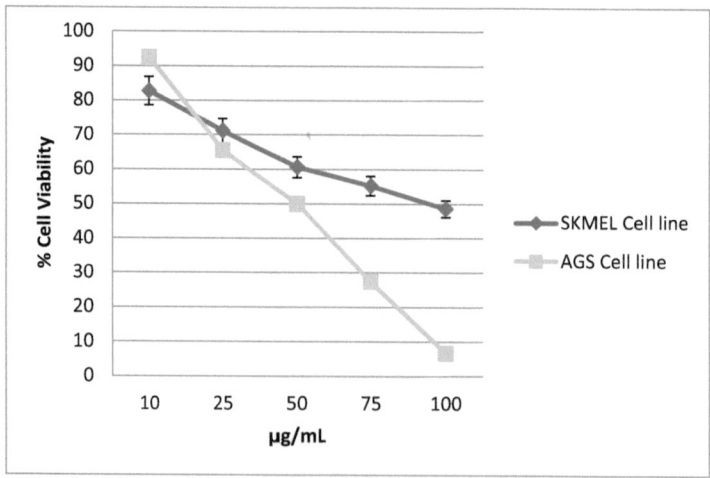

Fig. (10). Percentage viability of AgNPs against SKMEL and AGS cell line.

In general, the anti-cancer activity of AgNPs boosted by *Kalanchoe gastonis-bonnieri* aqueous leaf extract has excellent anti-cancer activity against AGS cells compared to SKMEL cell lines with a low IC_{50} value. This was determined by comparing the two types of cell lines [37 - 41]. To validate the mechanism of action of these chemicals, however, more research, such as an assay for apoptosis and necrosis, an assay for reactive oxygen species, a study of the cell cycle, and an assay for mitochondrial membrane damage, may be necessary.

Cytotoxicity

Because the newly produced AgNPs are more effective against the AGS cell line, the cytotoxic impact on a normal, non-malignant human gastrointestinal epithelial cell line (HGAEPC) was investigated. Fig. (**11** and **12**) provide direct microscopic observations of normal, non-malignant human gastrointestinal epithelial cells that were treated with AgNPs. Additionally, Figures 11 and 12 present the percentage of cell vitality that was observed. After an incubation period of 24 hours, the findings revealed that the AgNPs that were generated utilizing *Kalanchoe gastonis-bonnieri* were shown to be significant with higher IC_{50} values of 131.22 μg/mL against normal non-malignant gastric epithelial cells. This was determined by looking at the outcomes of the experiment. Therefore, the AgNPs that were generated utilizing the aqueous leaf extract of *Kalanchoe gastonis-bonnieri* were shown to be an anti-cancer drug. This was shown by the fact that they caused toxicity in gastric cancer cells while exhibiting no sign of toxicity in normal, non-malignant gastric epithelial cells.

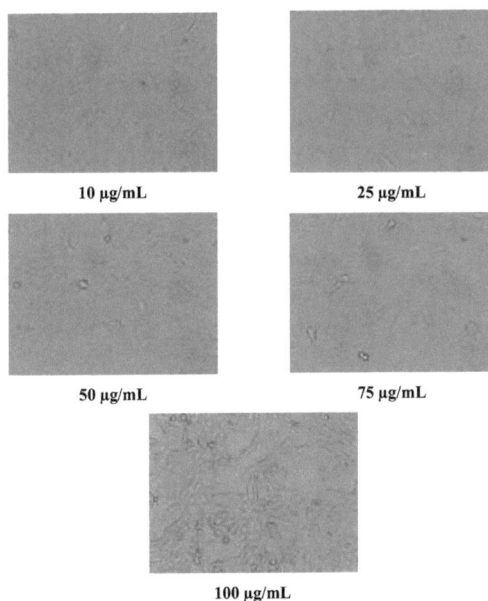

Fig. (11). Cytotoxicity of AgNPs against normal non-malignant human gastrointestinal epithelial cell at five different concentrations..

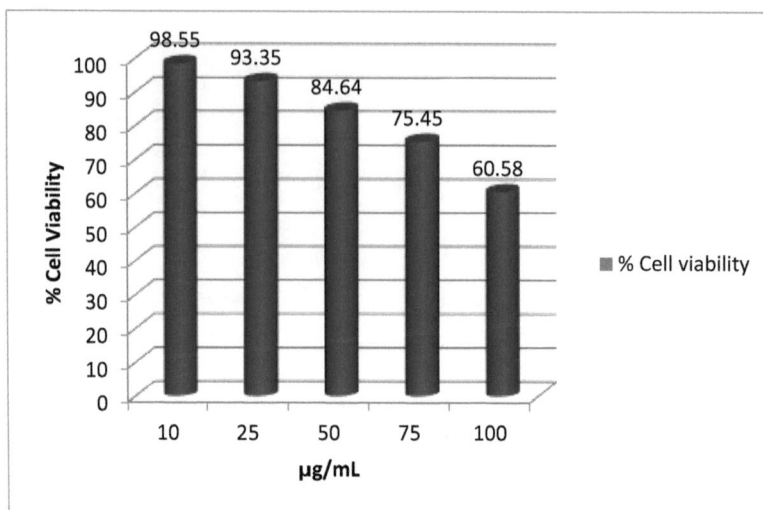

Fig. (12). Percentage viability of AgNPs against normal non-malignant human Gastrointestinal epithelial cell.

CONCLUSION

AgNPs were effectively produced at room temperature in the shortest amount of time possible by making use of the biomolecules that were present in the aqueous leaf extract of the plant *Kalanchoe gastonis-bonnieri*. This was accomplished without the use of any harmful chemical. The zeta potential value and HR-TEM pictures both indicated that the AgNPs produced had a high degree of stability. These nanoparticles were spherical and polydispersed. It has been shown that AgNPs augmented with *Kalanchoe gastonis-bonnieri* have effective anti-cancer action due to their high anti-oxidant activity at low concentrations. The anti-cancer activity against SKMEL and AGS cell lines was investigated as a result. The findings indicated that the AgNPs exhibited toxicity against the AGS cell line, which showed that the AgNPs have strong anti-cancer efficacy. The fact that it does not have any adverse effects on normal, non-malignant gastric epithelial cells demonstrates that it is an effective anti-cancer drug. In addition to these benefits, it also has powerful anti-inflammatory and anti-diabetic effects.

REFERENCES

[1] A. DeVierno Kreuder, T. House-Knight, J. Whitford, E. Ponnusamy, P. Miller, N. Jesse, R. Rodenborn, S. Sayag, M. Gebel, I. Aped, I. Sharfstein, E. Manaster, I. Ergaz, A. Harris, and L. Nelowet Grice, "A Method for Assessing Greener Alternatives between Chemical Products Following the 12 Principles of Green Chemistry", *ACS Sustain. Chem.& Eng.,* vol. 5, no. 4, pp. 2927-2935, 2017.
[http://dx.doi.org/10.1021/acssuschemeng.6b02399]

[2] S.P.K. Malhotra, and M.A. Alghuthaymi, "Biomolecule-assisted biogenic synthesis of metallic nanoparticles", In: *Agri-Waste and Microbes for Production of Sustainable Nanomaterials* Elsevier, 2022, pp. 139-163.
[http://dx.doi.org/10.1016/B978-0-12-823575-1.00011-1]

[3]   H. Rauscher, B. Sokull-Klüttgen, and H. Stamm, "The European Commission's recommendation on the definition of nanomaterial makes an impact", *Nanotoxicology,* vol. 7, no. 7, pp. 1195-1197, 2013.
[http://dx.doi.org/10.3109/17435390.2012.724724] [PMID: 22920756]

[4]   Á.I. López-Lorente, and M. Valcárcel, "The third way in analytical nanoscience and nanotechnology: Involvement of nanotools and nanoanalytes in the same analytical process", *Trends Analyt. Chem.,* vol. 75, pp. 1-9, 2016.
[http://dx.doi.org/10.1016/j.trac.2015.06.011]

[5]   M. Thabet Ahmed, "Design and Investment of High Voltage NanoDielectrics. IGI Global", *Publisher of Timely Knowledge,* no. Aug, p. 363, 2020.

[6]   A.T. Mohamed, "Emerging Nanotechnology Applications in Electrical Engineering. IGI Global", *Publisher of Timely Knowledge,* no. June, p. 318, 2021.

[7]   G. Miller, and F. Wickson, "Risk analysis of nanomaterials: Exposing nanotechnology's naked emperor: Risk analysis of nanomaterials", *Rev. Policy Res.,* vol. 32, no. 4, pp. 485-512, 2015.
[http://dx.doi.org/10.1111/ropr.12129]

[8]   A. Thabet, M. Allam, and S.A. Shaaban, "Investigation on enhancing breakdown voltages of transformer oil nanofluids using multi-nanoparticles technique", *IET Gener. Transm. Distrib.,* vol. 12, no. 5, pp. 1171-1176, 2018.
[http://dx.doi.org/10.1049/iet-gtd.2017.1183]

[9]   F.A. Ahmed Thabet Mohamed, "Al-Mufadi, and A. A. Ebnalwaled, "Synthesis and Measurement of Optical Light Characterization for Modern Cost-fewer Polyvinyl Chloride Nanocomposites Thin Films", *Transactions on Electrical and Electronic Materials,* vol. 25, pp. 98-109, 2023.
[http://dx.doi.org/10.1007/s42341-023-00489-x]

[10]   Y. Yan, X. Zhu, Y. Yu, C. Li, Z. Zhang, and F. Wang, "Nanotechnology strategies for plant genetic engineering", *Adv. Mater.,* vol. 34, no. 7, p. 2106945, 2022.
[http://dx.doi.org/10.1002/adma.202106945] [PMID: 34699644]

[11]   O. Akhavan, and E. Ghaderi, "Graphene nanomesh promises extremely efficient in vivo photothermal therapy", *Small,* vol. 9, no. 21, pp. 3593-3601, 2013.
[http://dx.doi.org/10.1002/smll.201203106] [PMID: 23625739]

[12]   D.K. Wang, M. Rahimi, and C.S. Filgueira, "Nanotechnology applications for cardiovascular disease treatment: Current and future perspectives", *Nanomedicine,* vol. 34, no. Mar, p. 102387, 2021.
[http://dx.doi.org/10.1016/j.nano.2021.102387] [PMID: 33753283]

[13]   M. Razavi, *Nanoengineering in Musculoskeletal Regeneration.* Academic Press, 2020, pp. 105-136.

[14]   K. Radhakrishnan, P. Senthil Kumar, G. Rangasamy, K. Ankitha, V. Niyathi, V. Manivasagan, and K. Saranya, "Recent advances in nanotechnology and its application for neuro-disease: a review", *Appl. Nanosci.,* vol. 13, no. 9, pp. 6631-6665, 2023.
[http://dx.doi.org/10.1007/s13204-023-02958-4]

[15]   M. Xie, M. Gao, Y. Yun, M. Malmsten, V.M. Rotello, R. Zboril, O. Akhavan, A. Kraskouski, J. Amalraj, X. Cai, J. Lu, H. Zheng, and R. Li, "Antibacterial Nanomaterials: Mechanisms, Impacts on Antimicrobial Resistance and Design Principles", *Angew. Chem. Int. Ed.,* vol. 62, no. 17, p. e202217345, 2023.
[http://dx.doi.org/10.1002/anie.202217345] [PMID: 36718001]

[16]   M. Ebrahimi, M. Asadi, and O. Akhavan, "Graphene-based Nanomaterials in Fighting the Most Challenging Viruses and Immunogenic Disorders", *ACS Biomater. Sci. Eng.,* vol. 8, no. 1, pp. 54-81, 2022.
[http://dx.doi.org/10.1021/acsbiomaterials.1c01184] [PMID: 34967216]

[17]   Y.V. Simos, K. Spyrou, M. Patila, N. Karouta, H. Stamatis, D. Gournis, E. Dounousi, and D. Peschos, "Trends of nanotechnology in type 2 diabetes mellitus treatment", *Asian J. Pharm. Sci.,* vol. 16, no. 1, pp. 62-76, 2021.

[http://dx.doi.org/10.1016/j.ajps.2020.05.001] [PMID: 33613730]

[18] H.Y. Mao, S. Laurent, W. Chen, O. Akhavan, M. Imani, A.A. Ashkarran, and M. Mahmoudi, "Graphene: promises, facts, opportunities, and challenges in nanomedicine", *Chem. Rev.,* vol. 113, no. 5, pp. 3407-3424, 2013.
[http://dx.doi.org/10.1021/cr300335p] [PMID: 23452512]

[19] Michal Wozniak, "Molecular Imaging and Nanotechnology—Emerging Tools in Diagnostics and Therapy", *Int J Mol Sci,* vol. 23, no. 5, pp. 2658-2658, 2022.

[20] D. Lombardo, M.A. Kiselev, and M.T. Caccamo, "Smart nanoparticles for drug delivery application: Development of versatile nanocarrier platforms in biotechnology and nanomedicine", *J. Nanomater.,* vol. 2019, pp. 1-26, 2019.
[http://dx.doi.org/10.1155/2019/3702518]

[21] Z. Zahed, R. Hadi, G. Imanzadeh, Z. Ahmadian, S. Shafiei, A.Z. Zadeh, H. Karimi, A. Akbarzadeh, M. Abbaszadeh, L.S. Ghadimi, H.S. Kafil, and F. Kazeminava, "Recent advances in fluorescence nanoparticles "quantum dots" as gene delivery system: A review", *Int. J. Biol. Macromol.,* vol. 254, no. Pt 2, pp. 127802-127802, 2024.
[http://dx.doi.org/10.1016/j.ijbiomac.2023.127802] [PMID: 37918598]

[22] H. Amani, E. Mostafavi, H. Arzaghi, S. Davaran, A. Akbarzadeh, O. Akhavan, H. Pazoki-Toroudi, and T.J. Webster, "Three-Dimensional Graphene Foams: Synthesis, Properties, Biocompatibility, Biodegradability, and Applications in Tissue Engineering", *ACS Biomater. Sci. Eng.,* vol. 5, no. 1, pp. 193-214, 2019.
[http://dx.doi.org/10.1021/acsbiomaterials.8b00658] [PMID: 33405863]

[23] M.T. Matter, L. Maliqi, K. Keevend, S. Guimond, J. Ng, E. Armagan, M. Rottmar, and I.K. Herrmann, "One-Step Synthesis of Versatile Antimicrobial Nano-Architected Implant Coatings for Hard and Soft Tissue Healing", *ACS Appl. Mater. Interfaces,* vol. 13, no. 28, pp. 33300-33310, 2021.
[http://dx.doi.org/10.1021/acsami.1c10121] [PMID: 34254508]

[24] H. Devalapally, A. Chakilam, and M.M. Amiji, "Role of nanotechnology in pharmaceutical product development", *J. Pharm. Sci.,* vol. 96, no. 10, pp. 2547-2565, 2007.
[http://dx.doi.org/10.1002/jps.20875] [PMID: 17688284]

[25] N. Jain, R. Jain, N. Thakur, B.P. Gupta, D.K. Jain, J. Banveer, and S. Jain, "Nanotechnology: a safe and effective drug delivery system", *Asian J. Pharm. Clin. Res.,* vol. 3, no. 3, pp. 159-165, 2010.

[26] C.L. Keat, A. Aziz, A.M. Eid, and N.A. Elmarzugi, "Biosynthesis of nanoparticles and silver nanoparticles", *Bioresour. Bioprocess.,* vol. 2, no. 1, p. 47, 2015.
[http://dx.doi.org/10.1186/s40643-015-0076-2]

[27] N. Jain, P. Jain, D. Rajput, and U. K. Patil, "Green synthesized plant-based silver nanoparticles: therapeutic prospective for anticancer and antiviral activity", In: *Micro Nano Syst. Lett* vol. 9. SpringerOpen, 2021no. 1, .
[http://dx.doi.org/10.1186/s40486-021-00131-6]

[28] C.P. Devatha, and A.K. Thalla, "Green Synthesis of Nanomaterials", In: *Synthesis of Inorganic Nanomaterials.* Elsevier, 2018, pp. 169-184.
[http://dx.doi.org/10.1016/B978-0-08-101975-7.00007-5]

[29] A. Palumbo, L.M. Casanova, M.F.P. Corrêa, N.M. Da Costa, L.E. Nasciutti, and S.S. Costa, "Potential therapeutic effects of underground parts of Kalanchoe gastonis-bonnieri on benign Prostatic Hyperplasia", *Evid. Based Complement. Alternat. Med.,* vol. 2019, pp. 1-10, 2019.
[http://dx.doi.org/10.1155/2019/6340757] [PMID: 30719063]

[30] S.B. Kedare, and R.P. Singh, "Genesis and development of DPPH method of antioxidant assay", *J. Food Sci. Technol.,* vol. 48, no. 4, pp. 412-422, 2011.
[http://dx.doi.org/10.1007/s13197-011-0251-1] [PMID: 23572765]

[31] Y. Mizushima, and M. Kobayashi, "Interaction of anti-inflammatory drugs with serum proteins,

especially with some biologically active proteins", *J. Pharm. Pharmacol.*, vol. 20, no. 3, pp. 169-173, 2011.
[http://dx.doi.org/10.1111/j.2042-7158.1968.tb09718.x] [PMID: 4385045]

[32]   K.E. Van Holde, and S.F. Sun, "Bovine serum albumin in water-dioxane mixtures", *J. Am. Chem. Soc.*, vol. 84, no. 1, pp. 66-72, 1962.
[http://dx.doi.org/10.1021/ja00860a017]

[33]   C.Y. Woumbo, D. Kuate, D.G. Metue Tamo, and H.M. Womeni, "Antioxidant and antidiabetic activities of a polyphenol rich extract obtained from Abelmoschus esculentus (okra) seeds using optimized conditions in microwave-assisted extraction (MAE)", *Front. Nutr.*, vol. 9, p. 1030385, 2022.
[http://dx.doi.org/10.3389/fnut.2022.1030385] [PMID: 36386938]

[34]   P. Tomasik, and D. Horton, "Enzymatic conversions of starch", *Adv. Carbohydr. Chem. Biochem.*, vol. 68, pp. 59-436, 2012.
[http://dx.doi.org/10.1016/B978-0-12-396523-3.00001-4] [PMID: 23218124]

[35]   A. Podsędek, I. Majewska, M. Redzynia, D. Sosnowska, and M. Koziołkiewicz, "In vitro inhibitory effect on digestive enzymes and antioxidant potential of commonly consumed fruits", *J. Agric. Food Chem.*, vol. 62, no. 20, pp. 4610-4617, 2014.
[http://dx.doi.org/10.1021/jf5008264] [PMID: 24785184]

[36]   S. Vinodhini, B.S.M. Vithiya, and T.A.A. Prasad, "Green synthesis of silver nanoparticles by employing the Allium fistulosum, Tabernaemontana divaricate and Basella alba leaf extracts for antimicrobial applications", *J. King Saud Univ. Sci.*, vol. 34, no. 4, p. 101939, 2022.
[http://dx.doi.org/10.1016/j.jksus.2022.101939]

[37]   K. Singh, V. Khanna, S. Sonu, S. Singh, S.A. Bansal, V. Chaudhary, and A. Khosla, "Paradigm of state-of-the-art CNT reinforced copper metal matrix composites: processing, characterizations, and applications", *J. Mater. Res. Technol.*, vol. 24, pp. 8572-8605, 2023.
[http://dx.doi.org/10.1016/j.jmrt.2023.05.083]

[38]   V. Khanna, V. Kumar, S.A. Bansal, C. Prakash, M. Ubaidullah, S.F. Shaikh, A. Pramanik, A. Basak, and S. Shankar, "Fabrication of efficient aluminium/graphene nanosheets (Al-GNP) composite by powder metallurgy for strength applications", *J. Mater. Res. Technol.*, vol. 22, pp. 3402-3412, 2023.
[http://dx.doi.org/10.1016/j.jmrt.2022.12.161]

[39]   M. Dahiya, V. Khanna, and S. Anil Bansal, "Effect of graphene size variation on mechanical properties of aluminium graphene nanocomposites: A modeling analysis", *Mater. Today Proc.*, vol. 73, no. 2, pp. 249-254, 2023.
[http://dx.doi.org/10.1016/j.matpr.2022.07.259]

[40]   P. Gupta, N. Ahamad, D. Kumar, N. Gupta, V. Chaudhary, S. Gupta, V. Khanna, and V. Chaudhary, "Synergetic Effect of CeO2 Doping on Structural and Tribological Behavior of Fe-Al2 O3 Metal Matrix Nanocomposites", *ECS J. Solid State Sci. Technol.*, vol. 11, no. 11, p. 117001, 2022.
[http://dx.doi.org/10.1149/2162-8777/ac9c92]

[41]   K. Singh, S.A. Bansal, V. Khanna, and S. Singh, "Effects of Performance Measures of Non-conventional Joining Processes on Mechanical Properties of Metal Matrix Composites", *Metal Matrix Composites*, no. Aug, pp. 135-165, 2022.
[http://dx.doi.org/10.1201/9781003194897-7]

CHAPTER 10

Recent Methods for Biogenic Synthesis of Metal Nanoparticles and their Applications

Giriraj Tailor[1,*], **Jyoti Chaudhary**[2], **Chesta Mehta**[2], **Saurabh Singh**[3] and **Deepshikha Verma**[4]

[1] *Department of Chemistry, Mewar University, Chittorgarh, Rajasthan, 31290, India*

[2] *Department of Chemistry, M.L. Sukhadia University, Udaipur, Rajasthan, 313001, India*

[3] *M.L.V. Government College, Bhilwara, Rajasthan, 311001, India*

[4] *Department of Chemistry, M.B.S. College of Engineering and Technology, 181101, Jammu and Kashmir, India*

Abstract: Nanoparticles are among the most important tools under investigation due to their application in optical, electrical, biological, sensing, and photocatalytic systems. Nanoparticles made by plants have a larger range of sizes and shapes and are far more stable. Investigators' fascination with producing metal-based nanoparticles, such as those of silver (Ag), platinum (Pt), gold (Au), zinc (Zn), copper (Cu), and cerium (Ce), has been aroused by the study of biological systems. In a manner analogous to this, microorganisms produce valuable substances like antibiotics, acids, and pigments as well as proteins and bioactive metabolites. The plant-based synthesis uses a variety of extracts, including fruit, leaves, roots, peel, bark, seeds, twigs, stems, shoots, and seedlings. The primary theme of the chapter is the synthesis of metallic nanoparticles mediated by plants. The potential applications of nanoparticles across a variety of fields have altered the research and industries that are briefly discussed in this chapter.

Keywords: Nanoparticles, Green synthesis, Photocatalytic, Thin film, Capping agent, Antibacterial, AFM, Solar cell, Silver NP.

INTRODUCTION

Nanotechnology provides methods for handling and using materials at nanoscales. Nanomaterials differ significantly in elements at larger scales in that they have a higher surface area. Additionally, there are several techniques to make nanoscale materials, but the green, eco-friendly approach is the best [1]. The application of biological processes has been viewed as an environment-friendly strategy and a

* **Corresponding author Giriraj Tailor:** Department of Chemistry, Mewar University, Chittorgarh, Rajasthan, 31290, India; E-mail: giriraj.tailor66@gmail.com

Virat Khanna, Suneev Anil Bansal, Vishal Chaudhary and Reddicherla Umapathi (Eds.)

credible technique for producing nanoparticles because of its ecological qualities [3, 4]. The synthesis of nanoparticles sustainably or biologically has several advantages, including rapid and eco-friendly production methods as well as inexpensive and biocompatible nanoparticle types. Pure nanoparticles can be made *via* physical and chemical methods, but they can be costly and even environmentally hazardous. A quick, simple, and affordable method for producing nanoparticles is known as "green synthesis." [19]. Furthermore, since microbes and plant components act as stabilisers, there is no need to apply extra stabilising agents [2]. The foundations for sustainable (biological) chemistry enable cleaner nanoparticle synthesis. The study of metabolism and chemical tolerance mechanisms are employed in the synthesis of biological material (photosynthesis) to produce nanoparticles [5]. Due to their wide variety, plants' ability to manufacture nanoparticles of metallic material is still unrealized. In nature, there exists a wide variety of plant species and a lot of them would have been suitable for producing nanoparticles [6].

This chapter discusses general methods for producing nanoparticles utilising a range of methods and uses. The biosynthesis of nanoparticles is currently chosen because of their acceptability towards toxicity, higher bioaccumulation, relative economies, simple synthesis procedure, uncomplicated post-synthesis processing, as well as simple biomass management [8]. The creative and advantageous qualities of nanoparticles are significantly responsible for the rapid growth in industry inventions and studies in the discipline of nanotechnology. New nanomaterials, a critical component of nanotechnology advancements, allow for the creation of novel goods and services [7]. NPs are produced using extracts of plants that have several phytoconstituents, which involve phenols, flavonoids, alkaloids, terpenoids, and acids [9]. The leaf, root, stem, seed, fruit, gum, flower, and other parts of plants are only a few of the materials used to make these organic products [10]. Extracts of plants interacting with metal salts result in the formation of NPs, which give the reaction solution a specific colour [11]. Many studies have been done in the context of this topic, and a variety of plant components, including leaves, seeds, twigs, flowers, and petals have been used to make these extracts. The synthesis of nanoparticles of noble metals is the only application of this viewpoint, even though various publications have looked at using plant-based substances in nanoparticle manufacturing [12], specifically gold (Au) [14], iron (Fe) [15], silver (Ag) [13] and palladium (Pd) [16]. Gold nanoparticles are among the more commonly utilised particles because they possess a broad range of therapeutic applications for numerous disorders, including cancer, diabetes mellitus, and cardiovascular diseases [17]. Due to the absence of harmful contaminants, the nanoparticles derived from conventional pharmaceuticals are appropriate to be utilised for healthcare studies and treatments [18, 13].

The current area of interest for the study is the identification of novel ecologically safe techniques for the bioproduction of metal-based nanoparticles. Most plants contain many molecules that can scavenge free radicals, reducing sugars, terpenoids, vitamins, phenolic compounds, nitrogen compounds, as well as additional metabolites with significant antioxidant activity [19]. The goal of this chapter is to give a broad review of recent advancements in environmentally benign ways to produce metallic nanoparticles and their diverse applications across a variety of fields. Fig. (**1**) provides the advantages of biogenic synthesis methods.

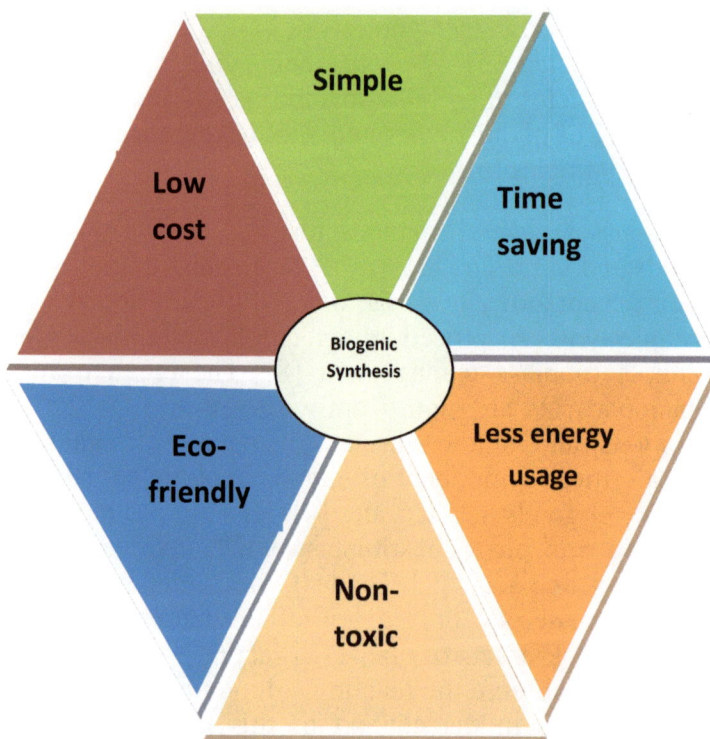

Fig. (1). Advantages of biogenic synthesis methods.

GREEN SYNTHESIS APPROACH FOR NANOPARTICLES

Plant-mediated Biosynthesis

Plant extract is employed for the production of nanoparticles from biomass wastes, such as leaves, peels, petals, fruits, roots, *etc.*, which is a cost-effective strategy. Furthermore, the dried part is favored; nevertheless, different plant tissues contain different amounts of water. [21] According to Mahanty *et al.*, transitioning from chemically produced nanoparticles to eco-friendly biosynthesized ones would help lessen the toxicity of chemicals in the environment [20] (Fig. **2**).

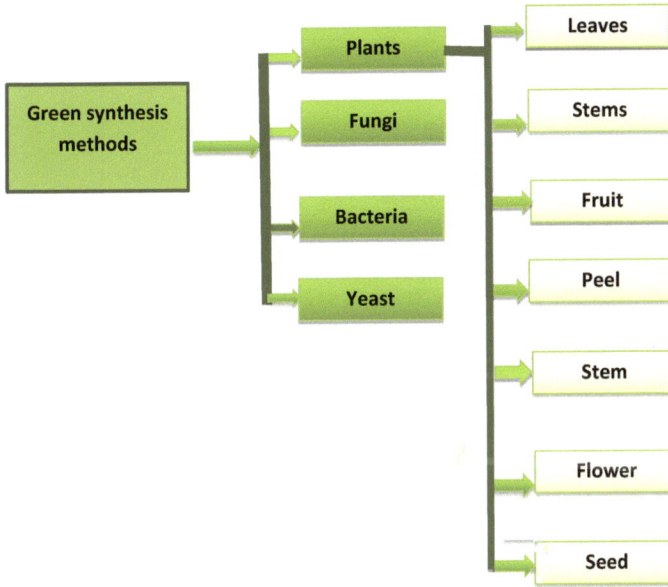

Fig. (2). Different green approaches for the preparation of nanoparticles.

It has been investigated whether certain plant species can create nanoparticles. Plant extracts as well as the leaves, stems, fruits, roots, and flowers have all been used to make metal nanoparticles (Fig. **3**). Table **1** shows the green synthesized metal-based nanoparticles using various parts of the medicinally relevant.

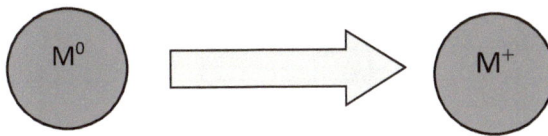

Fig. (3). Mechanisms of nanoparticle synthesis (M+-metal ion).

Table 1. Green synthesized metal-based nanoparticles using various parts of the medicinally relevant plants.

Common Name	Biological Name	Plant Parts	Metal NPs	Shape	Size	Applications	Refs.
Red spinach	*Aramanthus tricolor*	Leaves	Ag	Spherical	40.45 nm	Antibacterial	[97]
Aloe vera	Aloe barbadensis miller	Leaves	Au, Ag	Spherical, triangular	50–350 nm	Anticancer, Anti-inflammatory	[22]

Common Name	Biological Name	Plant Parts	Metal NPs	Shape	Size	Applications	Refs.
Mountain Grape	Barberry	Fruit	Cu, Fe	-	-	Catalyst	[36]
South Indian Olibanum	Boswellia ovalifoliolata	Bark	Ag	Spherical	-	Treats asthma	[43]
Thornapple	Datura metel	Leaves	Ag	Quasilinear superstructures	16–40 nm	Anti-Inflammatory	[23]
Orange	Citrus Aurantium Dulcis	Peel	Ag	Spherical	91 nm	Anti-bacterial	[38]
Dalchini	Cinnamon zeylanicum	Bark	Ag	spherical	31–40 nm	Antioxidant, Antibacterial, Anti-inflammatory	[41, 42]
Gum Tree	Eucalyptus hybrid	Leaves	Ag	-	50–150 nm	Antimicrobial	[24]
Fig	Ficus carica	Fruit	Ag	Spherical	54–89 nm	Antispasmodic, anti-inflammatory	[34]
Mango	Mangifera indica	Seed	Au	Spherical	46.8 nm	Antiseptic, Astringent	[31]
Mangosteen	*Garcinia mangostana*	Fruit	ZnO	-	-	Treats skin infections.	[35]
Neemaru	Memecylonedule	Leaves	Ag, Au	Triangular, Circular, Hexagon	20–50 nm	Treats Bruises, and conjunctivitis.	[26]
Drumstick	Moringa oleifera	Seed	Fe	Spherical	2.6 nm	Antioxidant	[30]
Banana	Musa sapientum	Peel	Ag	Spherical	23.7 nm	Antimicrobial activity	[37]
Frangipani	Plumeria alba	Flower	Ag	-	-	Antibacterial activity	[44]
Guava	Psidium guajava	Leaves	Au, Ag, ZnO	-	-	Anti-diabetic activity	[25]
Avocado	Persea americana	Seed	Cu	Spherical	42-90 nm	Antioxidant	[28, 29]
Pomegranate	Punica granatum	Seed	Fe	Semi Spherical	25-55 nm	Treats Jaundice and diarrhea.	[33]
Pomegranate	Punica granatum	Peel	Ag, Au	Spherical	5-10 nm	Immunity Booster	[39]
Sand Potato Bush	Phyllanthus pinnatus	Stem	-	Cubical	49 nm	Antimicrobial	[40]

Plant extracts had been recognised to reduce metal ions throughout the early 1900s, but the precise make-up of the restricting active components was not well known at the time. Due to their simplicity, use of living plants, complete extracts from plants, and the ability of plant tissue to turn metallic ions into nanoparticles, they have attracted a great deal of focus during the past few years.

Leaf

The antibacterial efficacy of gold nanoparticles was investigated for Brassica species by Kuppusamy *et al.* against a number of pathogenic infections [45]. It was reported that olive leaf extracts were used to produce silver nanoparticles (AgNPs), and their antibacterial potency to bacterial strains was assessed. The effects of pH, contact time, crude extract, and morphology of the Ag NPs were evaluated [47]. Ashwini *et al.* utilized *Cayratiapedata* immobilized with biogenically synthesized zinc oxide nanoparticles [46]. Shankar *et al.* used an extract of leaves of Cinnamomum tamala to quickly produce silver nanoparticles under mild circumstances and without any stabilizing agent [48]. The antibacterial efficacy, antioxidant potential, and cancer prevention potential were also evaluated for the synthesized Olea europaea-based silver nanoparticles [49].

Seeds

Antibacterial, antiviral, anticancer, and cardioprotective bioactivities have been observed in seed extracts. There may be industrial uses for certain bioactivities. Along with simple processing and accessibility, the bioactive components can be extracted from seeds without particular growth conditions or processors [50]. Among the intriguing and extremely useful phytochemicals present in seeds are caffeic acid, chlorogenic acid, epicatechin gallate, polyunsaturated fatty acids (PUFA), and cinnamic acid [51, 52]. Sharma *et al.* claim that nutmeg extract may aid in the transformation of $AgNO_3$ into silver nanoparticles by presenting the surface Plasmon peak at around 410 nm, confirming the reduction of Ag^+ to Ag^0. These biologically created Ag nanoparticles were found to be persistent for three months. This work demonstrated the fast and sustainable production of stable silver nanoparticles by biological processes. Whenever evaluated on test microorganisms, the nanoparticles a lso showed good antibacterial capabilities [53].

Stem

By using the stems of old banana plants, Dang H *et al.* were able to spontaneously manufacture silver (Ag) nanoparticles exhibiting antimicrobial properties. The antibacterial activities of the nano-Ag contrary to Staphylococcus epidermis and Escherichia coli were also assessed using the Kirby-Bauer sensitivity method [61]. Nickel oxide nanoparticles from Callicarpa maingayi stem bark extract [91] and

Balochistanica Berberis [92] were also made using the stem to study bioactivities. Their overall antioxidant capacity was 64.77 percent, their 2, 2-diphenyl-1-picrylhydrazyl content was 71.48 per cent, and they also had cytotoxic potential. There are many biological and agricultural applications for the metabolic potential of BBS-NiONPs, which are abundant in flavonoids, phenolics, and berberine [92].

Fruit

Fruit extracts contain naturally occurring bioactive substances that have significant therapeutic potential. Fruit extracts have significant reducing properties [54]. Microbes are essential to the biological synthesis of NP; these microorganisms must exist in pure strains maintained in an aseptic environment. Additionally, during downstream processing, it could be challenging to remove nanoparticles from biological broth cultures. Additionally, it takes time for the elemental oxide/elemental nanoparticles to replace the soluble metallic salts [55].

Peel

Nabi *et al.* utilised lemon peel extract to produce TiO_2 nanoparticles. The hydrolyzed compound contains hesperidin flavonoid, and a reducing and capping agent [56]. Due to its unique biological implications, including its antiallergic [58] and antioxidant [57] qualities, lemon peel is used. Citrus peel was used by Gao *et al.* [59] to create ZnO NPs, which was then characterized and compared to commercial ZnO NPs made using the solid-phase approach. Next, several combinations of carboxymethylcellulose sodium (CMC) were used to coat fresh strawberries to compare two types of ZnO NPs' preservation. Lee *et al.* used the fruit peel of G. mangostana and tetrachloroaurate to demonstrate the production and characterization of Au-NPs. Plant peels are a cheap, efficient, and safe way to harness the potential of unwanted waste materials [60].

Flower

ZnO nanoparticles were created by Dobrucka *et al.* utilising a T. pratense. The generated ZnO nanoparticles (EDX) were characterised by several spectroscopic techniques, which confirmed the formation of zinc oxide nanoparticles of 60 - 70 nm. The larger ZnO nanoparticles were created by the accumulation of smaller nanoparticles [64].

Root

Zingiber officinale (ginger) root extracts were employed for the formation of ZnO nanoparticles. The significant flavonoid concentration of the root extracts was revealed by the flavonoid assay. Characterization allowed the determination of the elemental composition and the size of the generated zinc oxide nanoparticles [62]. Coleus forskohlii root extract (RECo) was employed by Dhayalan *et al.* for reducing and capping Ag and Au metal nanoparticles. The manufactured gold and Ag nanoparticles were examined using a UV-Vis spectrophotometer and a high-resolution transmission electron microscope [63]. Fig. (**4**) gives the graphical illustration of biogenically synthesized metallic nanoparticles.

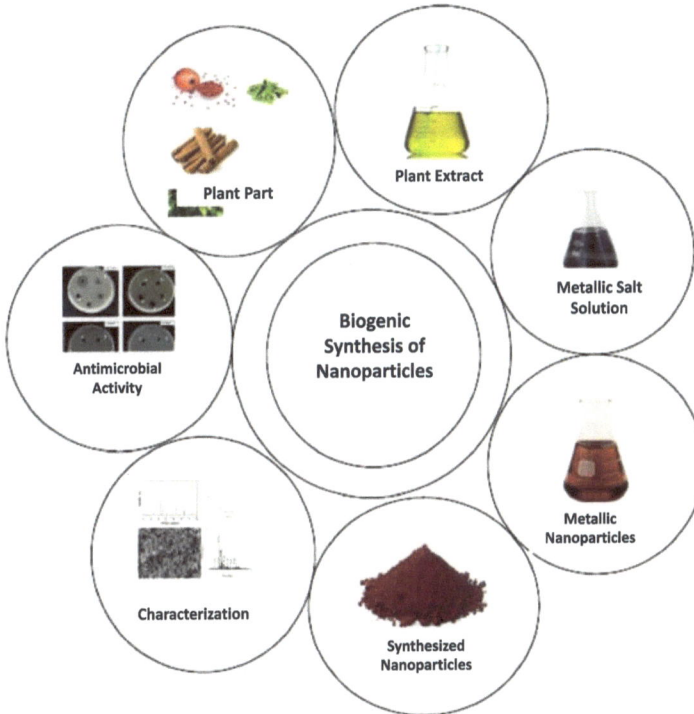

Fig. (4). Graphical illustration of biogenically synthesized metallic nanoparticles.

SPECTRAL ANALYSIS

UV-Visible

Dubey *et al.* produced nanosilver particles by extracting the leaves of Eucalyptus hybrid (safeda).To evaluate the bioreduction time, the generated silver nanoparticles were analysed by UV spectrophotometers [65] (Fig. **5**).

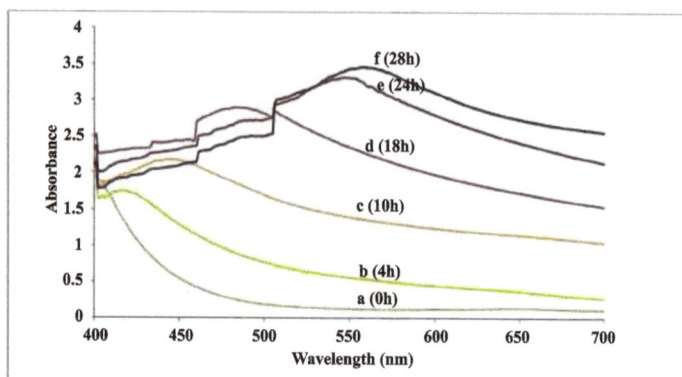

Fig. (5). Copper nanoparticles' UV-visible absorption spectra as a function of wavelength versus absorbance at different locations during the manufacturing process [95].

FTIR

Khalil *et al.* produced spherical, hexagonal, and triangular gold nanoparticles (NPs) using an environmentally friendly method that used olive leaf extract. The strongest influences on the size and form of gold NPs were pH and olive leaf fluid content. The FTIR spectra demonstrated that the proteins decreased and the nanoparticles stabilised by connecting to the fragments *via* available amine groups [66] (Fig. **6**).

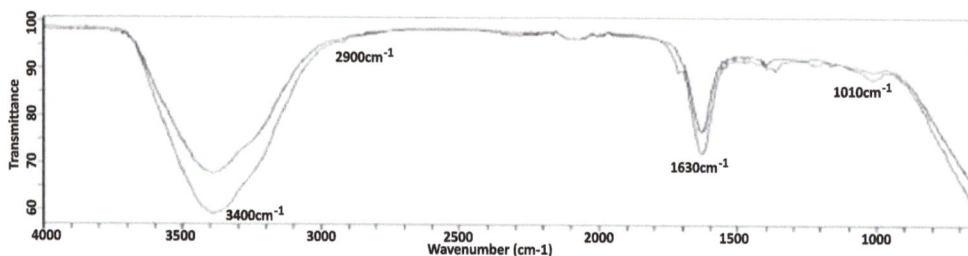

Fig. (6). FTIR scan of CoNP's with drumstick tree [96].

According to Sankar *et al.*, copper oxide nanoparticle formation may be aided by bioactive compounds found in the aqueous extract of the leaves of F. religiosa, which may be confirmed by FT-IR spectra. Copper oxide nanoparticles and F. religiosa leaf powder exhibit spectral peaks in the FT-IR study that point out bands associated with amide -NH stretching, nitro N=O bending amide, and CO stretching. The CuO vibrations have acknowledged an abundance of CuO nanoparticles with a bandwidth of 618 cm 1. The results show that the bioactive compounds in the leaf extract dominate copper oxide nanoparticle production [67]. The vibration caused by the C-H stretch with alkane, O-H stretches with carboxylic acid, C-H alkyl halides, C-N stretching with aromatic amines, and C-C stretch with aromatic compounds, in that order, led to the peaks in the IR absorption peak of the root extracts of P. hydropiper [68].

Scanning Electron Microscopy (SEM)

By using a standard SEM study, Barwant *et al.* studied the structural properties of the Ag-AgO-Ag$_2$O nanocomposite. Numerous nanoparticles were seen spherically in one location during the SEM scan. The EDX examination of Ag-AgO-Ag$_2$O nanocomposites showed simply a trace amount of Ag and O in the sample and no contamination peaks, proving the development of impurity-free Ag-AgO-Ag$_2$O nanocomposites. Ag (17.76%) and O (82.24%) were two of the relative percentages of elements that were revealed by the EDX study [69] (Fig. **7**).

Fig. (7). SEM photographs of the AgNPs biosynthesized from seed extract of Iklas, Irziz, and Shishi kinds of date palm. [Aldayel *et al.*, 2021] [94].

Transmission Electron Microscopy (TEM)

A stream of electrons is sent through an incredibly thin material transmission electron microscopy. R. nasutus nanoparticles were applied to the copper grid by Pasupuleti *et al.*, who then dried the nanoparticles under a lamp to create the ultra-thin film [70]. Here are some TEM pictures and absorption spectra of iron (Fe) nanoparticles that have been stabilised by tannin and made at room temperature [93] (Fig. **8**).

Mallikarjuna *et al.* analysed silver nanoparticles generated from Ocimum leaf extract using TEM. The Ag nanoparticles are clearly engulfed by a thin, coating of materials, which we presume to be an organic compound from Ocimum leaf broth as they are inspected more closely. Only a small portion of the generated nanoparticles are agglomerated, and they are between 3 and 20 nm in size. The estimated average particle size is 9.5 nm, indicating that particle sizes vary [71].

XRD

A crucial technique for analysing the structure of nanomaterials is X-ray diffraction (XRD), which uses X-rays' atomic-scale wavelengths. According to P. Kuppusamy *et al.*, Ag and Au nanoparticles were biosynthesized by using ethanol-based extract of Brassica oleracea L. The XRD patterns of Ag and Au nanoparticles showed four diffraction peaks, corresponding to the fcc structure of crystalline silver and gold. XRD planes provided more evidence that the particles are crystalline [45].

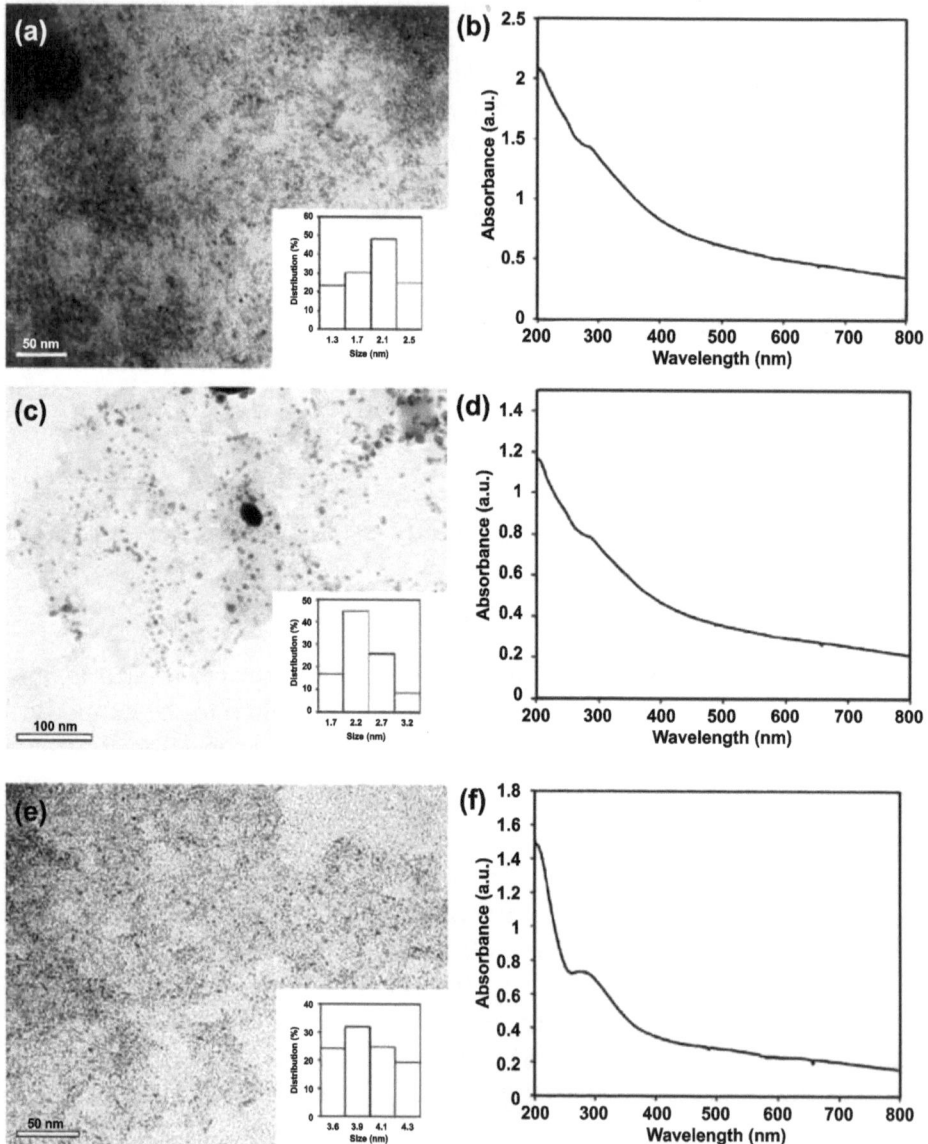

Fig. (8). Tannin absorption spectra and TEM photos. Fe nanoparticles developed at the average temperature for a range of time and precursor concentration, as shown in (a) and (b), 24 hours and 0.01 M, (c) and (d), 48 hours and 0.05 M, and (e) and (f), 24 hours and 0.05 M, respectively. (**Somchaidee *et al.*, 2018**) [93].

AFM

Atomic force microscopy (AFM) enables sub-nanometer 3D characterisation of nanoparticles. According to a study by Mossa*et al.*, AFM 3D images of the fabricated tea leaves extracted showed that silver nanoparticles had grown in a homogeneous dispersion. High nanoparticle diameter aggregation was observed at 7 mM and 9 mM. Granularity Accumulation Distribution (GAD) from AFM was used to display the particle size distribution of AgNPs. The generated Ag nanoparticles range in size from 66 - 99 nanometers on average. According to AFM, the generated silver nanoparticles had a particle diameter that varied from 70 to 180 nm. The nanoparticles' getting settled, which reduces their absorbance, may be the source of this huge diameter [72]. AFM (3D) images of the film created by Shinde *et al.* revealed the glass expanding over the surface [73].

APPLICATIONS

Antimicrobial/Antibacterial

Research has shown that the antibacterial activity of Cu_2ONPs derived from Ziziphus extract is superior to Escherichia coli against Staphylococcus aureus [75]. Cu_2ONPs produced using fruit extract of Capparis spinosa were observed to have a potent antibacterial effect against Bacillus cereus and S. aureus in another experiment, in contrast to Klebsiella pneumoniae and *E. coli* [76]. Dash *et al.* examined the antibacterial activity of green synthesised silver nanoparticles against Escherichia coli and Staphylococcus aureus. Treatment with AgNPs for each bacterial strain is all inhibited in a concentration-dependent manner. The concentration of resistance at which no detectable bacterial growth is visible was recorded as a specific MIC value [48, 97]. Fig. (**9**) shows various Applications of metallic nanoparticles.

Solar Cell Applications

Energy sources that are emission-free, renewable, and good for the environment must now take the place of dangerous fossil fuels. Due to their exceptional ability to convert incidental solar-powered light into electricity, dye sensitization has been identified to be one of the biggest fascinating options. (Das *et al.*, 2018, Jen *et al.*,) [77, 78]. Shashanka *et al.* used environmentally friendly ZnO nanoparticles to make dye-sensitized solar cells [79]. Sharma *et al.* produced NPs in an aqueous media using a green synthesis approach. The well-crystalline nature of the produced CuO NPs perfectly matched the monoclinic arrangement of bulk CuO [80]. Ahmed *et al.* examined how new SR/MWCNT nanocomposites' characteristics relate to nanotechnology approaches for creating novel dielectric properties in insulating materials [81].

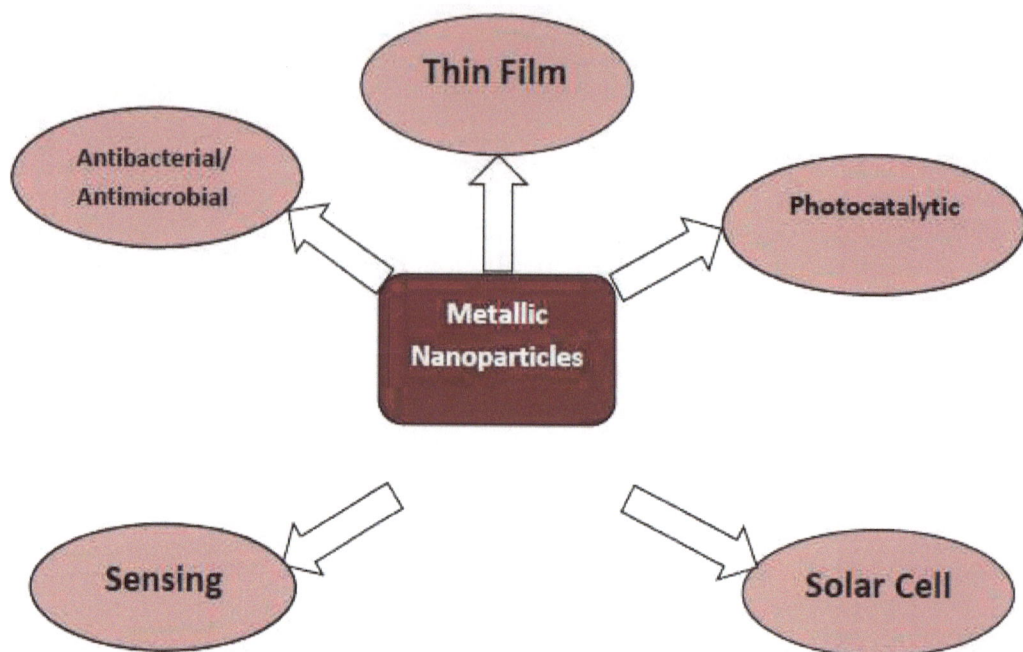

Fig. (9). Applications of metallic nanoparticles.

Thin Film

Silver nanostructured films were produced by Shinde *et al.* using guava leaf extract and silver nitrate (AgNO3) solution. Other guava tree parts, like the fruit, stem, and root extracts, include chloride, which produce a white precipitate that reduces Ag^+ to Ag^0 nanoparticles. To create silver (Ag^0) nanoparticles, the reduction process only uses the leaf of the guava tree [73]. Polypropylene nanocomposites were created by Ahmed *et al.* using the sol-gel process; the attraction forces of water droplets on the surfaces of the nanocomposites were adjusted by varying the concentrations (1, 5, and 10 wt%) and types of nanoparticles (clay, ZnO, SiO_2, and TiO_2) [74]. Without the use of a separate photo initiator, acrylic acid (AA) monomer was used as the crosslinker to create thin films of the poly(vinyl alcohol)/chitosan (PVA/CTS) based hydrogel that was around 4 m thick. The AgNP-loaded PVA/CTS hydrogel film may be used in medical applications as a biomaterial [88].

Sensing Application

In biosensing applications, metallic nanoparticles and a number of semiconductors' conductive features have been employed [85]. Due to significant quenching abilities, metal-based nanoparticles are an excellent choice for biosensors that are used to detect glutathione [86, 87]. Tamarind leaf extract was used by Ankamwar *et al*. to biologically create gold nanotriangles with possible use in vapour sensing. He looked into how different organic solvent vapours affected the conductivity of gold nanotriangles, which decreased by tamarind leaf extract [89]. Chelly *et al*. utilised Rumex roses (RR) plant extract as a reducing agent and developed new electrochemical sensors through the use of the produced Au and Ag nanoparticles as modifying agents of glassy carbon electrodes [90].

Photocatalytic Activity

Elango *et al*. produced lead nanoparticles (Pb-NPs) using cocos nucifera L extract. Lead acetate was mixed with Cocos nucifera coir extracts, and the result showed stable Pb-NPs. It was also discovered that malachite green dye generated at a brief UV wavelength of 254 nm was absorbed photo catalytically. UV spectrum investigation demonstrated the peak absorbance at 613 nm, especially in connection with the stimulation of surface plasmon vibration by Pb-NPs [82]. Alagesan *et al*. discovered that green Se NPs (selenium nanoparticles) successfully broke down the methylene blue dye when subjected to sunlight using withania somnifera (W. somnifera) leaf extract. The current results point to the advantages of creating Se NPs with potential activities utilising a green approach [83]. Iron oxide nanoparticles (Fe_2O_3 NPs) were produced environmentally friendly by Bibi *et al*. utilising pomegranate seed extract, which showed remarkable photocatalytic performance towards reactive blue was accomplished in fifty-six min. Fe_2O_3 nanomaterials can be made from pomegranate seed extract at a low cost and without harming the environment. They can be used to break down colour in wastewater [84].

CONCLUSION

In a more ecologically conscious way, plants can currently generate metallic nanoparticles successfully. This economical and useful route is also environmentally beneficial. The size and shape of the nanoparticles may be controlled in this novel, ecologically friendly technique by varying the temperature and volume of plant extract used. The green synthesis approach is simple, doable, economical, easily upgradeable, low energy-intensive, and ecologically sound, and minimises the use of potentially dangerous components while maximising process effectiveness. It is especially well adapted for creating non-polluting nanoparticles (NPs) required for medicinal and therapeutic purposes. Large-scale greener processes have a lot of potential because nanoparticles are used for so many different things. For instance, iron nanoparticles produced from easily accessible, affordable organic precursors for heavy metal removal or water purification may be a preferable substitute in environmental nanoscience and nanotechnology applications. A comprehensive study is also necessary to pinpoint a few of the reaction processes and reach more definitive conclusions.

Nanoparticles

Recent Advancements in Multidimensional Applications, Vol. 1 217

REFERENCES

[1] U. Kamran, H.N. Bhatti, M. Iqbal, and A. Nazir, "Green synthesis of metal nanoparticles and their applications in different fields: A review", *Z. Phys. Chem.*, vol. 233, no. 9, pp. 1325-1349, 2019.
[http://dx.doi.org/10.1515/zpch-2018-1238]

[2] V.V. Makarov, A.J. Love, O.V. Sinitsyna, S.S. Makarova, I.V. Yaminsky, M.E. Taliansky, and N.O. Kalinina, "Green nanotechnologies: synthesis of metal nanoparticles using plants", *Acta Nat. (Engl. Ed .)*, vol. 6, no. 1, pp. 35-44, 2014.
[http://dx.doi.org/10.32607/20758251-2014-6-1-35-44] [PMID: 24772325]

[3] D. Bhattacharya, and R.K. Gupta, "Nanotechnology and potential of microorganisms", *Crit. Rev. Biotechnol.*, vol. 25, no. 4, pp. 199-204, 2005.
[http://dx.doi.org/10.1080/07388550500361994] [PMID: 16419617]

[4] D. Mandal, M.E. Bolander, D. Mukhopadhyay, G. Sarkar, and P. Mukherjee, "The use of microorganisms for the formation of metal nanoparticles and their application", *Appl. Microbiol. Biotechnol.*, vol. 69, no. 5, pp. 485-492, 2006.
[http://dx.doi.org/10.1007/s00253-005-0179-3] [PMID: 16317546]

[5] P.K. Rai, V. Kumar, S. Lee, N. Raza, K.H. Kim, Y.S. Ok, and D.C.W. Tsang, "Nanoparticle-plant interaction: Implications in energy, environment, and agriculture", *Environ. Int.*, vol. 119, pp. 1-19, 2018.
[http://dx.doi.org/10.1016/j.envint.2018.06.012] [PMID: 29909166]

[6] M.S. Akhtar, J. Panwar, and Y-S. Yun, "Biogenic synthesis of metallicnanoparticles by plantextracts", *ACS Sustain. Chem.& Eng.*, vol. 1, no. 6, pp. 591-602, 2013.
[http://dx.doi.org/10.1021/sc300118u]

[7] A. Thabet, and A. Fahad, "Al mufadi and A. A. Ebnalwaled", In: *Synthesis and Measurement of Optical Light Characterization for Modern Cost-Fewer Polyvinyl Chloride Nanocomposites Thin Films, Transactions on Electrical and Electronic Materials Journal* vol. 24. Nature Springer, 2023, no. 6, pp. 1-12.

[8] G. Ingale, and A.N. Chaudhari, "Biogenic synthesis of nanoparticles and potentialapplications: An eco-friendlyapproach", *J. Nanomed. Nanotechnol.*, vol. 4, p. 165, 2013.
[http://dx.doi.org/10.4172/2157-7439.1000165]

[9] P. Kuppusamy, M.M. Yusoff, G.P. Maniam, and N. Govindan, "Biosynthesis of metallic nanoparticles using plant derivatives and their new avenues in pharmacological applications – An updated report", *Saudi Pharm. J.*, vol. 24, no. 4, pp. 473-484, 2016.
[http://dx.doi.org/10.1016/j.jsps.2014.11.013] [PMID: 27330378]

[10] D. Dhamecha, S. Jalalpure, and K. Jadhav, "Nepenthes khasiana mediated synthesis of stabilized gold nanoparticles: Characterization and biocompatibility studies", *J. Photochem. Photobiol. B*, vol. 154, pp. 108-117, 2016.
[http://dx.doi.org/10.1016/j.jphotobiol.2015.12.002] [PMID: 26716586]

[11] A.M. El Badawy, *Surface charge-dependent toxicity of silver nanoparticles*, 2015.
[http://dx.doi.org/10.1021/es1034188]

[12] S. Bhakya, S. Muthukrishnan, M. Sukumaran, and M. Muthukumar, "Biogenic synthesis of silver nanoparticles and their antioxidant and antibacterial activity", *Appl. Nanosci.*, vol. 6, no. 5, pp. 755-766, 2016.
[http://dx.doi.org/10.1007/s13204-015-0473-z]

[13] P.V. AshaRani, G. Low Kah Mun, M.P. Hande, and S. Valiyaveettil, "Cytotoxicity and genotoxicity of silver nanoparticles in human cells", *ACS Nano*, vol. 3, no. 2, pp. 279-290, 2009.
[http://dx.doi.org/10.1021/nn800596w] [PMID: 19236062]

[14] S. Sadhasivam, V. Vinayagam, and M. Balasubramaniyan, "Recent advancement in biogenic synthesis of iron nanoparticles", *J. Mol. Struct.*, vol. 1217, p. 128372, 2020.

[http://dx.doi.org/10.1016/j.molstruc.2020.128372]

[15]   R. Lakshmipathy, B. Palakshi Reddy, N.C. Sarada, K. Chidambaram, and S. Khadeer Pasha, "Watermelon rind-mediated green synthesis of noble palladium nanoparticles: catalytic application", *Appl. Nanosci.,* vol. 5, no. 2, pp. 223-228, 2015.
[http://dx.doi.org/10.1007/s13204-014-0309-2]

[16]   S. Menon, R. S, and V.K. S, "A review on biogenic synthesis of gold nanoparticles, characterization, and its applications", *Resource-Efficient Technologies,* vol. 3, no. 4, pp. 516-527, 2017.
[http://dx.doi.org/10.1016/j.reffit.2017.08.002]

[17]   P. K. Dikshit, "Green synthesis of metallicnanoparticles: Applications and limitations", *Catalysts,* vol. 11, no. 8, p. 902, 2021.
[http://dx.doi.org/10.3390/catal11080902]

[18]   H.A. al Salamet, "Plants: Green route for nanoparticle synthesis", *Int. Res. J. Biol. Sci.,* vol. 1, pp. 85-90, 2012.

[19]   A. Mahanty, S. Mishra, R. Bosu, U.K. Maurya, S.P. Netam, and B. Sarkar, "Phytoextracts-synthesized silver nanoparticles inhibit bacterial fish pathogen Aeromonas hydrophila", *Indian J. Microbiol.,* vol. 53, no. 4, pp. 438-446, 2013.
[http://dx.doi.org/10.1007/s12088-013-0409-9] [PMID: 24426148]

[20]   P. Tiwari, "Phytochemical screening and extraction: A review, Int", *Pharm. Sci.,* vol. 1, pp. 98-106, 2011.

[21]   S.P. Chandran, M. Chaudhary, R. Pasricha, A. Ahmad, and M. Sastry, "Synthesis of gold nanotriangles and silver nanoparticles using Aloe vera plant extract", *Biotechnol. Prog.,* vol. 22, no. 2, pp. 577-583, 2006.
[http://dx.doi.org/10.1021/bp0501423] [PMID: 16599579]

[22]   J. Kesharwani, K.Y. Yoon, J. Hwang, and M. Rai, "Phytofabrication of silver nanoparticles by leaf extract of Datura metel: Hypothetical mechanism involved in synthesis", *Journal of Bionanoscience,* vol. 3, no. 1, pp. 39-44, 2009.
[http://dx.doi.org/10.1166/jbns.2009.1008]

[23]   M. Dubey, "Green Synthesis of Nanosilver Particles From Extract of Eucalyptus hybrida (Safeda) Leaf", *Dig. J. Nanomater. Biostruct.,* vol. 4, pp. 537-543, 2009.

[24]   S.P. Patil, and P.M. Rane, "Psidium guajava leaves assisted green synthesis of metallic nanoparticles: a review", *Beni. Suef Univ. J. Basic Appl. Sci.,* vol. 9, no. 1, p. 60, 2020.
[http://dx.doi.org/10.1186/s43088-020-00088-2]

[25]   T. Elavazhagan, and T. Elavazhagan, "Memecylon edule leaf extract mediated green synthesis of silver and gold nanoparticles", *Int. J. Nanomedicine,* vol. 6, pp. 1265-1278, 2011.
[http://dx.doi.org/10.2147/IJN.S18347] [PMID: 21753878]

[26]   A. Nabikhan, K. Kandasamy, A. Raj, and N.M. Alikunhi, "Synthesis of antimicrobial silver nanoparticles by callus and leaf extracts from saltmarsh plant, Sesuvium portulacastrum L", *Colloids Surf. B Biointerfaces,* vol. 79, no. 2, pp. 488-493, 2010.
[http://dx.doi.org/10.1016/j.colsurfb.2010.05.018] [PMID: 20627485]

[27]   P. Jimenez, P. Garcia, V. Quitral, K. Vasquez, C. Parra-Ruiz, M. Reyes-Farias, D.F. Garcia-Diaz, P. Robert, C. Encina, and J. Soto-Covasich, "ppulp, leaf, peel and seed of avocado fruit: A review of bioactive compounds and healthy benefits", *Food Rev. Int.,* vol. 37, no. 6, pp. 619-655, 2021.
[http://dx.doi.org/10.1080/87559129.2020.1717520]

[28]   S. Rajeshkumar, and G. Rinitha, "Nanostructural characterization of antimicrobial and antioxidant copper nanoparticles synthesized using novel Persea americana seeds", *OpenNano,* vol. 3, pp. 18-27, 2018.
[http://dx.doi.org/10.1016/j.onano.2018.03.001]

[29]   L. Katata-Seru, T. Moremedi, O.S. Aremu, and I. Bahadur, "Green synthesis of iron nanoparticles

using Moringa oleifera extracts and their applications: Removal of nitrate from water and antibacterial activity against Escherichia coli", *J. Mol. Liq.,* vol. 256, pp. 296-304, 2018.
[http://dx.doi.org/10.1016/j.molliq.2017.11.093]

[30]   S. Vimalraj, T. Ashokkumar, and S. Saravanan, "Biogenic gold nanoparticles synthesis mediated by Mangifera indica seed aqueous extracts exhibits antibacterial, anticancer and anti-angiogenic properties", *Biomed. Pharmacother.,* vol. 105, pp. 440-448, 2018.
[http://dx.doi.org/10.1016/j.biopha.2018.05.151] [PMID: 29879628]

[31]   S.M. Roopan, R.S. Mathew, S.S. Mahesh, D. Titus, K. Aggarwal, N. Bhatia, K.I. Damodharan, K. Elumalai, and J.J. Samuel, "Environmental friendly synthesis of zinc oxide nanoparticles and estimation of its larvicidal activity against Aedes aegypti", *Int. J. Environ. Sci. Technol.,* vol. 16, no. 12, pp. 8053-8060, 2019.
[http://dx.doi.org/10.1007/s13762-018-2175-z]

[32]   I. Bibi, N. Nazar, S. Ata, M. Sultan, A. Ali, A. Abbas, K. Jilani, S. Kamal, F.M. Sarim, M.I. Khan, F. Jalal, and M. Iqbal, "Green synthesis of iron oxide nanoparticles using pomegranate seeds extract and photocatalytic activity evaluation for the degradation of textile dye", *J. Mater. Res. Technol.,* vol. 8, no. 6, pp. 6115-6124, 2019.
[http://dx.doi.org/10.1016/j.jmrt.2019.10.006]

[33]   S.J.P. Jacob, V.L.S. Prasad, S. Sivasankar, and P. Muralidharan, "Biosynthesis of silver nanoparticles using dried fruit extract of Ficus carica - Screening for its anticancer activity and toxicity in animal models", *Food Chem. Toxicol.,* vol. 109, no. Pt 2, pp. 951-956, 2017.
[http://dx.doi.org/10.1016/j.fct.2017.03.066] [PMID: 28377268]

[34]   M. Aminuzzaman, L.P. Ying, W-S. Goh, and A. Watanabe, "Green synthesis of zinc oxide nanoparticles using aqueous extract of Garcinia mangostana fruit pericarp and their photocatalytic activity", *Bull. Mater. Sci.,* vol. 41, no. 2, p. 50, 2018.
[http://dx.doi.org/10.1007/s12034-018-1568-4]

[35]   M. Nasrollahzadeh, M. Maham, A. Rostami-Vartooni, M. Bagherzadeh, and S.M. Sajadi, "Barberry fruit extract assisted in situ green synthesis of Cu nanoparticles supported on a reduced graphene oxide–Fe$_3$O$_4$ nanocomposite as a magnetically separable and reusable catalyst for the O-arylation of phenols with aryl halides under ligand-free conditions", *RSC Advances,* vol. 5, no. 79, pp. 64769-64780, 2015.
[http://dx.doi.org/10.1039/C5RA10037B]

[36]   H.M.M. Ibrahim, "Green synthesis and characterization of silver nanoparticles using banana peel extract and their antimicrobial activity against representative microorganisms", *Journal of Radiation Research and Applied Sciences,* vol. 8, no. 3, pp. 265-275, 2015.
[http://dx.doi.org/10.1016/j.jrras.2015.01.007]

[37]   A.A. Manal, "Silver nanoparticles biogenic synthesized using an orange peel extract and their use as an antibacterialagent", *Int. J. Phys. Sci.,* vol. 9, no. 3, pp. 34-40, 2014. Febr. 9
[http://dx.doi.org/10.5897/IJPS2013.4080]

[38]   N. Ahmad, and S. Sharma, "Rapid green synthesis of silver and gold nanoparticles using peels of Punica granatu", *Adv. Mater. Lett.,* vol. 3, no. 5, pp. 376-380, 2012.
[http://dx.doi.org/10.5185/amlett.2012.5357]

[39]   R. Balachandar, "Plant-Mediated Synthesis, characterization and bactericidalpotential of emergingsilvernanoparticlesusingstemextract of Phyllanthus pinnatus: A recentadvance in phytonanotechnology", *J. Clust. Sci.,* vol. 30, no. 6, pp. 1481-1488, 2019.

[40]   M. Sathishkumar, K. Sneha, I.S. Kwak, J. Mao, S.J. Tripathy, and Y.S. Yun, "Phyto-crystallization of palladium through reduction process using Cinnamom zeylanicum bark extract", *J. Hazard. Mater.,* vol. 171, no. 1-3, pp. 400-404, 2009. a
[http://dx.doi.org/10.1016/j.jhazmat.2009.06.014] [PMID: 19576689]

[41]   M. Sathishkumar, K. Sneha, S.W. Won, C.W. Cho, S. Kim, and Y.S. Yun, "Cinnamon zeylanicum

bark extract and powder mediated green synthesis of nano-crystalline silver particles and its bactericidal activity", *Colloids Surf. B Biointerfaces,* vol. 73, no. 2, pp. 332-338, 2009. b
[http://dx.doi.org/10.1016/j.colsurfb.2009.06.005] [PMID: 19576733]

[42]   N. Savithramm, M.L. Rao, and P.S. Devi, "Evaluation of antibacterial efficacy of biologically synthesized silver nanoparticles using stem bark of BoswelliaovalifoliolataBal. And Henry and ShoreatumbuggaiaRoxb", *J. Biol. Sci. (Faisalabad, Pak.),* vol. 11, no. 1, pp. 39-45, 2010.
[http://dx.doi.org/10.3923/jbs.2011.39.45]

[43]   B.M. Patil, and A.A. Hooli, "Evaluation of antibacterial activities of environmental benign synthesis of silver nanoparticles using the flower extracts of Plumeria albaLinn", *J. Nanosci. Nanoeng. Appl.,* vol. 3, pp. 13-20, 2013.

[44]   P. Kuppusamy, "Intracellular biosynthesis of Au and Ag nanoparticles using ethanolic extract of Brassica oleraceaL", *J. Environ. Sci. (China),* vol. 29, pp. 151-157, 2015.
[http://dx.doi.org/10.1016/j.jes.2014.06.050] [PMID: 25766024]

[45]   A. Jayachandran, A. T R, and A.S. Nair, "Green synthesis and characterization of zinc oxide nanoparticles using Cayratia pedata leaf extract", *Biochem. Biophys. Rep.,* vol. 26, p. 100995, 2021.
[http://dx.doi.org/10.1016/j.bbrep.2021.100995] [PMID: 33898767]

[46]   M.M.H. Khalil, E.H. Ismail, K.Z. El-Baghdady, and D. Mohamed, "Green synthesis of silver nanoparticles using olive leaf extract and its antibacterial activity", *Arab. J. Chem.,* vol. 7, no. 6, pp. 1131-1139, 2014.
[http://dx.doi.org/10.1016/j.arabjc.2013.04.007]

[47]   S.S. Dash, S. Samanta, S. Dey, B. Giri, and S.K. Dash, "Rapid Green Synthesis of Biogenic Silver Nanoparticles Using Cinnamomum tamala Leaf Extract and its Potential Antimicrobial Application Against Clinically Isolated Multidrug-Resistant Bacterial Strains", *Biol. Trace Elem. Res.,* vol. 198, no. 2, pp. 681-696, 2020.
[http://dx.doi.org/10.1007/s12011-020-02107-w] [PMID: 32180127]

[48]   H. Sellami, S.A. Khan, I. Ahmad, A.A. Alarfaj, A.H. Hirad, and A.E. Al-Sabri, "Green synthesis of silvernanoparticlesusingOleaeuropaeaLeafextract for theirenhancedantibacterial, antioxidant, cytotoxic and biocompatibilityapplications", *Int. J. Mol. Sci.,* vol. 22, no. 22, p. 12562, 2021.
[http://dx.doi.org/10.3390/ijms222212562] [PMID: 34830442]

[49]   C. Chakraborty, "Utilization of various seeds: Areview", *Pharm Innov. J.,* vol. 6, no. 11, pp. 93-101, 2017.

[50]   C.V.S. Prakash, and I. Prakash, "Bioactive chemical constituents from pomegranate (Punica granatum) juice, seed and peel-a review", *Int. J. Res. Chem. Environ.,* vol. 1, no. 1, pp. 1-18, 2011.

[51]   N. Nazar, I. Bibi, S. Kamal, M. Iqbal, S. Nouren, K. Jilani, M. Umair, and S. Ata, "Cu nanoparticles synthesis using biological molecule of P. granatum seeds extract as reducing and capping agent: Growth mechanism and photo-catalytic activity", *Int. J. Biol. Macromol.,* vol. 106, pp. 1203-1210, 2018.
[http://dx.doi.org/10.1016/j.ijbiomac.2017.08.126] [PMID: 28851642]

[52]   G. Sharma, "S.S. lee, green synthesis of silver nanoparticle using Myristica fragrans (nutmeg) seed extract and its biological activity", *J. Nanomater. Biostructures,* vol. 9, no. 1, pp. 325-332, 2014. [Di g.].

[53]   H. Kumar, K. Bhardwaj, D.S. Dhanjal, E. Nepovimova, F. Şen, H. Regassa, R. Singh, R. Verma, V. Kumar, D. Kumar, S.K. Bhatia, and K. Kuča, "Fruit extractmediatedgreensynthesis of metallicnanoparticles: A New Avenue in pomologyapplications", *Int. J. Mol. Sci.,* vol. 21, no. 22, p. 8458, 2020.
[http://dx.doi.org/10.3390/ijms21228458] [PMID: 33187086]

[54]   H. Kumar, K. Bhardwaj, K. Kuča, A. Kalia, E. Nepovimova, R. Verma, and D. Kumar, "Flower-based green synthesis of metallic nanoparticles: Applications beyond fragrance", *Nanomaterials (Basel),* vol. 10, no. 4, p. 766, 2020.

[http://dx.doi.org/10.3390/nano10040766] [PMID: 32316212]

[55]   G. Nabi, Q-U. Ain, M.B. Tahir, K. Nadeem Riaz, T. Iqbal, M. Rafique, S. Hussain, W. Raza, I. Aslam, and M. Rizwan, "Green synthesis of TiO 2 nanoparticles using lemon peel extract: their optical and photocatalytic properties", *Int. J. Environ. Anal. Chem.,* vol. 102, no. 2, pp. 434-442, 2022.
[http://dx.doi.org/10.1080/03067319.2020.1722816]

[56]   V. Elangovan, *Cancer Lett.,* vol. 3835, no. 94, pp. 90416-90412, 1994.
[http://dx.doi.org/10.1016/0304]

[57]   O. Keleş, *Turk. J. Vet. Anim. Sci.,* vol. 25, p. 559, 2001.

[58]   Y. Gao, D. Xu, D. Ren, K. Zeng, and X. Wu, "Green synthesis of zinc oxide nanoparticles using Citrus sinensis peel extract and application to strawberry preservation: A comparison study", *Lebensm. Wiss. Technol.,* vol. 126, p. 109297, 2020.
[http://dx.doi.org/10.1016/j.lwt.2020.109297]

[59]   K. Xin Lee, K. Shameli, M. Miyake, N. Kuwano, N.B. Bt Ahmad Khairudin, S.E. Bt Mohamad, and Y.P. Yew, "Green synthesis of gold nanoparticles using aqueous extract of Garcinia mangostana Fruit Peels", *J. Nanomater.,* vol. 2016, pp. 1-7, 2016.
[http://dx.doi.org/10.1155/2016/8489094]

[60]   H. Dang, D. Fawcett, and G.E.J. Poinern, "Biogenic synthesis of silver nanoparticles from waste banana plant stems and their antibacterial activity against Escherichia coli and Staphylococcus Epidermis", *International Journal of Research in Medical Sciences,* vol. 5, no. 9, pp. 3769-3775, 2017.
[http://dx.doi.org/10.18203/2320-6012.ijrms20173947]

[61]   L. F. A. A. Raj, and E. Jayalakshmy, "Biosynthesis and Characterization of Zinc Oxide Nanoparticles Using Root Extract of Zingiber officinale", *OJCHEG,* vol. 31, no. 1, pp. 51-56, 2015.

[62]   M. Dhayalan, M.I.J. Denison, M. Ayyar, N.N. Gandhi, K. Krishnan, and B. Abdulhadi, "Biogenic synthesis, characterization of gold and silver nanoparticles from Coleus forskohlii and their clinical importance", *J. Photochem. Photobiol. B,* vol. 183, pp. 251-257, 2018.
[http://dx.doi.org/10.1016/j.jphotobiol.2018.04.042] [PMID: 29734113]

[63]   R. Dobrucka, and J. Długaszewska, "Biosynthesis and antibacterial activity of ZnO nanoparticles using Trifolium pratense flower extract", *Saudi J. Biol. Sci.,* vol. 23, no. 4, pp. 517-523, 2016.
[http://dx.doi.org/10.1016/j.sjbs.2015.05.016] [PMID: 27298586]

[64]   M. Dubey, "Green Synthesis of Nanosilver Particles From Extract of Eucalyptus hybrida (Safeda) Leaf", *Dig. J. Nanomater. Biostruct.,* vol. 4, pp. 537-543, 2009.

[65]   M.M.H. Khalil, E.H. Ismail, and F. El-Magdoub, "Biosynthesis of Au nanoparticles using olive leaf extract", *Arab. J. Chem.,* vol. 5, no. 4, pp. 431-437, 2012.
[http://dx.doi.org/10.1016/j.arabjc.2010.11.011]

[66]   R. al Sankaret, "Anticancer activity of Ficus religiosa engineered copper oxide nanoparticles", *Mater. Sci. Eng. C Mater. Biol. Appl.,* vol. 44, pp. 234-239, 2014.
[http://dx.doi.org/10.1016/j.msec.2014.08.030]

[67]   G. Ali, A. Khan, A. Shahzad, A. Alhodaib, M. Qasim, I. Naz, and A. Rehman, "Phytogenic-mediated silver nanoparticles using Persicaria hydropiper extracts and its catalytic activity against multidrug resistant bacteria", *Arab. J. Chem.,* vol. 15, no. 9, p. 104053, 2022.
[http://dx.doi.org/10.1016/j.arabjc.2022.104053]

[68]   M. al Barwantet, *Plant-Mediated Biological Synthesis of Ag-Ago-Ag2O Nanocomposites Using Leaf Extracts of Solanum Elaeagnifolium for Antioxidant.* Anticancer, and DNA Cleavage Activities, 2021.

[69]   V.R. Pasupuleti, Ganapathi Narasimhulu, C. Reddy, I. Rahman, G. Narasimhulu, C.S. Reddy, I. Ab Rahman, and S.H. Gan, "Biogenic silver nanoparticles using Rhinacanthus nasutus leaf extract: synthesis, spectral analysis, and antimicrobial studies", *Int. J. Nanomedicine,* vol. 8, pp. 3355-3364, 2013.
[http://dx.doi.org/10.2147/IJN.S49000] [PMID: 24039419]

[70] K. Mallikarjuna, "Green synthesis of silver nanoparticles using ocimum leaf extract and their characterization", *J. Nanomater. Biostructures,* vol. 6, no. 1, pp. 181-186, 2011. [Dig.].

[71] A.A. Moosa, "Green synthesis of silver nanoparticles using spent tea leaves extract with atomic force microscopy", *Int. J. Curr. Eng. Technol.,* vol. 5, no. 5, pp. 3233-3241, 2015.

[72] N.M. Shinde, A.C. Lokhande, and C.D. Lokhande, "A green synthesis method for large area silver thin film containing nanoparticles", *J. Photochem. Photobiol. B,* vol. 136, pp. 19-25, 2014.
[http://dx.doi.org/10.1016/j.jphotobiol.2014.04.011] [PMID: 24836517]

[73] A.T. Mohamed, and K.E.A. Ahmed, "Controlling on attraction forces of water droplets on surfaces of polypropylene nanocomposites coatings", *Transactions on Electrical and Electronic Materials,* vol. 19, no. 5, pp. 387-395, 2018.
[http://dx.doi.org/10.1007/s42341-018-0054-4]

[74] R. Khani, B. Roostaei, G. Bagherzade, and M. Moudi, "Green synthesis of copper nanoparticles by fruit extract of Ziziphus spina-christi (L.) Willd.: Application for adsorption of triphenylmethane dye and antibacterial assay", *J. Mol. Liq.,* vol. 255, pp. 541-549, 2018.
[http://dx.doi.org/10.1016/j.molliq.2018.02.010]

[75] K. Ebrahimi, S. Shiravand, and H. Mahmoudvand, "Biosynthesis of copper nanoparticles using aqueous extract of Capparis spinosa fruit and investigation of its antibacterial activity", *Marmara Pharm. J.,* vol. 21, no. 4, pp. 866-871, 2017.
[http://dx.doi.org/10.12991/mpj.2017.31]

[76] T.K. Das, "Whispering gallery mode assisted enhancement in the power conversion efficiency of DSSc and QDSSC devices using TiO_2 microsphere photoanodes ACSAppl", *Energy Mater.,* vol. 1, pp. 765-774, 2018.

[77] J.S. Shaikh, N.S. Shaikh, S.S. Mali, J.V. Patil, K.K. Pawar, P. Kanjanaboos, C.K. Hong, J.H. Kim, and P.S. Patil, "Nanoarchitectures in dye-sensitized solar cells: metal oxides, oxide perovskites and carbon-based materials", *Nanoscale,* vol. 10, no. 11, pp. 4987-5034, 2018.
[http://dx.doi.org/10.1039/C7NR08350E] [PMID: 29488524]

[78] R. Shashanka, H. Esgin, V.M. Yilmaz, and Y. Caglar, "Fabrication and characterization of green synthesized ZnO nanoparticle based dye-sensitized solar cells", *J. Sci. Adv. Mater. Devices,* vol. 5, no. 2, pp. 185-191, 2020.
[http://dx.doi.org/10.1016/j.jsamd.2020.04.005]

[79] J.K. Sharma, M.S. Akhtar, S. Ameen, P. Srivastava, and G. Singh, "Green synthesis of CuO nanoparticles with leaf extract of Calotropis gigantea and its dye-sensitized solar cells applications", *J. Alloys Compd.,* vol. 632, pp. 321-325, 2015.
[http://dx.doi.org/10.1016/j.jallcom.2015.01.172]

[80] A. T. Mohamed, "Emerging Nanotechnology Applications in Electrical Engineering' IGI Global, publisher of timely knowledge", *Softcover: 9781799885375, ISBN10: 1799885364, EISBN13, ISBN13, ISBN-13: 9781799885368: 9781799885382, Jun.,* p. 318, 2021.
[http://dx.doi.org/10.4018/978-1-7998-8536-8]

[81] G. Elango, and S.M. Roopan, "Green synthesis, spectroscopic investigation and photocatalytic activity of lead nanoparticles", *Spectrochim. Acta A Mol. Biomol. Spectrosc.,* vol. 139, pp. 367-373, 2015.
[http://dx.doi.org/10.1016/j.saa.2014.12.066] [PMID: 25574657]

[82] V. Alagesan, and S. Venugopal, "Green synthesis of selenium nanoparticle using leaves extract of Withaniasomnifera and its biological applications and photocatalytic activities", *Bionanoscience,* vol. 9, no. 1, pp. 105-116, 2019.
[http://dx.doi.org/10.1007/s12668-018-0566-8]

[83] I. Bibi, N. Nazar, S. Ata, M. Sultan, A. Ali, A. Abbas, K. Jilani, S. Kamal, F.M. Sarim, M.I. Khan, F. Jalal, and M. Iqbal, "Green synthesis of iron oxide nanoparticles using pomegranate seeds extract and photocatalytic activity evaluation for the degradation of textile dye", *J. Mater. Res. Technol.,* vol. 8, no.

6, pp. 6115-6124, 2019.
[http://dx.doi.org/10.1016/j.jmrt.2019.10.006]

[84]   G. Evtugyn, and T. Hianik, "Electrochemical immuno- and aptasensors for mycotoxin determination", *Chemosensors (Basel), vol.* 7, no. 1, p. 10, 2019.
[http://dx.doi.org/10.3390/chemosensors7010010]

[85]   W. Dong, R. Wang, X. Gong, and C. Dong, "An efficient turn-on fluorescence biosensor for the detection of glutathione based on FRET between N,S dual-doped carbon dots and gold nanoparticles", *Anal. Bioanal. Chem.,* vol. 411, no. 25, pp. 6687-6695, 2019.
[http://dx.doi.org/10.1007/s00216-019-02042-3] [PMID: 31407048]

[86]   A. T. Mohamed, "Design and Investment of High Voltage NanoDielectrics' IGI Global, publisher of timely knowledge", *ISBN10: 1799838293, EISBN13: 9781799838302, ISBN-13: 9781799838296, Aug.,* p. 363, 2020. n
[http://dx.doi.org/10.4018/978-1-7998-3829-6]

[87]   N.T. Nguyen, and J.H. Liu, "A green method for in situ synthesis of poly(vinyl alcohol)/chitosan hydrogel thin films with entrapped silver nanoparticles", *J. Taiwan Inst. Chem. Eng.,* vol. 45, no. 5, pp. 2827-2833, 2014.
[http://dx.doi.org/10.1016/j.jtice.2014.06.017]

[88]   B. Ankamwar, M. Chaudhary, and M. Sastry, "Gold nanotriangles biologically synthesized using tamarind leaf extract and potential application in vapor sensing", *Synth. React. Inorg. Met.-Org. Nano-Met. Chem.,* vol. 35, no. 1, pp. 19-26, 2005.
[http://dx.doi.org/10.1081/SIM-200047527]

[89]   M. Chelly, S. Chelly, R. Zribi, H. Bouaziz-Ketata, R. Gdoura, N. Lavanya, G. Veerapandi, C. Sekar, and G. Neri, "Synthesis of silver and gold nanoparticles from Rumex roseus plant extract and their application in electrochemical sensors", *Nanomaterials (Basel),* vol. 11, no. 3, p. 739, 2021.
[http://dx.doi.org/10.3390/nano11030739] [PMID: 33804238]

[90]   K. Shameli, M. Bin Ahmad, E.A. Jaffar Al-Mulla, N.A. Ibrahim, P. Shabanzadeh, A. Rustaiyan, Y. Abdollahi, S. Bagheri, S. Abdolmohammadi, M.S. Usman, and M. Zidan, "Green biosynthesis of silver nanoparticles using Callicarpa maingayi stem bark extraction", *Molecules,* vol. 17, no. 7, pp. 8506-8517, 2012.
[http://dx.doi.org/10.3390/molecules17078506] [PMID: 22801364]

[91]   S. Uddin, L.B. Safdar, S. Anwar, J. Iqbal, S. Laila, B.A. Abbasi, M.S. Saif, M. Ali, A. Rehman, A. Basit, Y. Wang, and U.M. Quraishi, "Green synthesis of nickel oxide nanoparticles from Berberisbalochistanica stem for investigating bioactivities", *Molecules,* vol. 26, no. 6, p. 1548, 2021.
[http://dx.doi.org/10.3390/molecules26061548] [PMID: 33799864]

[92]   P. Somchaidee, and K. Tedsree, "Green synthesis of high dispersion and narrow size distribution of zero-valent iron nanoparticles using guava leaf (Psidium guajava L) extract", *Advances in Natural Sciences: Nanoscience and Nanotechnology,* vol. 9, no. 3, p. 035006, 2018.
[http://dx.doi.org/10.1088/2043-6254/aad5d7]

[93]   F.M. Aldayel, M.S. Alsobeg, and A. Khalifa, "In vitro antibacterial activities of silver nanoparticles synthesised using the seed extracts of three varieties of Phoenix dactylifera", *Braz. J. Biol.,* vol. 82, p. e242301, 2022.
[http://dx.doi.org/10.1590/1519-6984.242301] [PMID: 34346959]

[94]   N. Nagar, and V. Devra, "Green synthesis and characterization of copper nanoparticles using Azadirachta indica leaves", *Mater. Chem. Phys.,* vol. 213, pp. 44-51, 2018.
[http://dx.doi.org/10.1016/j.matchemphys.2018.04.007]

[95]   S. Younis, "Synthesis, characterization & bacterial evaluation of cobalt nanoparticles using drumstick leaf extract via green route", *Easychair,* 2020.

[96]   I. Fatimah, and Z.H.V.I. Aftrid, "Characteristics and antibacterial activity of green synthesized silver nanoparticles using red spinach (Amaranthus Tricolor L.) leaf extract", *Green Chem. Lett. Rev.,* vol.

12, no. 1, pp. 25-30, 2019.
[http://dx.doi.org/10.1080/17518253.2019.1569729]

[97] Y.M. Long, L.G. Hu, X.T. Yan, X.C. Zhao, Q.F. Zhou, Y. Cai, and G.B. Jiang, "Surface ligand controls silver ion release of nanosilver and its antibacterial activity against Escherichia coli", *Int. J. Nanomedicine,* vol. 12, pp. 3193-3206, 2017.
[http://dx.doi.org/10.2147/IJN.S132327] [PMID: 28458540]

CHAPTER 11

Performance Benchmarking of Different Convolutional Neural Network Architectures on Covid-19 Dataset

Harsh Kumar Mishra[1], Anand Singh[1] and Ayushi Rastogi[2,3,*]

[1] *DST – Centre of Interdisciplinary Mathematical Sciences, Institute of Science, Banaras Hindu University (BHU)Varanasi, Varanasi, Uttar Pradesh, India*

[2] *Scitechesy Research and Technology Private Limited, Central Discovery Centre, BioNEST BHU, Banaras Hindu University, Varanasi – 2210035, India*

[3] *Department of Humanities and Applied Sciences, School of Management Sciences, College of Engineering, Lucknow – 226001, Uttar Pradesh, India*

Abstract: The utilization of chest X-rays could offer valuable assistance in the initial screening of patients before undergoing RT-PCR testing. This potential approach holds promise within hospital environments grappling with the challenge of categorizing patients for either general ward placement or isolation within designated COVID-19 zones. This study investigates the use of chest X-rays as a preliminary screening technique for suspected COVID-19 cases in hospital settings, given the limited testing capacity and probable delays for RT-PCR testing. We assess how well several neural network architectures perform in automated COVID-19 identification in X-rays with the goal of locating a model that has the highest levels of sensitivity, low latency, and accuracy. The results reveal that InceptionV3 exhibits better robustness while MobileNet obtains the maximum accuracy. This strategy may help healthcare organisations better manage patients and allocate resources optimally, especially when radiologists are hard to come by. This will help in choosing an architecture that has better accuracy, sensitivity, and lower latency. The chosen models are pre-trained using the technique of transfer learning to save computation power and time. After the training and testing of the model, we observed that while MobileNet gave the best accuracy among all the models (VGG16, VGG19, MobileNet and InceptionV3), IncpetionV3 was still better when it comes to robustness.

Keywords: Chest X-ray, RT-PCR testing, Neural Network, Transfer learning.

---

* **Corresponding author Ayushi Rastogi:** Scitechesy Research and Technology Private Limited, Central Discovery Centre, BioNEST BHU, Banaras Hindu University, Varanasi – 2210035, India; E-mail: sweetayushi19@gmail.com

Virat Khanna, Suneev Anil Bansal, Vishal Chaudhary and Reddicherla Umapathi (Eds.)

INTRODUCTION

Machine learning and computational-based methods have excited many attempts for versatile purposes including water purification [1], energy storage systems [2], antibiotic generation [3], prediction of toxicity [4], weather forecasting [5], and designing effective antigen/antibody biosensors [6]. These tools are transforming how we analyze data, solve problems, and interact with the world. This paper contributes to this exciting landscape by exploring the application of CNNs for COVID-19 detection in chest X-rays, showcasing the potential of these methods to improve healthcare through data-driven solutions.

Background: The emergence of COVID-19, a novel respiratory virus, has led to an unprecedented surge in patient caseloads, placing an immense strain on global healthcare infrastructures. Numerous countries are grappling with healthcare systems that are already stretched thin, encompassing finite resources such as diagnostic instruments, hospital capacities, personal protective gear for medical personnel, and respiratory support devices. In order to optimize the utilization of these scarce resources, it becomes imperative to discern whether individuals presenting with severe acute respiratory illness (SARI) potentially harbor COVID-19 infections.

Methodology: Within the ambit of this research, we propose the utilization of chest X-rays to detect COVID-19 infections among individuals exhibiting symptoms of SARI. Our choice of utilizing X-rays for dataset preparation stems from their inherent advantages vis-à-vis traditional diagnostic modalities. Several key advantages make chest X-rays a compelling choice:

1. Widespread availability and affordability: Compared to other diagnostic imaging tools, chest X-rays are more readily available and cost-effective in many healthcare settings.
2. Rapid analysis: Digital X-ray images allow for swift transmission and analysis, potentially accelerating diagnosis and triage decisions.
3. Enhanced safety and convenience: Portable X-ray scanners enable bedside examinations within isolation wards, minimizing staff exposure and patient movement, and reducing the risk of nosocomial infections.
4. Abundant data accessibility: Open-source repositories and readily available X-ray data facilitate research and development of robust machine learning models for automated analysis.

Leveraging these advantages, this paper investigates the performance of various Convolutional Neural Network (CNN) architectures for automatic COVID-19 detection in chest X-rays. Our aim is to identify the CNN architecture that offers the optimal balance of accuracy, sensitivity, and computational efficiency for practical clinical application. By exploring chest X-ray-based triage as a complementary tool to current diagnostic methods, we hope to contribute to optimizing resource allocation and improving patient management during the ongoing COVID-19 pandemic.

Problem Statement

Recognizing features from images is in high demand for various purposes ranging from medical diagnosis to security. The work currently done in this field includes the development of deep learning models for the classification of chest X-rays [7]. The primary goal of this project is to compare different CNN architectures on the prepared Covid-19 dataset and based on their performance find the best model that can be used to diagnose Covid19. This will help in saving precious time and resources during this period of pandemic.

Motivation

Since there has been a significant increase in COVID-19 infections worldwide, numerous different screening techniques have been created to find potential COVID-19 cases. However, there are not many open-source programs that employ chest X-ray pictures that are currently accessible [8]. Publicly accessible datasets containing chest X-rays for COVID-19 remain limited [9]. Given this context, the need arises to consolidate information dispersed across online sources and curate a bespoke dataset. This dataset is then fed into various models, enabling a comparative analysis of outcomes across distinct parameters. The objective is to discern the most appropriate architecture.

This will also help in making a baseline for a custom Neural Network architecture that predicts results specific to this problem.

Contribution

The main contribution of this work is in subjecting the same dataset to different Neural Network architectures and comparing the results to see which one is the most efficient when it comes to X-ray images. In the contemporary environment, non-radiologists often interpret radiographs. Furthermore, many radiologists may not be familiar with all the intricacies of the infection due to the virus' novelty, and they may not have the necessary competence to produce a diagnosis that is as accurate as possible. As a result, people leading this analysis can use this automated program as a guide. It is important to emphasize that our intention is not to endorse the substitution of any particular model for the established COVID-19 diagnostic tests. Instead, we propose its application as a triage mechanism. This approach aims to assist in the assessment of whether a patient displaying symptoms of SARI should proceed with the COVID-19 diagnostic test.

We introduce a performance benchmarking study of four different CNN architectures on a Covid-19 dataset prepared from different available open-source image datasets on the internet. The experiment was performed on Google Colab which provides free access to GPUs for research and classroom purposes. This study uniquely compares the results which can be further used to form a basis for the development of different models for the detection of COVID-19 or other thoracic diseases using chest X-rays and deep learning algorithms.

Research Questions

The following Table **1** lists the questions that this experiment addresses and also the motivation behind it:

Table 1. List of research questions.

Research Question	Motivation
Which model performs the best when subjected to the same kind of images under identical conditions?	Investigate the possible reasons for different results of models that are trained on identical weights and are subjected to identical sets of images.
What is the effect of model complexity on accuracy?	Investigate the threshold of model complexity for accuracy above which the model starts overfitting or performs worse than comparatively simpler models.
Is the transfer learning technique good for radiological image datasets?	Investigate if transfer learning can be used for a certain kind of dataset that is not generally used. This will increase the affectivity and acceptance of this technique as it not only saves time and computation power but works well for a wide range of datasets
Can accuracy be trusted as the metric for the comparison of different models?	Investigate the results of each model and find if more metrics are required to compare the efficacy and accuracy of models. A model needs to be not only accurate but also robust when subjected to different images.
What are the results of Grad-Cam on the best-performing network?	Investigate the results of the Grad-Cam-based heat maps generated. It was applied on just two networks MobileNet and InceptionV3.

The remainder of the essay is structured as follows. A brief history is provided in Background section. The relevant work in this area is described in related works section. A description of the dataset and the models used for the experiment is provided in Research Framework section. The experimental information on the models' performance benchmarking activity is elaborated in Experiment details section. The outcome is discussed in Results and Discussion section. Our study project is concluded with future directions in Conclusion and Future Scope section.

BACKGROUND

In addition to being a potent tool for minimizing physician fatigue, artificial intelligence in medical diagnostics also offers radiology professionals tremendous help in managing workloads that are only expected to grow. Radiologists are required to work at previously unheard-of speeds while managing many and increasing image volumes. Today, they must prioritize the most urgent cases while still managing patient care and navigating through reams of images. As we are using a deep learning model, we will use the transfer learning concept by using the pre-trained VGG16, VGG19, InceptionV3, and MobileNet. The next subsections explain transfer learning and its designs (Table **2**).

Table 2. Traditional ML *vs.* Transfer learning [8].

Isolated, discrete learning: No knowledge is gained or retained. Learning is done without taking into account previously acquired information in other openings.	A new task's learning is dependent on previously learned tasks: The learning process may be quicker, more precise, or need less training data.

Transfer Learning

A machine learning technique called transfer learning [7] uses a model created for one task as the foundation for a model on another.

Pre-trained models are frequently used as the foundation for deep learning tasks in computer vision and natural language processing because they save both time and money compared to developing neural network models from scratch and because they perform vastly better on related tasks. In transfer learning, the learned features are first applied to a base network that is trained on a base dataset and task, and then the features are transferred to a second target network that is trained on a target dataset and task. The distinction between Traditional ML and Transfer learning is shown in the (Figs. **1** and **2**).

If the traits are general—that is, applicable to both the base task and the target task—rather than task-specific, this procedure is more likely to succeed. When utilizing transfer learning, you can see these three advantages [8]:

1. Greater start. On the source model, the initial competence is more than it would otherwise be (prior to model refinement).
2. Greater slope. As compared to how it would otherwise be, the rate of skill growth during source model training is steeper.

3. Increased asymptote. The taught model's converged skill is superior to what it would be without training.

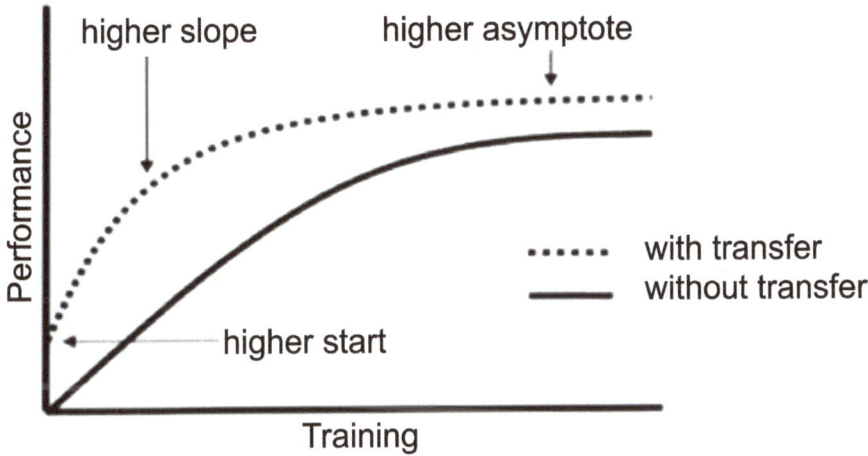

Fig. (1). Performance comparison in case of transfer learning [9].

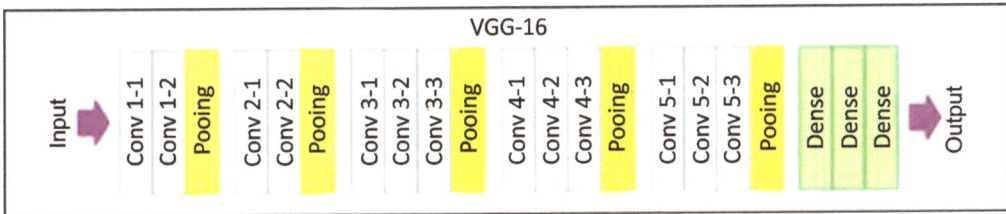

Fig. (2). VGGNet16 architecture [11].

VGG16 Architecture

In the study, "Very Deep Convolutional Networks for Large-Scale Image Recognition," K. Simonyan and A. Zisserman from the University of Oxford introduced the convolutional neural network model known as VGG16 [10]. In the top five tests, ImageNet, a dataset of more than 14 million images separated into 1000 classes, shows that the model performs 92.7% accurately. The model that was presented to ILSVRC-2014 was well-known. It outperforms AlexNet by successively replacing multiple huge kernel-sized filters (11 and 5, respectively, in the first and second convolutional layers) with 33 kernel-sized filters. The NVIDIA Titan Black GPUs were used for the weeks-long training of VGG16.

The RGB picture with a fixed size of 224 by 224 is the input to the cov1 layer. The picture is run through a stack of convolutional layers with a very narrow receptive field—3×3 (the smallest size to capture the concepts of left/right, up/down, and center)—used by the filters. It also uses 1by 1 convolution filters in one of the configurations, which may be thought of as a linear transformation of the input channels (followed by non-linearity). The convolution layer's input undergoes spatial padding, ensuring the preservation of spatial resolution post-convolution. To elaborate, for 3x3 convolution layers, a pixel padding of 1 is employed. Additionally, a consistent convolution stride of 1 pixel is employed.

Following a stack of convolutional layers (which varies in depth between designs), three Fully Connected (FC) layers are used: the first two have 4096 channels each, while the third uses 1000-way ILSVRC classification and so has 1000 channels (one for each class). The soft-max layer is the last one.

Across all networks, the fully connected layers share identical configurations. Refer to Fig. (**3a** and **3b**) for a comprehensive illustration of the VGG16 architecture [12]. Fig. (**2**) outlines the configurations for the ConvNet, denoted by labels (A-E). While adhering to the foundational architectural blueprint, the only discernible distinction among these configurations lies in their depth. This variation extends from networks with 11 weight layers. In the network E (16 conv. and 3 FC layers), there are 19 weight layers ranging from A (8 conv. and 3 FC layers). Convolutional layers have a relatively small width (number of channels), starting from 64 in the first layer and rising by a factor of 2 after each maximum pooling layer until it reaches 512.

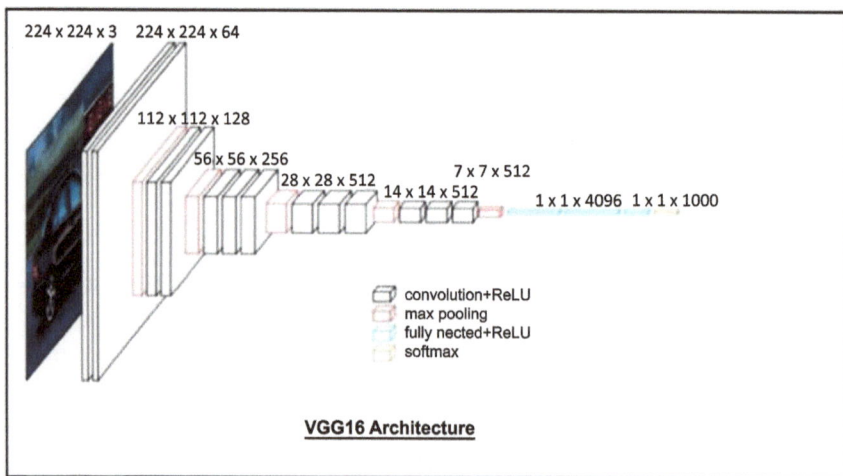

Fig. (3a). VGG16 architecture detailed representation [12].

Herein, we will perform a comparative study of different CNN models using the transfer learning model with VGG16 and other architectures and compare the results to show which model is a better fit for predicting Covid-19. We will also discuss the basic structure of other architectures.

VGG19 Architecture

A variation of the VGG model called VGG19 [13] has 19 layers in total: 16 convolutional layers, 3 fully connected layers, 5 MaxPool layers, and 1 SoftMax layer. There are further VGG variations, including VGG11, VGG16, and others. **19.6 billion FLOPs** make up VGG19.

• This network received an RGB picture with a fixed size of (224 * 224), indicating that the matrix has the form of (224, 224, 3).

• The mean RGB value of each pixel, calculated throughout the whole training set, was the sole preprocessing carried out.

• They were able to cover the entirety of the image by using kernels that were (3 * 3) in size with a stride size of 1 pixel.

• To maintain the image's spatial resolution, spatial padding was applied.

• Max pooling was carried out with stride 2 across a 2 by 2-pixel frame.

• Rectified linear unit (ReLu) was then employed to add non-linearity to the model in order to enhance classification accuracy and computation time. As opposed to earlier models that used tanh or sigmoid functions, this one performed far better.

• Three completely linked layers were used, the first two of which had a size of 4096, followed by a layer with 1000 channels for classification using a 1000-way ILSVRC, and the third layer is a softmax function.

MobileNet Architecture

Real-world applications employ MobileNet [15], a CNN architecture that is effective and portable. In order to create lighter models, MobileNets essentially replace the typical convolutions employed in older designs with **depthwise separable convolutions**. According to their needs, model creators can trade off latency or accuracy for speed and small size using the two new global hyperparameters introduced by MobileNets (width multiplier and resolution multiplier).

On depth-wise separable convolution layers, MobileNets are constructed. A depthwise convolution and a pointwise convolution make up each depthwise separable convolution layer. A MobileNet contains 28 layers if depthwise and pointwise convolutions are counted separately. By properly adjusting the width of multiplehyperparameters, it is possible to further minimize the 4.2 million parameters that make up a conventional MobileNet. The supplied picture is 224 by 224 by 3, in size (Fig. **4a**).

ConvNet Configuration					
A	A-LRN	B	C	D	E
11 weight layers	11 weight layers	13 weight layers	16 weight layers	16 weight layers	19 weight layers
input (224 × 224 RGB image)					
conv3-64	conv3-64 LRN	conv3-64 **conv3-64**	conv3-64 conv3-64	conv3-64 conv3-64	conv3-64 conv3-64
maxpool					
conv3-128	conv3-128	conv3-128 **conv3-128**	conv3-128 conv3-128	conv3-128 conv3-128	conv3-128 conv3-128
maxpool					
conv3-256 conv3-256	conv3-256 conv3-256	conv3-256 conv3-256	conv3-256 conv3-256 **conv1-256**	conv3-256 conv3-256 **conv3-256**	conv3-256 conv3-256 conv3-256 **conv3-256**
maxpool					
conv3-512 conv3-512	conv3-512 conv3-512	conv3-512 conv3-512	conv3-512 conv3-512 **conv1-512**	conv3-512 conv3-512 **conv3-512**	conv3-512 conv3-512 conv3-512 **conv3-512**
maxpool					
conv3-512 conv3-512	conv3-512 conv3-512	conv3-512 conv3-512	conv3-512 conv3-512 **conv1-512**	conv3-512 conv3-512 **conv3-512**	conv3-512 conv3-512 conv3-512 **conv3-512**
maxpool					
FC-4096					
FC-4096					
FC-1000					
soft-max					

Fig. (3b). VGG19 Architecture [14].

Recap - MobileNetV1

Table 1. MobileNet Body Architecture

Type/Stride	Filter Shape	Input Size
Conv / s2	3 x 3 x 3 x 32	224 x 224 x 3
Conv dw / s1	3 x 3 x 32 dw	112 x 112 x 32
Conv / s1	1 x 1 x 32 x 64	112 x 112 x 32
Conv dw / s2	3 x 3 x 64 dw	112 x 112 x 64
Conv / s1	1 x 1 x 64 x 128	56 x 56 x 64
Conv dw / s1	3 x 3 x 128 dw	56 x 56 x 128
Conv / s1	1 x 1 x 128 x 128	56 x 56 x 128
Conv dw / s2	3 x 3 x 128 dw	56 x 56 x 128
Conv / s1	1 x 1 x 128 x 256	28 x 28 x 128
Conv dw / s1	3 x 3 x 256 dw	28 x 28 x 256
Conv / s1	1 x 1 x 256 x 256	28 x 28 x 256
Conv dw / s2	3 x 3 x 256 dw	28 x 28 x 256
Conv / s1	1 x 1 x 256 x 512	14 x 14 x 256
5x Conv dw / s1	1 x 1 x 512 dw	14 x 14 x 512
Conv / s1	1 x 1 x 512 x 512	14 x 14 x 512
Conv dw / s2	3 x 3 x 512 dw	14 x 14 x 512
Conv / s1	1 x 1 x 512 x 1024	7 x 7 x 512
Conv dw / s2	3 x 3 x 1024 dw	7 x 7 x 1024
Conv / s1	1 x 1 x 1024 x 1024	7 x 7 x 1024
Avg Pool / s1	Pool 7 x 7	7 x 7 x 1024
FC / s1	1024 x 1000	1 x 1 x 1024
Softmax / s1	Classifier	1 x 1 x 1000

Figure 3. Left: Standard convolutional layer with batchnorm and ReLU. Right: Depthwise Separable convolutions with Depthwise and Pointwise layers followed by batchnorm and ReLU.

Fig. (4a). MobileNet Architecture [16].

InceptionV3 Architecture

The fundamental goal of InceptionV3 [17] is to utilize fewer computing resources by altering the Inception structures from the first two iterations. The proposal presented in 2015's work titled 'Rethinking the Inception Architecture for Computer Vision' was put forth by a collaborative effort involving Christian Szegedy, Vincent Vanhoucke, Sergey Ioffe, and Jonathon Shlens. This insightful article highlighted the virtues of Inception Networks, often referred to as GoogLeNet or Inception v1. In comparison to the VGGNet, Inception Networks have shown enhanced computational efficiency. This advantage extends to both the parameter count within the network and the overall economic expenditure, encompassing memory usage and other associated resources. It is important to take care not to lose the computational benefits while making changes to an Inception Network. Due to the unknown effectiveness of the new network, it becomes difficult to modify an Inception network for various use cases. In an Inception v3 model, a number of network optimization strategies have been proposed to relax the restrictions and make model adaption simpler. The methods include regularization, dimension reduction, factorized convolutions, and parallelized calculations.

An Inception v3 network's architecture is developed gradually and methodically, as follows:

- **Factorized Convolutions:** Through the reduction of network parameters, this technique effectively enhances computational efficiency while also gauging the network's performance.
- **Smaller convolutions:** Employing smaller convolutions significantly expedites training. For instance, substituting a 5x5 filter, with 25 parameters, with two 3x3 filters leads to mere 18 parameters (3x3 + 3x3).
- **Asymmetric convolutions:** The utilization of asymmetric convolutions is another strategy. This involves the sequence of a 1x3 convolution, followed by a 3x1 convolution. While slightly increasing the parameter count compared to the suggested asymmetric convolution, this approach remains more efficient than replacing a 3x3 convolution with a 2x2 convolution.
- **Auxiliary classifier:** A noteworthy addition is the auxiliary classifier, strategically inserted between layers during the training process. The loss incurred by this auxiliary network contributes to the overall loss of the primary network. Notably, GoogLeNet employs auxiliary classifiers to foster depth within the network, differing from the role they play in Inception v3, which is mainly a regularizer.

- **Grid size reduction:** The process of grid size reduction conventionally involves pooling techniques. However, a more impactful alternative is proposed to address the computational constraints.

Grad-Cam Technique

Grad-CAM [18], a class-discriminative localization method, is able to produce visual explanations from any CNN-based network without the need for architectural modifications or new training. Grad-CAM is evaluated for localisation and model fidelity, where it performs better than baselines. Grad-CAM generates a rudimentary localization map, emphasizing crucial regions within an image to aid in concept prediction. This is achieved by leveraging gradients associated with a chosen target concept (such as the 'dog' logits or even a caption). These gradients are subsequently directed into the ultimate convolutional layer.

Grad-CAM allows us to visually confirm where our network is looking, ensuring that it is actually focusing on the right picture patterns and activating around them. If the appropriate patterns or items in the picture are not where the network should be activating, then we know:

• The network has not fully assimilated the dataset's fundamental patterns.
• The training process should be reviewed.
• More information must be acquired.
• The deployment of the model is not yet ready.

RELATED WORKS

To recognize various thoracic disorders, including pneumonia, many deep learning-based methods have been created. The primary source of reference for this study is the CovidAID research [19]. Building upon the foundational work of CheXNet [20], which notably outperformed experienced radiologists in detecting pneumonia from chest X-rays, this current research is rooted. CheXNet [20] stands out for its superior performance compared to preceding systems, coupled with a simplified architecture compared to subsequent methodologies. The model is trained on the ChestX-ray14 dataset [21], the largest publicly accessible repository of chest X-ray data. CheXNet [20], a 121-layer DenseNet-based model, was trained on 112,120 frontal view chest X-ray images from the ChestX-ray14 dataset [20]. This model has been adeptly trained to categorize thoracic CT scans into 14 distinct disease groups, including pneumonia.

Notably, given the visual coherence of the input data, it emerges as a robust pre-trained backbone for constructing a COVID-19 pneumonia identification model.

In light of the recent global surge in infections, numerous screening techniques have been developed to identify potential COVID-19 cases [19]. However, the availability of open-source programs employing chest X-ray imagery remains limited [21]. Similarly, there is a scarcity of publicly accessible chest X-ray data specific to COVID-19 [21].

Addressing this, the sole approach featuring an open-source and consistently maintained tool capable of effectively distinguishing COVID-19 from other forms of pneumonia, while displaying a heightened sensitivity towards COVID-19 detection, is COVID-Net [22].

COVID-Net [22] learns the architectural design starting from the original design prototype and requirements through machine-driven design exploration. It uses a chest X-ray picture as input and produces one of three predictions: normal, pneumonia, or COVID-19. We use this model as our reference point and contrast our findings with it.

RESEARCH FRAMEWORK

Without a doubt, creating a visual recognition dataset is a really difficult undertaking. To extract clips, annotate, validate, and create a benchmark dataset for computer vision algorithms, it takes a lot of work. Each stage in the production of a bespoke dataset must be completed. To prepare our custom dataset we went to various online sources from where X-ray images can be downloaded. These images were public and could be used for research purposes.

Images play a crucial role in training any custom model. Deep learning heavily relies on these pictures. A model's accuracy is determined by the training set of photos. Therefore, one has to prepare and gather photos before training a custom model. Below is a description of how the whole dataset was created.

As there was nowhere a particular dataset present in the open-source environment that could be used for doing this comparative study and given the few X-ray samples available, we decided to combine different datasets available on the internet to form a new custom dataset. We kept the participation of each class as equal as possible and also performed careful cleaning of the dataset in order to remove any truncated images. This led to a decrease in the size of the formed dataset but at the same time, we were confident that performing these techniques would eventually help our models to become more robust and perform better.

The dataset utilized in this study is sourced from IEEE8023 [23], encompassing COVID-19 frontal-view chest X-ray images, as well as the chest-X-ray pneumonia dataset containing frontal-view chest X-ray images depicting normal lungs. Additionally, we incorporated a Kaggle dataset [24] to compile the comprehensive dataset.

Given the absence of predefined data divisions within the Covid-chest X-ray dataset, we took the initiative to establish our partitioning. Specifically, we allocated 90% of the data for training purposes and reserved 10% for testing. To bolster the validation phase, we drew from various publicly accessible sources [23, 24]. Since the dataset was comparatively small for a deep learning project, we used an Image data generator [25] for pre-processing of the data to make the model more robust to outliers and other errors. Keras Image data generator [25] is a tool that lets you augment images in real-time while the model is still training. This not only makes the model more robust but also saves overhead memory.

The distribution of the dataset is given as follows:

Classes	Train Set	Test Set
Normal	2406	332
Pneumonia	2396	220
Covid-19	2506	174

Once, we prepared the dataset and filtered out the bad images we then, moved on to feeding them into our pretrained neural network models. These models use Imagenet weights for learning features in an image. The upper portion of the models has been removed and instead of them, we added our layers that suit our purpose. This development is consistent with all the models used in the experiment.

EXPERIMENT DETAILS

Three categories—Normal, Pneumonia (clubbing bacterial and viral as indicated by the datasets from which it was built), and Covid—can be applied to the dataset. Our model has been taught to operate in the three-class arrangement. To better clarify any potential misunderstanding between common pneumonia and COVID, the three-class design was chosen. A particular frontal view chest X-ray picture will be categorized into the following categories: normal, COVID-19, and pneumonia.

Refer to Fig. (3) for a selection of frontal view chest X-ray samples, these were systematically grouped according to their respective classes. Notable observations include the consolidation present in the right lower lobe of a patient diagnosed with bacterial pneumonia, and the radiographic illustration of patchy consolidation across the right middle and lower zones, characteristic of viral pneumonia. Furthermore, a patient afflicted with COVID-19 pneumonia exhibits a final image depicting sporadic ground glass opacity in the left lower zone.

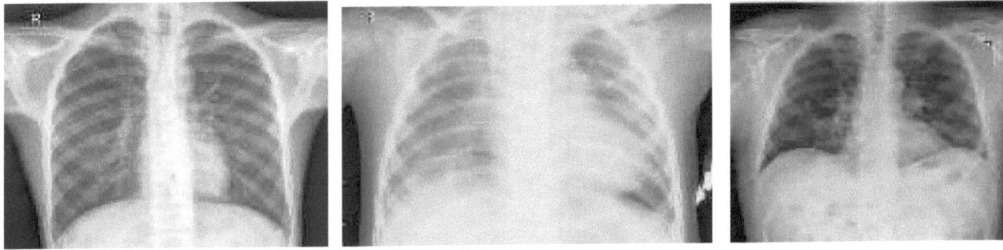

(a) Illustrating a healthy lung   (b) Pneumonia (Bacterial+Viral)   (c) Covid-19

Machine Setup

On the dataset Google Colab single GPU for training and model creation, we presented baseline results. We employed the same machine for inference. We train our models with an RMSProp optimizer for 10 epochs and 228 iterations for each epoch. We start the models with ImageNet weights pre-trained on the dataset. We use a single GPU with a 32-batch size.

Model Architecture

In background section of this paper, we have already discussed the model architectures that we have used for the purpose of performance benchmarking. However, all these models have their top layer removed, and instead of that, a Global Average Pooling layer and over it a Dropout layer were added. We used softmax cross-entropy as the activation function and kept the dropout percentage at 0.4. The loss function used was categorical cross-entropy. The purpose of adding these layers was to train our model for the specific purpose of determining chest x-rays as these when trained on the imagenet weights [26] serve a general purpose of categorizing over 1000 different objects.

RESULTS AND DISCUSSION

Based on our study, we can now perform performance benchmarking of different neural network architectures on the Covid-19 dataset.

Analysis of Results

Before going into the details of the result, first, we present the analysis of RQs as follows:

- MobileNet model gave the best result.
- It was found that the complexity of a model is not the only factor that affects the accuracy of the results. It depends on other conditions as well.
- Transfer learning can also be used for radiological images.

We did a comparative analysis of different CNN architectures. As we trained our models, we got different results for every model and the result is shown as follows:

Model Architecture	Validation Accuracy	Validation Loss
VGG16	0.6690	0.6839
VGG19	0.6903	0.6972
MobileNet	0.9446	0.1840
InceptionV3	0.8679	0.3518

VGG16 Model

The VGG16 model gives us an accuracy of 0.6690 and a validation loss of 0.6839.

The appropriate graphs are shown in below Fig. (**4b**).

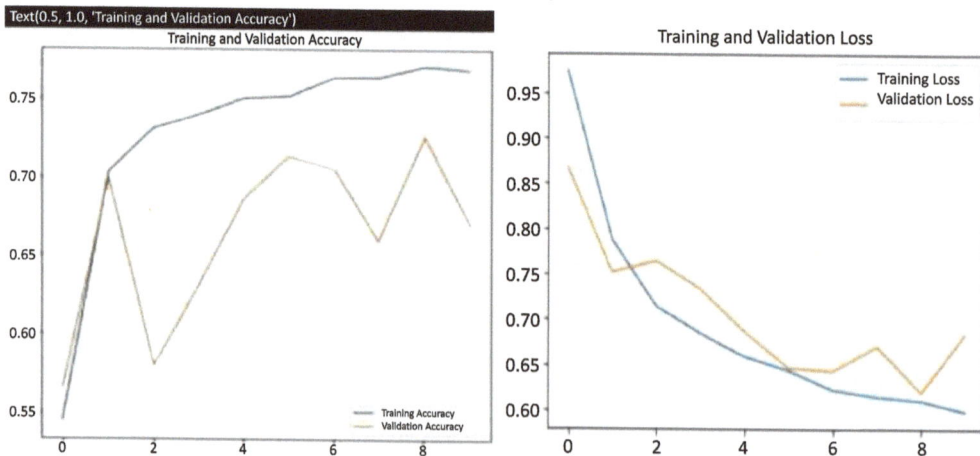

Fig. (4b). (**a**)Training and Validation Accuracy (**b**) Training and Validation Loss.

The training accuracy of a model shows its accuracy on examples. According to the theory, both these accuracies tend to increase when the number of examples in the dataset increases. This is the general trend we should observe in our example as well.

As the above graphs are investigated, it is seen that the training accuracy remains comparatively smooth and follows an upward trend but the testing accuracy is not just rough but also falls and jumps abruptly for some values. However, in broader terms, one can say that it too follows an upward trend. The fall of the values remains a matter of further study.

We see similar trends in training and validation loss as well. Both of them follow a general downward trend with the training loss curve being much smoother as compared to the validation loss curve.

VGG19 Model

The VGG19 model gives us an accuracy of 0.6903 and a validation loss of 0.6972.

The same trend was found as in the previous case.

The appropriate graphs are shown below (Figs. **5** and **6**).

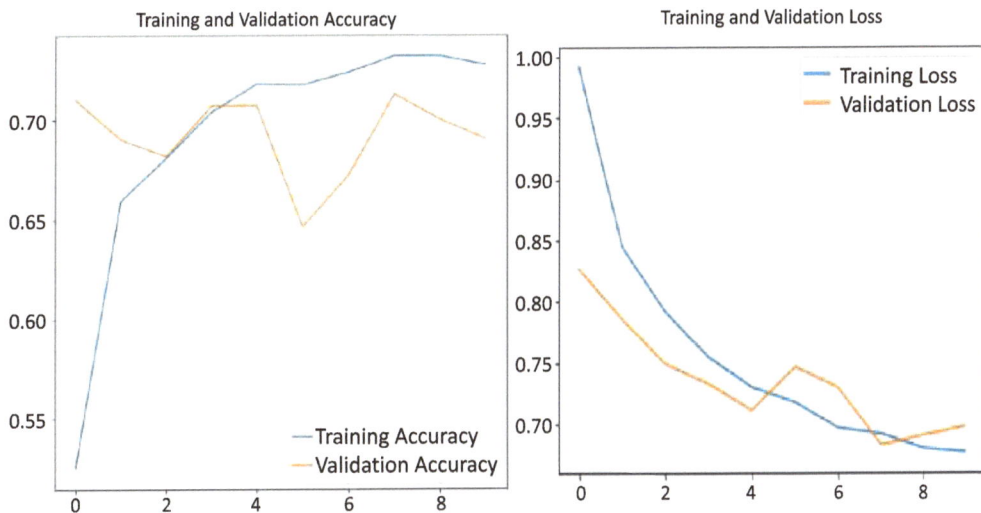

Fig. (5). (a) Training and Validation accuracy (b) Training and validation loss.

Mobile Net Model

The MobileNet model gives us an accuracy of 0.9446 and a validation loss of 0.1840

In this case, the graph did perform better than the above two but the curve was rather rough so, the results cannot be curtained to be true. However, we can be certain of the fact that this architecture will perform better than the above two ones, and with certain modifications, it has a lot of potential.

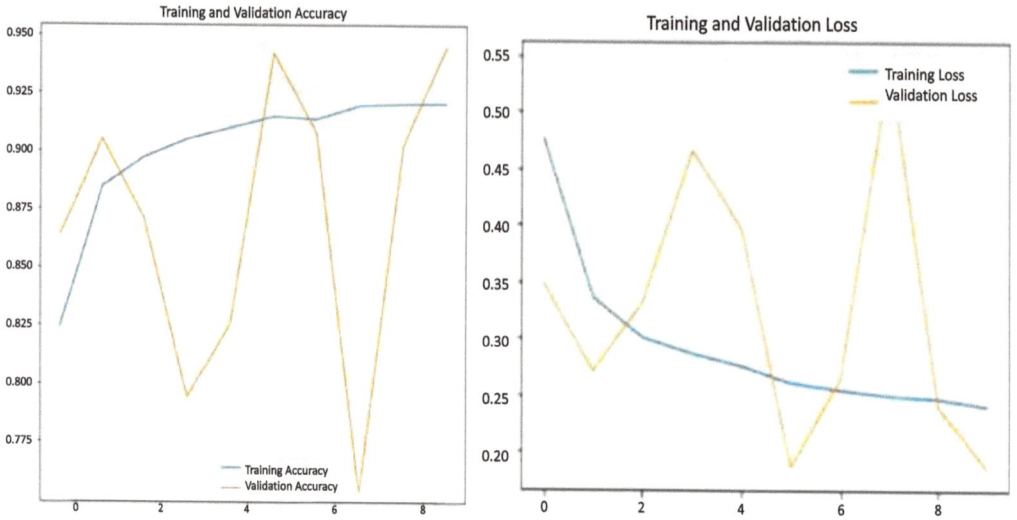

Fig. (6). (**a**)Training and Validation accuracy (**b**) Training and validation loss.

The appropriate graphs are shown below (Fig. **7**).

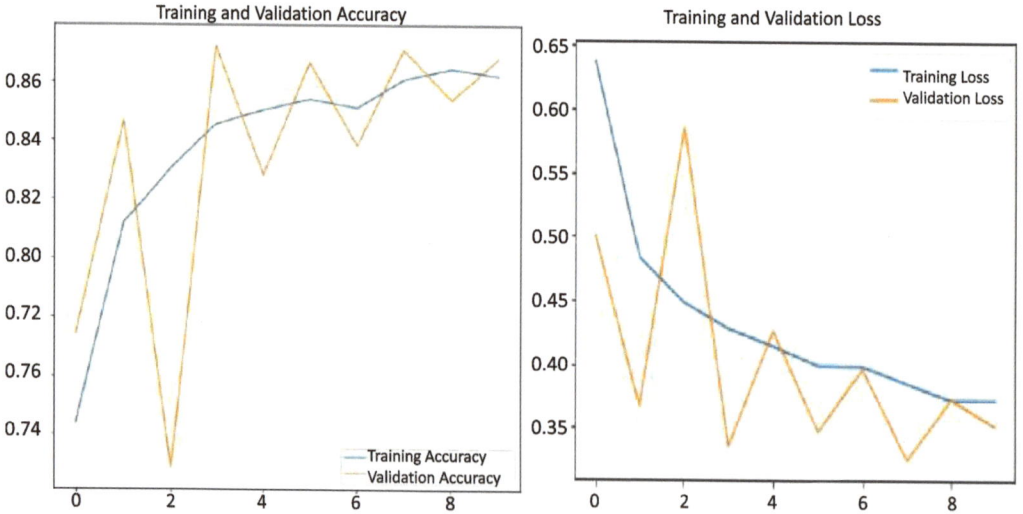

Fig. (7). (**a**) Training and Validation accuracy (**b**) Training and validation loss.

InceptionV3 Model

The MobileNet model gives us an accuracy of 0.8679 and a validation loss of 0.3518.

The curve is certainly the most consistent among all four but it did perform worse when compared to the results with the MobileNet architecture. However, it is not always about the accuracy of the result as sometimes we want our models to be consistent as well as accurate and if there comes a situation of barter then, giving away some accuracy for some consistency can be considered.

The appropriate graphs are shown below:

Grad-Cam Results

After all the models were trained and tested, we chose the two most promising models for further investigation as only accuracy and validation loss were not enough to compare them. A more visual approach was indeed the need of the hour. Therefore, an image from the validation test was given to both models, and the results obtained were as follows in Fig. (**8**).

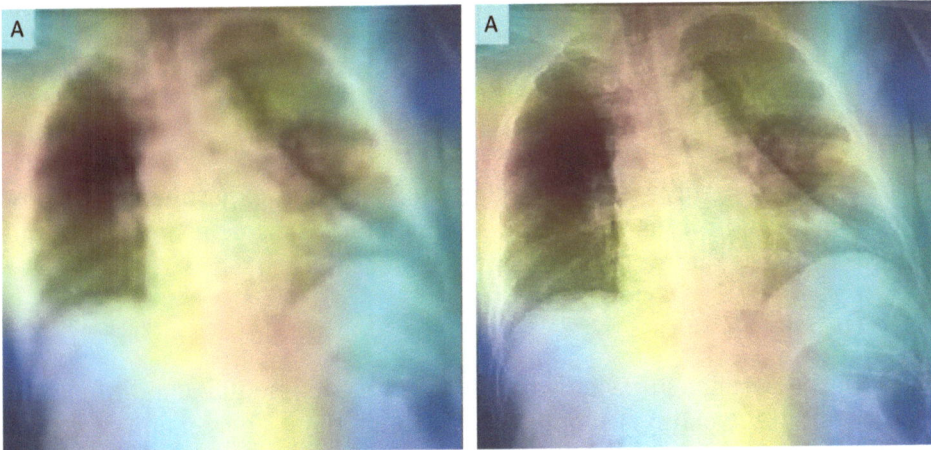

Fig. (8). (**a**) InceptionV3 (**b**) MobileNet.

On looking at the results, not much difference can be observed . However, a radiologist might have a different opinion of the results. Although the accuracies of the models were different on applying Grad-Cam we see that not much difference is present so, we can say that for further research, any one of them will perform well.

Discussions

From a bird's eye view, it seems that the MobileNet model outperforms its successors InceptionV3 and VGG16 /VGG19 in terms of accuracy but looking at the training and validation accuracy and loss curves we cannot be sure of whether it is the correct thing to do. Presently, most of the models prepared focus predominantly on increasing the overall accuracy which seems the very logical thing to do but after doing these tests, we can say that a model should not be judged only on one metric but should be carefully tested on a number of metrics.

The amount of effort that is presently put in to collect high-quality and large quantities of data is similar if not more effort is needed to check the validity of the result produced by the deep learning models. We know that a deep learning model is very data-hungry but we should also not forget that these models if want to perform well in the medical domain where chances to make mistakes are numerous and the cost of these mistakes is significantly high, must be tested heavily over various parameters.

On comparing our results with the CheXNet [20], we see that our model being smaller in size and having been trained on a much smaller dataset still performed satisfactorily and tells a lot about the use of other techniques like transfer learning which can be used over existing model architectures to update them and make them more robust and accurate.

Risks to Validity

In this part, we go through potential dangers to our experiment and how we mitigated them. Validity is the degree to which an experiment's results accurately reflect the variables that were intended to be measured.

Risks to Internal Validity

It particularly pertains to whether or not an experimental condition makes a difference and if the claim is supported by enough data. The version of the models that we employed in the experiment poses the biggest hazard in this case. An improved version of the same architecture may allow the model to predict data more accurately.

Risks to External Validity

It alludes to how universal the results of the trial were. Since we take the top layer from our networks and fine-tune it with three classes, we employ transfer learning in our networks. The experiment faced significant computational power constraints, which prevented the model from being trained from scratch to more effectively learn the dataset's class labels. It may not be possible to extrapolate the experiment's findings from this X-ray image to another real-time X-ray image in a lab.

Construct Validity

It assesses if an experimental variable's operational definition accurately captures the concept's initially actual theoretical meaning. Here we used models of increasing complexity for performance benchmarking and considered that as the number of layers will increase the overall accuracy of the model will increase. However, the most complex model did not perform the best instead a relatively simpler model gave better results.

Conclusion Validity

The degree to which our findings are trustworthy and credible is known as conclusion validity. We trained different models under identical conditions and found that the results improved as the number of convolutional layers increased but after a certain point, it started degrading the results. We have also observed that these neural networks can help in diagnosis even when the subject to be found is new and not seen before by humans like in the case of this novel virus.

CONCLUSION AND FUTURE SCOPE

Since there has been a significant increase in COVID-19 infections all over the world, several different screening techniques [19] have been created to find potential COVID-19 cases. There is a need for a safe and accurate detection method that can produce results at a faster rate than any other methods available now.

The whole process of this project concludes that the transfer learning technique using pre-trained models gives better performance in predicting COVID-19 in the chest X-ray images. The performance will be improved even more when the size of the data set is much bigger than that of the dataset that we used in our project.

Similarly, to COVID-19 detection, the transfer learning model can also be used in various applications that involve image detection in CCTV or in the domain of NLP as well. It reduces some of the effort to initialize the weights as it borrows from a pre-trained model and performs well. It tackles problems like having little or almost no labeled data availability.

Advancements in Machine Learning (ML) and Deep Learning (DL) are reshaping drug discovery and disease diagnosis. ML algorithms facilitate target identification, drug screening, and toxicity assessment, accelerating drug development. In disease diagnosis, ML aids in medical imaging interpretation and in predicting disease risks, and also enables personalized treatment plans. Genomic medicine benefits from ML-driven analysis, enhancing understanding of individualized drug responses. Challenges include data privacy and model interpretability. Future trends focus on overcoming ethical concerns and expanding applications in personalized medicine, promising transformative impacts on healthcare.

In addition to the application of machine learning in the analysis of X-ray pictures of COVID-19 patients, machine learning utility can be applied in various other fields such as drug discovery [27], predictions of antiviral [28] as well as anticancer [29] peptides, and neural network-based diagnosis [30].

REFERENCES

[1]   L. Li, S. Rong, R. Wang, and S. Yu, "Recent advances in artificial intelligence and machine learning for non linear relationship analysis and process control in drinking water treatment: A review", *Chem. Eng. J.,* vol. 405, no. 1, p. 126673, 2021.

[2]   M. Kazeminejad, and M. Karamifard, "Optimal and economic design of a stand-alone hybrid renewable energy system integrated with battery storage using an artificial electric field algorithm", *Przegląd Elektrotechniczny,* vol. 1, no. 11, pp. 315-321, 2022.
[http://dx.doi.org/10.15199/48.2022.11.63]

[3]   G. Liu, and J.M. Stokes, "A brief guide to machine learning for antibiotic discovery", *Curr. Opin. Microbiol.,* vol. 69, no. 102190, p. 102190, 2022.
[http://dx.doi.org/10.1016/j.mib.2022.102190] [PMID: 35963098]

[4]   A.H. Vo, T.R. Van Vleet, R.R. Gupta, M.J. Liguori, and M.S. Rao, "An Overview of Machine Learning and Big Data for Drug Toxicity Evaluation", *Chem. Res. Toxicol.,* vol. 33, no. 1, pp. 20-37, 2020.
[http://dx.doi.org/10.1021/acs.chemrestox.9b00227] [PMID: 31625725]

[5]   B. Choubin, M. Borji, A. Mosavi, F. Sajedi-Hosseini, V. P. Singh, and S. Shamshirband, "Snow avalanche hazard prediction using machine learning methods", *Journal of Hydrology,* vol. 577, p. 123929, 2019.
[http://dx.doi.org/10.1016/j.jhydrol.2019.123929]

[6]   Maryam Rad, "Neisseria meningitidis detection by coupling bacterial factor H onto Au/scFv antibody nanohybrids", *Applied Physics A,* vol. 129.6, p. 401, 2023.
[http://dx.doi.org/10.1007/s00339-023-06620-2]

[7]   K. Weiss, T.M. Khoshgoftaar, and D. Wang, "A survey of transfer learning", *J. Big Data,* vol. 3, no. 1, p. 9, 2016.
[http://dx.doi.org/10.1186/s40537-016-0043-6]

[8]   D. Sarkar, "A Comprehensive Hands-on Guide to Transfer Learning with Real-World Applications in Deep Learning Towards Data Science", Available from: https://miro.medium.com/max/2400/1*9GTEzcO8KxxrfutmtsPs3Q.png (online)

[9]   "The benefits of transfer learning. The three types of performance... | Download Scientific Diagram (researchgate.net)", Available from: https://cdn-images1.medium.com/max/800/1*tECctD6W1FCYDvCHbAWD-A.jpeg

[10]   Simonyan, Karen, Zisserman, and Andrew, "Very Deep Convolutional Networks for Large-Scale Image Recognition", *arXiv,* .1409.1556

[11]   R. Prajapati, "Face Recognition using Transfer Learning! | by Rahul Prajapati | Medium", Available from: https://neurohive.io/wp-content/uploads/2018/11/vgg16.png (online)

[12]   G. Surma, "Style Transfer - Styling Images with Convolutional Neural Networks | by Greg Surma |

Medium", Available from: https://neurohive.io/wp-content/uploads/2018/11/vgg16-1-e1542731207177.png (online)

[13] R. Sudha, "A Convolutional Neural Network Classifier VGG-19 Architecture for Lesion Detection and Grading in Diabetic Retinopathy Based on Deep Learning", *Comput. Mater. Continua,* vol. 66, pp. 827-842, 2020.
[http://dx.doi.org/10.32604/cmc.2020.012008]

[14] A. Hossein, "Implementing VGG13 for MNIST dataset in TensorFlow | by Amir Hossein | Medium", Available from: https://iq.opengenus.org/content/images/2020/02/Screenshot-from-2020-02-22-16-25-15.png (online)

[15] A. Howard, M. Zhu, B. Chen, D. Kalenichenko, W. Wang, T. Weyand, and M. Andreetto, "MobileNets: Efficient Convolutional Neural Networks for Mobile Vision Applications", *arXiv,* 1704. 04861.

[16] "MobileNets: Efficient Convolutional Neural Networks for Mobile Vision Applications (2)_yp532的博客CSDN博客", Available from: https://miro.medium.com/max/410/1*TJAjuueT9_pk2Nlv1zmb4A.png (online)

[17] C. Szegedy, V. Vanhoucke, S. Ioffe, J. Shlens, and Z. Wojna, "Rethinking the Inception Architecture for Computer Vision", *arXiv,* .1512.00567
[http://dx.doi.org/10.1109/CVPR.2016.308]

[18] R.R. Selvaraju, M. Cogswell, A. Das, R. Vedantam, D. Parikh, and D. Batra, "Grad-CAM: Visual Explanations from Deep Networks *via* Gradient-Based Localization", *2017 IEEE International Conference on Computer Vision (ICCV),* 2017pp. 618-626 Venice, Italy
[http://dx.doi.org/10.1109/ICCV.2017.74]

[19] A. Mangal, S. Kalia, H. Rajgopal, K. Rangarajan, V. Namboodiri, S. Banerjee, and C. Arora, "Detection Using ChestX-Ray", *arXiv,* .2004.09803

[20] P. Rajpurkar, J. Irvin, K. Zhu, B. Yang, H. Mehta, T. Duan, D. Ding, A. Bagul, C. Langlotz, K. Shpanskaya, M.P. Lungren, and A.Y. Ng, "CheXNet: Radiologist-Level Pneumonia Detection on Chest X-Rays with Deep Learning", *arXiv,* .1711.05225

[21] X. Wang, and Y. Peng, *Le Lu, Zhiyong Lu, MohammadhadiBagheri, Ronald M. Summers.ChestX--: Hospital-scale Chest X-ray Database and Benchmarks on Weakly- SupervisedClassification and Localization of Common Thorax Diseases.* IEEE CVPR, 2017, pp. 3462-3471.

[22] L. Wang, Z.Q. Lin, and A. Wong, "COVID-Net: a tailored deep convolutional neural network design for detection of COVID-19 cases from chest X-ray images", *Sci. Rep.,* vol. 10, no. 1, p. 19549, 2020.
[http://dx.doi.org/10.1038/s41598-020-76550-z] [PMID: 33177550]

[23] J. P. Cohen, P. Morrison, and L. Dao, "COVID-19 Image Data Collection", *arXiv,* .2003.11597

[24] D. Kermany, K. Zhang, and M. Goldbaum, *Labeled Optical Coherence Tomography (OCT) and Chest X-Ray Images for Classification* Mendeley Data, 2018.
[http://dx.doi.org/10.17632/rscbjbr9sj.2]

[25] "tf.keras.preprocessing.image.ImageDataGenerator (tensorflow.org)", Available from: https://www.tensorflow.org/api_docs/python/tf/keras/preprocessing/image/ImageDataGenerator (online)

[26] J. Deng, W. Dong, R. Socher, L-J. Li, K. Li, and F-F. Li, "ImageNet: A large-scale hierarchical image database", *2009 IEEE Conference on Computer Vision and Pattern Recognition,* 2009 pp. 248-255 Miami, FL, USA
[http://dx.doi.org/10.1109/CVPR.2009.5206848]

[27] D. Selwood, *Chem. Biol. Drug Des.,* vol. 100, pp. 699-721, 2022.
[http://dx.doi.org/10.1111/cbdd.14136] [PMID: 36002440]

[28] C.A. Kieslich, F. Alimirzaei, H. Song, M. Do, and P. Hall, "Data-driven prediction of antiviral

peptides based on periodicities of amino acid properties", *Proceedings of the 31st European Symposium on Computer Aided Process Engineering,* vol. 50, pp. 2019-2024, 2021.
[http://dx.doi.org/10.1016/B978-0-323-88506-5.50312-0]

[29]   F. Alimirzaei, and C. A. Kieslich, "Machine learning models for predicting membranolytic anticancer peptides", *Computer Aided Chemical Engineering,* vol. 52, pp. 2691-2696, 2023.
[http://dx.doi.org/10.1016/B978-0-443-15274-0.50428-5]

[30]   Wanqi Lai, "Skin cancer diagnosis (SCD) using Artificial Neural Network (ANN) and Improved Gray Wolf Optimization (IGWO)", *Sci Rep,* vol. 13, p. 19377, 2023.
[http://dx.doi.org/10.1038/s41598-023-45039-w]

CHAPTER 12

Application of Novel Nanotherapeutic Strategies in Treatment Using Herbal Medicines

Sumanta Bhattacharya[1,*]

[1] *Maulana Abul Kalam Azad University of Technology, West Bengal, India*

Abstract: Herbal remedies are gaining popularity as an alternative to allopathic medicine because of how much better they are at curing modern health problems. By facilitating the efficient distribution of medicinal molecules to both targeted and non-targeted regions, nanotherapeutic approaches enhance the pharmacokinetic efficacy of herbal remedies. Active and system-based nanostructures have the potential to utterly transform herbal therapy. Nanomedicine may benefit from third-generation nanotechnology, namely system-based nanostructures, due to their self-healing properties. Research and Market predicts that the pharmaceutical market's use of nanotechnology will increase by 15.3% by 2026. The effectiveness of dual therapy treatment is enhanced by nanotechnology. The creation of cell-penetrating peptides, which allow the transport of drug molecules to the afflicted cells, is made possible by nanotechnology. The rate of medication metabolism is accelerated by nanomaterials. The use of nanotechnology to enhance histidine activity has significant implications for the treatment of cancer and acute genetic disorders. Acute illnesses such as cancer, genetic disorders, neurological disorders, behavioural disorders, cardiovascular disorders, and bone fractures can all benefit from a nanotherapeutic approach to treatment. Nanomedicines' market share is growing at an exponential rate because of their superior therapeutic efficacy. Increased access to Ayurvedic treatment will result from nanotechnology's ability to boost the efficacy of herbal remedies. Waste management is further supported by the use of nanotechnology, which enhances the ability to extract bioactive components from plant-based waste products. Due to the dynamic nature of infectious illnesses, nano vaccines work more effectively than traditional vaccinations. This chapter will describe research on the use of nanotechnology in various ayurvedic practices, which will broaden the use of herbal remedies for the treatment of long-term health problems. Additionally, it will investigate the potential of nanomaterials to enhance the efficacy of herbal remedies, which can aid in the development of novel ayurvedic treatment approaches.

Keywords: Active nanostructure, Bioactivity, Drug delivery, Genomic vaccine, Herbal medicine, Immunotherapy, Molecular pathogenesis, Nanobiotechnology.

---

* **Corresponding author Sumanta Bhattacharya:** Maulana Abul Kalam Azad University of Technology, West Bengal, India; E-mail: sumanta.21394@gmail.com

Virat Khanna, Suneev Anil Bansal, Vishal Chaudhary and Reddicherla Umapathi (Eds.)

INTRODUCTION

Allopathic approaches to medicinal treatment based on the remediation of a target disease are becoming an outdated idea in today's scenario as the prevalence of chronic diseases is increasing due to the changing nature of the environment. The improvement of overall body function and the boosting of the immune system become the basis of modern medicinal treatment, just like in ancient times. Phytomedicines are gaining importance in the effective treatment of chronic diseases due to their potential to improve the efficiency of removing the root cause of disease by keeping the balance among the physiological, mental, and emotional health of human beings. The nervous system is one of the most important and sensitive physiological systems, as it controls other body functions. The allopathic dosage of neurological diseases increases the risk because of side effects leading to major damage to healthy neurons. On the other hand, herbal medicines focus on the repair of the overall nervous system, reducing the risk of side effects. The advancement of modern technologies like nanotechnology, biotechnology, neuroheormosis, *etc.* improves the efficiency of phytomedicines. The rapid development of the field of bioinformatics led to the emergence of neuroinformatics, which enables the detailed study of the mechanisms of the nervous system. In this way, the field of herbal medicine is developing through the application of advanced techniques. The adoption of ayurvedic medicine instead of conventional allopathic medicines promotes the healthy growth of the human population under the current challenges of global warming and climate change. The increasing use of phytomedicines also facilitates the egalitarian economic development of society by expanding the scope of the primary sector. The improvement of tribal populations is also another component of the promotion of Ayurveda. The emergence of neurophytomedicine plays a significant role in the achievement of sustainable development in the health sector. It also promotes economic growth and the development of the economy.

Therapeutic compounds are commonly carried by nanoparticles because of their enhanced chemical activity and their capacity to penetrate tissue barriers. This is due to the fact that medicinal compounds may be delivered more efficiently via nanoparticles. Nanoparticles are built with precise instructions to carry various substances into the human body and influence specific cells. These substances include proteins, plasmids, antibodies, oligonucleotides, fluorophores, ligands, polymers, radioisotopes, tissue-engineered products, and more. Another possible target for nanoparticles is a specific cell of interest. Dendrimers, nanopolymeric dendritic structures that carry drug molecules, liposomes, quantum dots, and nanoshells are all examples of nanomechanical systems used in biosensors. Quantum dots are tiny light-emitting nanocrystals that can enter small human cells and aid in magnetic resonance imaging of the body when activated. Additionally, the existence of unique membrane features allows them to readily distinguish target cells from other cells. The regulated release of drugs and efficient target selection make these nanoparticles useful chemotherapeutic agents. These traits and talents aid in avoiding the chemotherapy-induced death of healthy cells. Neural stem cell treatment for neurological disorders also makes substantial use of nanotechnology. In addition to assisting with stem cell monitoring, these designed polymeric nanostructures aid in neural stem cell generation, regulation, and targeted distribution. The created nanocomposites boost the efficacy of scaffolds for therapeutic purposes by making tissue engineering more effective.

ROLES OF NANOTECHNOLOGY IN HERBAL MEDICINES

The evolution of active nanostructures revolutionizes the application of phytomedicines in the treatment process. The active nanostructure enables the integration of different technologies into micro- or nanostructures, which are able to fit into any structure and make desired changes. The nanocarriers are framed in this way. The microstructures of the drug carrier molecules facilitate the delivery of drug molecules effectively to the desired target cells. The improved structures of the nanomaterials enable the interaction of drug molecules with the affected cell more effectively than with normal molecules. The nanocarriers act as effective carriers of the bioactive components present in the herbal plants to the target cells through effective tissue-specific delivery characteristics, increased surface charge density, controllable surface chemistry, and increased bioavailability of the components in the affected area. The herbal plant extracts are proven to be more effective than the synthetic chemical extracts, as they are capable of remediating the root cause of the health issues. Nanotechnology induces the immune response of the human body by accelerating the functions of lymphocytes. It also plays significant roles in the treatment of cancerous diseases by facilitating the successful delivery of chemotherapeutic plant extracts into the tumor cells while mitigating the side effects. It also induces an anti-tumor response mechanism during tumorigenesis and increases the tumor's immune defense activities by controlling the immune suppression mechanism in the tumor microenvironment [1]. Nanoparticles are extensively used in in situ diagnostic imaging for their better biodistribution, tissue permeability, cellular uptake, and targeting efficiency. The introduction of a capsule endoscopy camera along with the nanocarriers enables us to monitor the phytokinetic activities of the bioactive components of the plant extracts [2].

The cellular internalization process of nanomaterials depends on their structural characteristics and interactions with cellular membranes. The lipid structures of the cell membranes and the cellular organelles restrict the entry of nanocarriers into the cellular environment. As a result, the pharmacokinetic and pharmacodynamic characteristics of the drug molecules have extensive effects, leading to a reduction in the bioavailability of the drug molecules in the targeted regions. The endocytosis process affects the physiological and biochemical processes, viz., apoptosis, immune surveillance activities, regulation of cell surface receptors and transporters, membrane remodeling abilities, neurotransmission, and intra- and intercellular communications in the cells. The size of the nanoparticles is less than 200 nm, and the moderate negative surface charge density of the nanoparticles facilitates their entry into the mitochondria, endoplasmic reticulum, and Golgi bodies (Fig. **1**) below shows the Cellular internalization of nanomaterials.

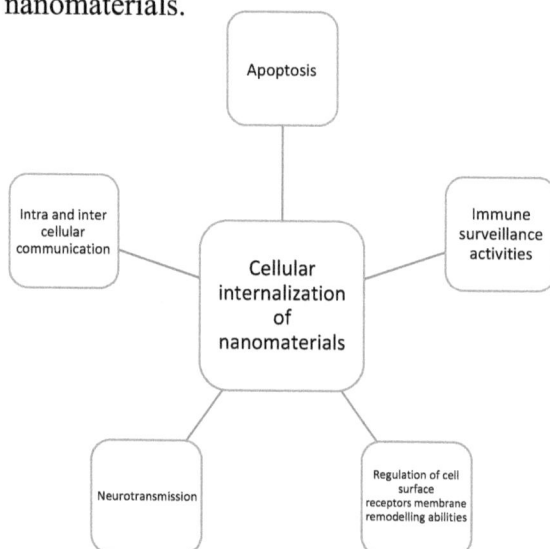

Fig. (1). Cellular internalization of nanomaterials.

Most of the components of plant extracts are destroyed in the higher acidic pH of the stomach and also metabolized in the liver. It disables their entry into the blood, resulting in a lack of bioactive components in the affected cells. The therapeutic efficacy of herbal plants can be increased by the use of nanomaterials that can easily break the blood-brain barrier and enter the affected cells. The nanostructured lipid carrier, lipid drug conjugate, parenteral emulsions, solid lipid nanoparticles, *etc.* are some important classes of nanomaterials that ensure the successful delivery of drug molecules in the affected cells [3]. The polymeric and lipid nanoparticles enable deep skin penetration and controlled delivery of phytochemical molecules that exhibit a more effective outcome in the treatment of acute dermatological diseases. Moreover, the lower toxicity of the nanomaterials reduces the possibility of allergic reactions in the skin [4]. Nanotechnology increases the bioavailability of bioactive components in the target by ensuring effective targeted delivery Below Given (Figs. **2** and **3**) explains the application of nanocarriers in drug delivery system and application of nanotechnology in immunization.

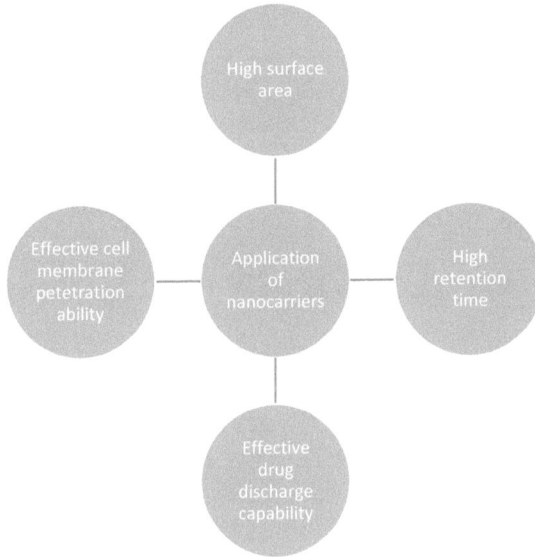

Fig. (2). Application of nanocarriers in drug delivery system.

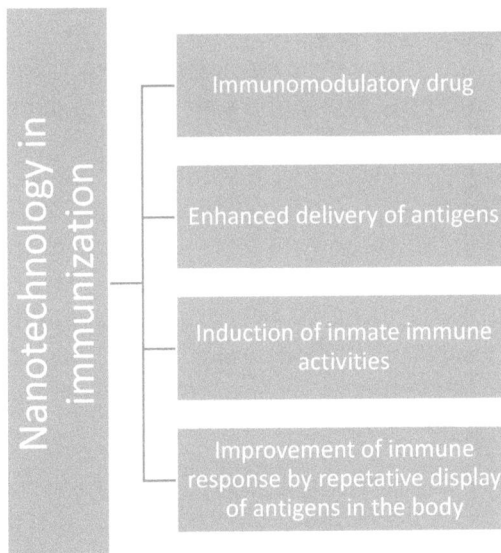

Fig. (3). Application of nanotechnology in immunization.

Nanomaterials induce the immune responses caused by herbal medicines. The immunomodulatory drugs synthesized from herbal extracts ensure greater efficiency when integrated with nanomaterial [5]. Nanogels and cationic liposomes are potential examples of carriers of immunomodulatory drugs. These exhibit low bioavailability and rapid clearance when exposed to the oral cavity. That's why these drugs can be directly injected into the gut or liver when integrated with nanomaterials. Nanomaterials induce the immune response of the body by facilitating the enhanced delivery of antigen plant extracts. Moreover, it activates the innate immune system via Toll-like receptors and the repetitive display of antigens in the body [6].

APPLICATION OF NANOTECHNOLOGY IN HERBAL MEDICINE

The disruption in the mechanism of the transformation of signals from the senses to the central nervous system and from the central nervous system to body muscles causes the prevalence of neurodevelopmental diseases. The phytoconstituents present in the plant extracts form stable chemical bonds with the receptors and signals to facilitate the steady flow of signals from the nervous system to different parts of the human body. The interaction between the bioactive constituent and different biomolecules responsible for neurological activities becomes the key to understanding the bioactivity of phytoconstituents in the treatment of different neurological diseases. Phytochemicals facilitate the maintenance of overall chemical balance in the brain instead of the target-based approach of allopathic medicines. Flavonoids, phenols, alkaloids, fatty acids, terpenes, saponins, *etc.* are some important secondary metabolites used for the treatment of several neurodegenerative, neuropsychiatric, and cognitive disorders. Phytochemicals control the function of receptors for the major inhibitory neurotransmitters. Several bioactive constituents, like polyphenols, activate specific transcription factors that facilitate the synthesis of RNAs. These RNAs activate specific signal transduction pathways by the translation of favorable protein molecules, which results in the resistance of neurons to various stress factors, leading to the prevention of neuropsychiatric and neurodegenerative diseases. The neuroprotective activities of phytochemicals make the phytomedicines more effective towards the overall improvement of the nervous system, mitigating the risk of neurological diseases [7].

The two main metabolic pathways that turn polysaccharides made by photosynthesis in plants into bioactive substances are the aromatic amino acid pathway and the mevalonic acid pathway. These bioactive substances do not contribute significantly to the process of plant growth because they are by-products of the metabolic process. However, these bioactive substances help the plant bodies fight off diseases and herbivorous animals that might try to eat them.

Also, a number of bioactive elements help plants recover from nitrogenous wastes and survive in harsh weather conditions like drought and flooding. Typically, bioactive substances fall into one of the following groups:

With the exception of amides and amino acids, alkaloids are heterocyclic nitrogen-containing chemicals that are generated from plant extracts. It has alkaline properties and is mostly produced from amino acids. Decarboxylation of amino acids yields amines, which interact with amine oxides to produce aldehydes, which are used to make alkaloids. Aldehydes and amine groups condense in a Mannich-type reaction to produce the distinctive heterocyclic rings of alkaloids. It demonstrates a wide spectrum of phytochemical properties, such as anti-malarial, anti-asthmatic, anti-cancer, cholinomimetic, analgesic, antibacterial, anti-hyperglycemic, vasodilatory, and anti-arrhythmic activity, among others.

The polyphenolic substances known as flavonoids are what give flowers their color and scent. Due to the highly conjugated heterocyclic structures present, it functions as a pigment. Due to their anti-oxidative, anti-mutagenic, anti-carcinogenic, and anti-inflammatory properties, they are frequently used as therapeutic medicines. They can also control how human cells' enzymatic functions are carried out. Cancer, Alzheimer's disease, atherosclerosis, and other major disorders are all treated with flavonoids' anti-oxidant abilities. The low cardiovascular mortality rate of flavonoids, which has been demonstrated in recent studies on their chemical characteristics, makes a wide range of applications for them as a preventative measure for cardiac disorders possible. Flavonoids are essential for the growth and development of plants because their colorful compounds make pollination easier [8].

Anthocyanins are water-soluble plant pigments that are primarily responsible for giving fruits and vegetables their colors. These are oxygen-containing heterocyclic compounds. The colors of the plants' vegetables and fruits were caused by the oxygen atoms' resonance effect under different pH and temperature conditions. A wide range of pharmacological benefits, including antioxidant, anti-cancer, anti-inflammatory, antibacterial, and anti-obesity actions, are also exhibited by these pigments. By scavenging free radicals, such as reactive oxygen and reactive nitrogen spices, the colored ingredients immediately lower the risks of the onset of chronic diseases and oxidative stress. Anthocyanins decrease the likelihood of cardiovascular illnesses by enhancing blood lipid profiles and biomarkers, inhibiting the growth of malignant cells by downregulating cyclooxygenase enzyme activity, and inducing apoptosis through oxidative stress reduction and lipid peroxidation. By preventing the activation of the mitogen-activated protein kinase pathway, it also slows the growth of malignancies [9].

Glycosides of triterpenes and steroids are the source of saponins, which are produced by the mevalonic acid pathway. The term "saponin" is also used to describe steroidal glycosides. The structural diversity of saponins' chemical properties, which depends on their amphipathicity, causes the bioactivity of various compounds to vary. Different saponin components that are isolated from plants are being employed to create a variety of medicinal medicines. Saponins have anti-inflammatory, anti-fungal, anti-microbial, anti-cancer, anti-viral, and anti-parasitic phytochemical effects. The penetration of plasma membranes and the creation of complexes with sterols are evidence of saponins' bioactivity.

Nanophytomedicine is an emerging branch of medical science that deals with the application of nanotechnology in herbal medicine, also known as phytotherapy or phytopharmacology. Nanotechnology plays an important role in the effective targeted and non-targeted delivery of drug molecules in medical science. The nano adsorbents lead to the effective adsorption of drug molecules, which exhibit more therapeutic activities due to an increase in bioavailability. The lower in vivo efficacy of phytomedicines is due to their poor permeability, low systematic availability, instability, extensive first-pass metabolism, and low aqueous solubility. The interaction of herbal drugs with nanoparticles increases their bioavailability in the human body and improves the potential for delivery of drug molecules at the target. Nanoparticles are widely used as carriers of drug molecules for their increased chemical activity and ability to cross the tissue barrier. Specially engineered nanomaterials are prepared that are able to carry drug molecules, proteins, plasmids, antibodies, oligonucleotides, fluorophores, ligands, polymers, radioisotopes, tissue-engineered products, *etc.* to the human body and affect desired cells. Some nanomaterials, such as dendrimers, liposomes, polymeric nanoparticles, polymeric micelles, carbon nanotubes, mesoporous silicon, quantum dots, nanocrystals, nanospheres, phytosomes, *etc.*, participate in drug delivery mechanisms. The aqueous solubility and permeability of the blood-brain barrier of the phytomedicines can be improved by reducing the size of the drug molecules to the nanoscale. The flavonoids and lignins present in the Cucscuta chinensis plant exhibit improved bioavailability and adsorption capability in the human body after being reduced to nanosize by the nanosuspension method. The rate of diffusion of the suspended nanomedicines is higher than that of phytomedicines into the gastrointestinal medium. The drug delivery mechanism of herbal drugs can be increased by increasing their solubility, enhancing their stability, reducing their toxicity, improving their macrophage distribution mechanism, and enhancing their pharmacological activities. Nanotechnology ensures the targeted delivery of phytomedicines in a cell- or tissue-specific way through the development of effective drug carriers. The increase in solubility and permeability of the phytomedicines leads to the transcytosis of neuro phytomedicine across tight endothelial and epithelial cells.

Nanotechnology also facilitates the therapeutic modality of combination therapy by enabling the co-delivery of two or more phytomedicines in the human body. The integration of phytomedicines with the nanopolymers increases the systemic drug concentration, leading to an improvement in anticancer efficacy.

Nanobioremediation is the application of nanotechnology to prevent environmental degradation, where the nanoparticles, having more chemical reactivity due to having more surface area per unit mass, react with the pollutants to form less harmful products or adsorb the pollutants to make them immobile, preventing the pollutants from contaminating the environment. Nanomaterials developed in different forms like nanotubes, nanowires, films, quantum dots, colloids, *etc.* are placed in the contaminated groundwater and soil in the form of fertilizers, pesticides, fungicides, composts, *etc.* that also intensify the activity of plants in the uptake of the toxicants from the soil and groundwater, facilitating the process of purification. Nanoparticles prevent the contamination of toxicants in the soil, groundwater, and surface water by reducing or oxidizing the contaminants, immobilizing the contaminants by combining with them, forming a wall between polluted and pure water to prevent the spread of pollutants, *etc.* Nanomaterials are instructed to be more economical, efficient, and eco-friendly than the prevailing materials in each resource conservation and protection setting.

By adding nanosensors to phytomedicines, scientists can learn more about how they work in terms of pharmacokinetics. Smart pills are being prepared by the introduction of nano-based electronic devices to the pharmaceutical pills that perform advanced imaging, sensing, and delivery of drugs. This enables medical practitioners to track the delivery pathway of drugs in the human body and study the mechanism of drug activity. 'Atmo Gas Capsule' is an advanced smart pill that examines the activity of gaseous substances in the human body. 'Smart Sensor Capsules' are prepared with advanced nanotechnology that is used along with the vaccines to monitor the activity of vaccines in the human body. It is also able to be consumed orally instead of by injection. This is a groundbreaking innovation in the field of research in medical science. Nanopatch vaccines use nanoparticles to deposit the vaccines on immune cells present in the skin and lower the risk of infection. It is an easier vaccination process and eliminates the need for vaccine refrigeration.

Nanoflares are specially engineered nanomaterials in the field of cancer treatment. It effectively detects the presence of cancer cells in the bloodstream. It can bind with the genetic target in cancer cells and generate light when the target genetic sequence is found.

Nanobots, the advanced robots developed with nanotechnology and robotics, act as miniature surgeons when inserted into the human body. It effectively repairs the intracellular structures, resulting in the healing of particular diseases within the human body. The DNA-based nanobots are also prepared for the eradication of genetic diseases by modifying the target genome sequence. Nanomaterials also act as effective carriers of genetic material in the body.

Biopharmaceutical technology focuses on the synthesis of pharmaceutical products from biological resources instead of synthetic chemicals, which increases the efficiency of medicinal products by reducing the side effects. Biotechnology plays an important role in the treatment of acute diseases like cancer and other genetic diseases through targeted and non-targeted therapy, immunotherapy, hormonal therapy, targeted drug delivery, gene therapy, the manufacture of vaccines, *etc*. The plant extract can become a potential component of chemotherapy that reduces the possibility of effects on healthy cells during the time of chemotherapy. Bio-based nanomaterials like cantilevers (a biosensor-based nanomechanical system) and dendrimers (a nanopolymeric dendritic structure that carries drug molecules in the core) are potential carriers of drug molecules that ensure the targeted delivery of drugs with higher efficiency. Nowadays, vaccines are developed based on advanced biotechnology like genomics, proteomics, transcriptomics, metabolomics, *etc*., which is likely to reduce the possibility of genetic diseases in the near future. Moreover, gene therapy is proven to be more effective in healing the carcinogenic effects in the human body. The advancement of biopolymer technology is one of the latest additions to the pharmaceutical industry. Biotechnology has recently improved to the point where advanced genetic engineering can help agriculture expand sustainably and environmentally. Through modern recombinant DNA technology and transgenesis, the phenotypic traits of crops can be adjusted, resulting in the cultivation of advanced, healthy herbal crops. Plant genome modification can also be used to grow pest-resistant crops. The contaminants can be successfully remediated from the soil and water by the application of nanoparticles with fertilizers, which leads to the improvement of soil health followed by the boosting of agricultural productivity. The development of key medicinal crops through the 4th generation technology revolution will make it easier to achieve sustainable growth in the medical sciences as well as the agricultural sector.

Neurohormesis characteristics of several phytochemicals facilitate the preparation of optimum doses for the effective treatment of several neurodegenerative diseases like Alzheimer's, Parkinson's disease, epilepsy, *etc*. The hormesis effects of the apparently toxic phytochemicals increase the immune response and mitigate the risk of neurological diseases. The hermetic responses of several phytochemicals like resveratrol, catechin, sulforaphane, hypericin, allicin, *etc*.

induce the adaptive stress response signaling pathways, resulting in the amelioration of the resistive potential towards injury and disease. The hormesis characteristics of the phytochemicals increase the hemodynamic characteristics of the body, leading to the effective management of mental and physical stresses caused by aging. The neuron cells are more susceptible to damage with age due to the rapid rate of DNA damage. Neurohormesis enables the repair of damaged DNA, resulting in a decrease in the rate of cell death through radical scavenging and increased antioxidant activities of the phytochemicals. Mitohormesis is the process of exhibiting hormesis activities by interacting with the mitochondria, which play a central role in bioenergy production and nutrient metabolism. Berberine, an alkaloid extracted from Captidis rhizoma and Hydrastis canadensis, directly interacts with mitochondria through the electron transport chain and reduces the oxidative stresses generated in the neuronal cells. The consumption of epicatechin, an important bioactive component belonging to the flavonol group, leads to the improvement of the cognitive function of the human brain by modifying neuronal spine concentration, hippocampal angiogenesis, and memory function [10].

The ongoing environmental issues related to climate change lead to serious health problems in humans. The conventional allopathic medication process becomes incapable of dealing with these ongoing challenges as it is unable to cure the root causes of the diseases. Under this scenario, traditional medicinal practices like Ayurveda are gaining importance. The importance of phytomedicine is increasing day by day in order to achieve sustainability in the healthcare sector. Under these circumstances, the advancement of phytomedicine with the help of modern technology becomes a necessary step in order to improve the health of humans. The application of modern technologies like nanotechnology, biotechnology, and genetic engineering improves the quality and efficiency of phytomedicines. The genomic, proteomic, and transcriptomic study of medicinal plants leads to the advancement of phytomedicines. Biotechnology will improve both the productivity and efficiency of medicinal plants, leading to the advancement of the health sector as well as the boosting of the agricultural sector. The application of nanotechnology enables sustainable and eco-friendly growth of medicinal plants by increasing productivity through the application of nanofertilizers and nanoremediators. The in vivo efficacy of neurophytomedicines will be increased by the application of nanocarriers that will lead to effective targeted and non-targeted drug delivery. The production of bioactive secondary metabolites will be increased in the plant body by the application of specially engineered genes through nanocarriers. The phytokinetic study of the neurophytomedicines can be improved through the application of nanosensors, which will facilitate the detailed in vivo study of the mechanism of neurophytomedicines. The neuroinformatic study reveals the detailed biological mechanisms of the nervous system, whereas the bioinformatic study of the plant extracts leads us to study the bioactive mechanisms of phytoconstituents. It results in the development of more effective neurophytomedicines for the sustainable treatment of neurological diseases [11].

The detailed study of neuroinformatics facilitates the preparation of databases for the biological functions of the nervous system. The neuroinformatic study of the activities of all the sensory, motor, and interneurons needs to be improved in order to develop effective phytotherapeutic techniques for the treatment of dysfunctions in the nervous system and also psychological disorders.

Also, the phytochemicals might be able to act as biomarkers by interacting with DNA, proteins, and other biomolecules in a useful way. It will help to study the pharmacokinetic mechanism of medicinal products within the nervous system, leading to the development of research in the fields of neurology, pharmacology, and herbal medicine. Modern biotechnological innovations enable the development of new herbal plants that produce more medicinal secondary metabolites. It will facilitate the development of the field of herbal medicine and also improve the pharmacokinetic efficiency of the phytochemicals in the nervous system. The emergence of nanophytomedicine improves the in vivo efficacy of phytomedicines by ameliorating the bioavailability and delivery mechanisms of phytochemicals in the body [12].

NOVEL NANOTHERAPEUTIC APPROACHES

Modern therapeutic practices are shifting towards combination therapy and polypharmacy due to their better therapeutic outcomes. It deals with the use of more than one medicinal drug together to facilitate the treatment of more than one disease effectively. The combination of herbal drugs with allopathic drugs increases the therapeutic outcome by increasing bioavailability, pharmacokinetic efficiency, and reducing side effects. The treatment using combination therapy is based on the principles of reinforcement, potentiation, restraint, detoxification, and counteraction, depending on the compatibility of the drugs. The nanotechnology reinforces the therapeutic efficacy of combination therapy by acting as drug payloads, facilitating the prolonged circulation of drug molecules in the bloodstream, and ensuring uniform and sustained release of drugs. It also optimizes the requirement for drugs by reducing the frequency of dosage and ensuring the effective release of drug molecules in the affected cells. The combined therapy plays an important role in the treatment of acute diseases like cancer, AIDS, hypertension, infectious diseases, *etc.* The therapeutic treatments of cancerous diseases can be improved by nanotechnology through preferential accumulation of drug molecules in the tumor sites via increased permeability of the blood-brain barrier and improved retention effects, decreased side effects, and an increased therapeutic index. Polymeric nanoparticles, nanodiamonds, fullerenes, graphene oxide nanocomposites, polymeric micelles, carbon nanoparticles, dendrimers, liposomes, *etc.* are used in the enhancement of combination therapy (Fig. **4**) below shows the application of nanotechnology in combination therapy.

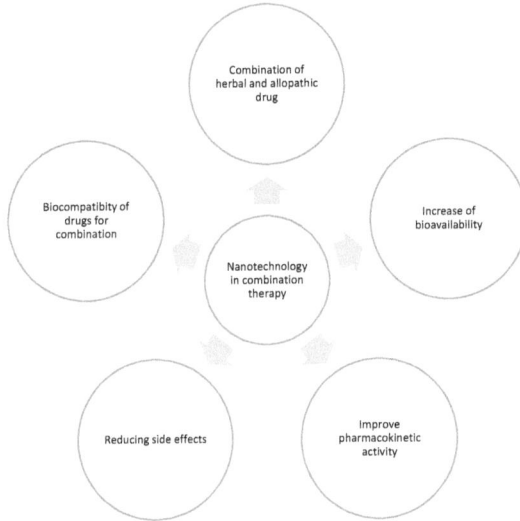

Fig. (4). Application of nanotechnology in combination therapy.

One recent development in nanotechnology is the use of nanobubbles. Cavity is produced at hydrophobic surfaces of water that contains gas and lasts for a long time. Because of its gas solubility, high surface charge density, low buoyancy, and capacity to create free radicals, it may find utility in the production of biomedicines. There are three main parts to it: a gaseous core, a polymer or liquid outer layer, and a liquid phase that can include water or other inorganic or organic solutions. There are a number of cutting-edge methods for synthesising nanobubbles, including mechanical stirring, the nanoscale pore membrane method, microfluidic flow regulation, acoustic cavitation (inducing negative pressure in the liquid through high-speed propeller rotation via high-speed sound waves), hydrodynamic cavitation (altering the flow velocity of liquid), the dissolved gas method, periodic pressure variation, hydraulic air compression, and many more [13].

The nanobubble water curcumin extract has potential application in the treatment of musculoskeletal injuries by lowering the levels of alanine aminotransferase, triglycerides, and alkaline phosphatase and increasing the density of high-density lipoprotein in the affected muscles. The nanobubbles have effective drug delivery capacity due to their smaller size and longer circulation time in the blood. It is also preferable as a potential drug delivery agent due to its lesser toxic effects on the body. It facilitates stimuli-responsive delivery of herbal drugs by reducing drug concentration fluctuations and drug toxicity. Nanobubbles increase the therapeutic efficiency of cancerous diseases through photodynamic therapy. The higher solubility of nanobubbles in gaseous solution enables their interaction with dissolved oxygen molecules and the conversion of those into reactive oxygen species, which successfully destroy the nearby tumor cells. It also does not have any side effects due to the absence of any toxic chemicals within it. The integration of chemotherapeutic plant extracts with nanobubbles is one of the potential therapeutic strategies for the treatment of cancer. The tumor theranostics approach with nanobubbles facilitates real-time monitoring of the cancer tumor. The nanobubbles also enable the delivery of herbal drugs through indirect drug delivery, codelivery, target drug delivery, and the therapeutic gas delivery method. Cancer and vascular disease theranostics are the two potential fields of application of nanobubble technology, along with herbal medicines. Modern bacteria are more contagious than ever before because they are more resistant to antiseptics and disinfectants. Antiseptics like Dettol are more effective against Staphylococcus aureus infections thanks to the development of nanobubble-mediated molecular liquid antiseptic agents [14]. In addition, nano shuttles made of graphene and copper oxide have antimicrobial properties because they can emit oxygen nanobubbles under control, preventing bacteria from interfering [15] (Fig. 5) below shows the Application of nanocarriers in drug delivery system.

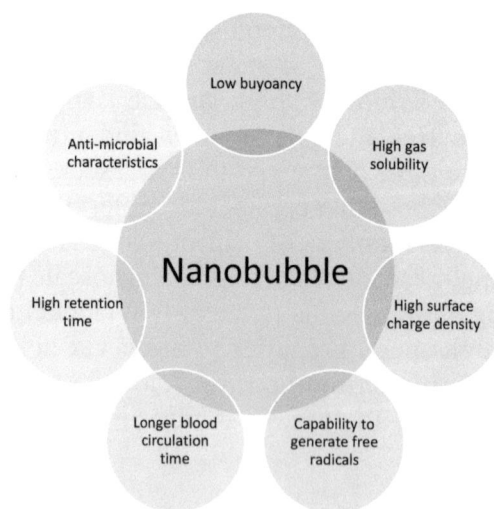

Fig. (5). The characteristics of nanobubbles in the nanopharmaceutical approach.

The electrospinning technology enables the extraction of nanofibers from organic polymers. The high porosity of these electrospinning nanofibers with their small porous structures enables the easy accumulation of drug molecules and the delivery of the same in the targeted cell. The bioactive herbal extracts can easily combine with the porous structure of the nanofibers, as these fibers are usually derived from organic polymers. It facilitates the delivery of ofloxacin antibiotics extracted from straw plants through the ocular route. These electrospinning nanofibers act as potential nanocarriers of herbal drugs due to their large surface area to volume ratio, small pore sizes, high porosity, and composite internal structure. The nanofibers have an effective drug-loading capacity due to their efficient internal composition. The drug-loading capacity of the nanofibers can be improved by bringing innovation to the electrospinning technique. The dual extrusion electrospinning technology enables the manufacture of multilayered, 3D scaffold nanofiber materials. It facilitates the preparation of nanofiber composites from different feed materials in an alternative way, leading to the formation of nano-mixed mixtures by the formation of multiple layers of different nanofibers. The dual-layer mucoadhesive patch nanofibers prepared in this manner enable the successful delivery of protein molecules through the oral mucosa. The porous structure of the nanofiber composites can be improved by melt electrospinning, in which polymers are heated to a high temperature before spinning. The drug release capacity of the nanofibers can be improved by making desired changes in the nanofiber structures. The core-shell nanofiber structures exhibit instant drug release by increasing the rate of drug delivery and their effective interaction with the target cells. It plays an important role in the increase in the therapeutic efficiency of herbal drugs. In this structure, the number of drug-loaded layers increases, which enhances the possibilities of interaction with the drug molecules within a short period of time. The outer polymeric layers of core-shell nanofibers act as a rate-controlling barrier that enables the easy crossing of the blood-brain barrier. The porous composite structure provides a high surface area and great mechanical strength in the nanofiber structures. The bio-based polymeric structures exhibit effective biocompatibility and less toxic effects, which also mitigate the risks of side effects. The composite porous structures mimic the extracellular matrix of the cell, which facilitates easy penetration of the barrier of the cells.

The newly synthesized inorganic and organic hybrid nanomolecules act as potential drug delivery systems by encapsulating drug molecules in the porous structures. The porous and hydrophobic characteristics of hybrid nanomolecules control the diffusion rate of the encapsulated drug molecules and allow them to cross the blood-brain barrier effectively [16]. It also enables the delivery of genes and protein molecules into human cells. The low density of the nanomolecules facilitates their steady flow in the gastrointestinal tract. The hybrid structure increases the retention time of drug molecules in the affected cells [17]. –Fluroxamide hybrid is one of the potential hybrid drug carriers. The integration of both inorganic and organic molecules increases the dissolution rate of all types of drug molecules [18].

The biocompatibility of the nanocarriers can be improved by their interaction with cell-penetrating peptides like penetratin, Tat, sC18, VP22, polyarginine, *etc.* These peptides combine with nanomaterials like liposomes, albumin, silica nanoparticles, gold nanoparticles, poly (lactic-co-glycolic acid) nanoparticles, insulin, dendrimers, polypeptide micelles, gold magnetite nanoparticles, *etc.* through electrostatic or covalent bonding. These cell-penetrating peptides effectively interact with the lipid-based components in cell membranes. The lipid-lipid interactions between cell membranes and cell-penetrating peptides facilitate the chemical transformation in the morphological structures of cell membranes that leads to the entry of nanoparticle-mediated drug molecules in the affected cells. The cell-penetrating peptides enter the cells through direct penetration or endocytosis processes. The lipid core components of cell membranes allow for easy penetration of the nanocarrier when combined with cell-penetrating peptides. Certain cell-penetrating peptides known as tumor-homing peptides facilitate the entry of the drug molecules into the cancerous cells by penetrating the membranes of tumor cells. These peptides are largely used in the synthesis of chemotherapeutic agents from plant extracts [19 - 21] (Fig. **6**) below shows the applications of electospinning nanofiber composites.

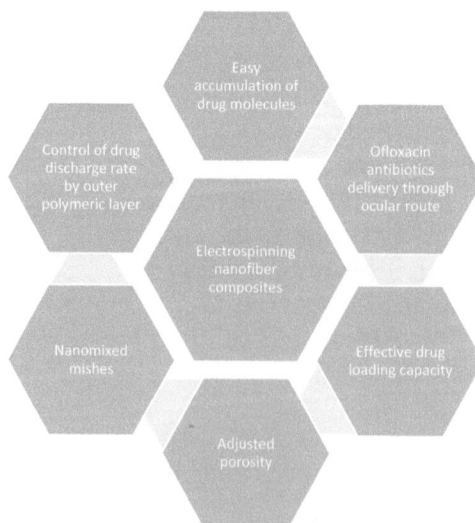

Fig. (6). Applications of electrospinning nanofiber composites.

Biologically derived nanoparticles contain pathogen-associated molecular patterns for the activation of B cells and T cells and repetitive high-density motifs that enable the preparation of efficient vaccines. Moreover, the nanomaterials are also capable of bringing the vaccines to the targeted cells effectively. The nanoparticles' ability to self-assemble facilitates their ability to protect against antigen attack. The biodegradable characteristics of biologically derived nanoparticles also provide a potential platform for the carrying and delivery of the vaccines to the desired target. The cell-penetrating structures of the nanomaterials enable them to penetrate lymphatic nodes via direct transport. The biologically derived nanoparticles also exhibit less toxic characteristics than others. Genetic manipulation or chemical linkage enables the surface display of many heterologous antigens, and novel coupling strategies such as sortase-mediated conjugation and Addvax provide additional flexibility to further modify these highly versatile platforms for use against various diseases. The virus-like particles, other membrane vaccines, protein nanocages, hybrid nanostructures, *etc.* are some biologically based nanomaterials that are used to synthesize vaccines for different infectious diseases [22, 23].

The characteristics of phagocytosis of the human immune cells, especially lymphocytes and neutrophil granulocytes, restrict the entry of external nanoparticles into the cells, as these cells treat the nanomaterials as external pathogens. However, the cell-penetrating membrane and other innovative strategies enable it to penetrate the cell membranes and enter the herbal bioactive components into the cells. The specifically engineered nanoparticles facilitate the in vivo interaction with the cells with high affinity and specificity by avoiding the non-specific pathways of macrophages' phagocytosis actions [24, 25].

By enhancing the mechanical, electrical, and biological properties of scaffolds, various nanocomposites play a significant role in tissue engineering. The biocompatibility of synthetic tissues and scaffolds is improved. The polymeric nanocomposites made of gold speed up cell multiplication, which in turn speeds up the mending of tissues including cartilage, bone, and muscle. The gold nanoparticle acts as a substitute for bone morphogenic proteins, which can lead to the development of abnormal bone structures and trigger localised inflammatory responses. The use of hybrid scaffolds, such as those containing both gelatin and gold nanoparticles, can hasten the healing process by increasing the proliferation rate of osteoblasts. Another important factor in bone tissue damage is nanoparticles derived from polymeric titanium oxide. Scaffolds' electrical characteristics are enhanced by incorporating nanoparticles into them. By changing the bone morphology, silver nanoparticles show improved striation behaviour, which enhances myocardial infarction therapy. The antibacterial properties of the tissue scaffolds are enhanced by the silver-based polymeric nanocomposites, further increasing their therapeutic efficacy [26].

When used to treat neurological disorders, nanotechnology increases the output and regulation of neural stem cells, leading to better therapeutic outcomes. Production of brain stem cells on a massive scale is made easier with nano substrates. By increasing immunomodulatory responses against defective neurological responses, nano-enabled tissue engineering improves therapeutic output by facilitating the synthesis of effective patient-specific stem cells and tissues. Nanomaterials, when added directly to a culture medium, speed up the process of neural stem cell generation. The morphology of the coating of the culture system is also improved by the specifically developed nanocomposites. Transfecting human embryonic stem cells is made possible by the creation of a biodegradable nanostructure of poly (β-amino esters) that self-assembles. Neuronal stem cells may be safely transplanted into damaged neurons using well-designed nanostructures, such as lipid nanocapsules. Biomarkers such as Amyloid β can be more effectively detected using MRI imaging of neuroinflammation when supermagnetic iron oxide nanoparticles are added. Gold nanoparticles are useful for imaging because their gold cores have unique optical properties, such as plasmonic capabilities [27].

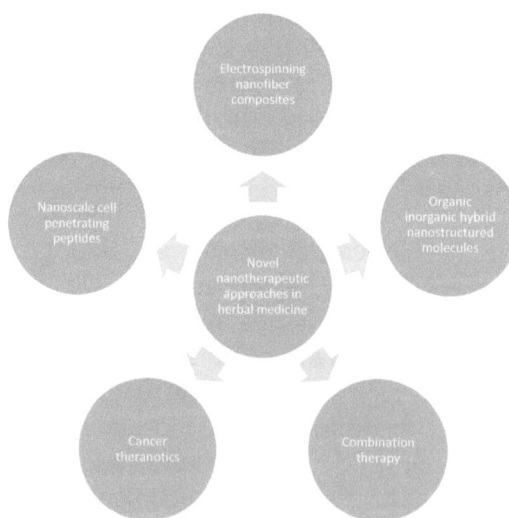

Fig. (7). Different novel nanotherapeutic approaches in herbal medicine.

The zero-dimensional, one-dimensional, two-dimensional, and three-dimensional nanomaterials marked with radioactive particles increase the efficiency of the radiotherapeutic approach to the treatment of cancer. Passive and active nano-targeting of radionuclides enables the effective imaging of cancerous tumors and chemotherapy. Nanostructured radiopharmaceutical products like radium-223, lutetium-177, copper-64, technetium-99, *etc.* facilitate molecular imaging and radionuclide therapy of cancerous diseases. The advancement of nuclear nanomedicine facilitates multimodal imaging and opens up a new era in the tumor theranostics approach to cancer treatment [26, 27] (Fig. **7**) below shows the different novel nanotherapeutic approaches in herbal medicine.

CULMINATION

The pharmacokinetic properties of drug molecules are enhanced by the use of nanotherapeutic strategies, which increase bioavailability and biodistribution. By enhancing drug metabolism, tissue retention, and cellular absorption capability of the bioactive components in drug molecules, nanotechnology enhances the therapeutic activity of drug molecules through precision drug delivery. By facilitating their efficient absorption in intracellular components, nanoparticle coatings alter the pharmacology of medicinal compounds. In order to effectively treat cancer and genetic illnesses, medication molecules can even interact with protein, DNA, messenger RNA, and small interfering RNA molecules. It makes it possible to fix health problems at their source. By improving the medication's release, distribution to specific targets, and elimination capacity, it lessens the likelihood of adverse drug reactions. Drug molecules and bioactive substances are able to more effectively access tumour microenvironments with the help of nanocarriers. Additionally, it facilitates the active transport of cancer stem cells to the target, which is useful in stem cell treatment. Nanotherapy enhances cancer treatment options through integrin targeting, self-assembly prodrug strategies, exosome-based targeting, non-coding RNA-based targeting, and other methods. Nanovaccines, which are made from biomimetic and bioengineered nanoparticles derived from bacterial, fungal, and viral pathogens, have outperformed chemical-based vaccines in preventing infectious diseases. This is because nano vaccines have several advantages over chemical-based vaccines, such as higher drug solubility, longer circulation time, the ability to overcome biological carriers, improved bioavailability, targeting of infection sites, and modulation of drug release profile. Nanoparticles, due to their self-assembling nature, have better immunological properties in the body. Novel vaccinations against viral infections are being developed utilising self-assembling nanoparticles carrying antigens. Nanomaterials produced by bacteria have several promising vaccine-making qualities, including immunogenicity, enhanced cellular absorption capacity, self-adjuvant capacity, and the ability to deliver foreign antigens. Nanomaterials' inherent healing properties allow for accelerated wound healing. By controlling molecules and the pathophysiology of chronic wounds and permitting cell type specificity, it speeds up the natural wound healing process. By combining their roles as biomarkers and biosensors, nanoparticles make it possible to research medication pharmacokinetics and treatment protocols in greater details. Effective biosensors for illness detection and therapeutic action monitoring are integrated microfluidics, transduction devices, automated sampling, and nanomaterials on a single chip. Compared to electrical and chemical sensors, it is small, easy to use, and inexpensive. Adverseeffects of reduced medication are possible thanks to its enhanced pharmacokinetic ability and focused drug delivery capabilities. Instead of interacting with the target, nano biomarkers efficiently collect its chemical, biological, optical, and electronic signals, minimising the likelihood of unwanted effects. Unlike chemical biosensors, which need extensive manipulation and treatment, nanobiosensors allow for the noninvasive transfer of clinically active biomarkers in various biofluids, including saliva, tears, and perspiration. By interacting with antigens, nanosensors based on antibody fragments make it easier to detect antibodies. The creation of new anti-cancer drugs is being facilitated by nanotechnology. It improves the efficacy of radiation treatment that targets hypoxia. When it comes to treating myocardial leukaemia, new nanomaterials are crucial. By making it easier for medications to penetrate the blood-brain barrier, it helps with the resolution of many haematological disorders. Genomic vaccines derived from several therapeutic plants are also developed using nanotechnology. By enhancing the immunological response, nanotechnology further boosts the efficacy of phytomedicines. It also makes macrophages better at phagocytosis. Improved radiopharmaceutical efficacy is a result of nanotechnology. Some new nanomaterials with potential medicinal uses include electrospun nanofiber composites, nanoinhibitors, nano-based hydroxypropyl beta-cyclodextrin, and in-house herbal nanoparticles derived from rDNA.

FUTURE PERSPECTIVES

The advancement in the field of nanotechnology enables its application in increasing the efficiency of herbal medicines. The application of different technologies enables the increased inefficiency of the nanotechnology-enabled drug delivery system. There is ample scope for advancement in the field. The integration of biotechnology enables the successful delivery of genes and gene-mediated pharmaceutical products, including gene-based vaccines, into targeted cells. The plant extracts also act as a potential resource for the synthesis of nanomaterials. The application of microbial engineering facilitates the synthesis of effective cell-penetrating peptides that increase the penetrating capacity of nanoparticles in the affected cells and increase the bioavailability of herbal medicines. The production of biopolymers through microbial action acts as a potential source for the nanofibers. An improvement in the electrospinning technology is required to prepare more effective nanofibers. The preparation of a composite nano mixture mesh with more than one fibrous structure has the potential to improve drug delivery systems. The nanoimmunotherapy techniques enable the entry of external pathogens into the body. The histidine proteins have potential applications in cancer treatment. The nanostructures tagged with histidine proteins can be a potential way of treating cancer. Nanotechnology also facilitates the targeting of drugs into the affected cell by avoiding the metabolism in the stomach and liver. To improve the efficiency of nanocarriers, the retention time and rate of release of drug molecules need to be increased. The biosynthesis of nanomaterials from plant extracts is a sustainable and eco-friendly strategy that has effective biomedical applications.

CONCLUSION

The application of nanotechnology increases the efficiency and widens the scope of application of herbal medicines. The novel nanopharmaceutical approaches increase the scope of the application of nanotechnology in herbal medicines by enhancing the capabilities of nanomaterials to cross the blood-brain barrier and interact with affected cells. The interaction between nanomaterials and cell membranes can be improved by developing the structural characteristics of the nanomaterials. The physical and chemical characteristics of nanomaterials can be improved through different innovative approaches. The knowledge from different advanced fields like biotechnology, microbiology, surface chemistry, *etc.* can be incorporated with nanotechnology to improve its efficiency. The diagnosis of different diseases like cancer and other genetic diseases can be improved by the application of nanotechnology. The nanotechnology-enabled radiopharmaceutical approaches facilitate multimodal imaging of cancer cells. The smart pills derived through nanostructured imaging enable the monitoring of the pharmacokinetic and pharmcodynamic activities of drug molecules. The specially engineered nanofibers also revolutionize the drug delivery system by enabling effective encapsulation of drug molecules in composite nanofiber structures. The knowledge gained from different advanced technologies will upgrade this field. The third generation of nanotechnology enables the introduction of self-healing characteristics in the materials, which have a wide range of applications in the fields of medical science and pharmacology.

REFERENCES

[1]   S. Gao, X. Yang, J. Xu, N. Qiu, and G. Zhai, "Nanotechnology for Boosting Cancer Immunotherapy and Remodeling Tumor Microenvironment: The Horizons in Cancer Treatment", *ACS Nano,* vol. 15, no. 8, pp. 12567-12603, 2021.
[http://dx.doi.org/10.1021/acsnano.1c02103] [PMID: 34339170]

[2]   S. Sim, and N. Wong, "Nanotechnology and its use in imaging and drug delivery (Review)", *Biomed. Rep.,* vol. 14, no. 5, p. 42, 2021.
[http://dx.doi.org/10.3892/br.2021.1418] [PMID: 33728048]

[3]   S.H. Ansari, M. Sameem, and F. Islam, "Influence of nanotechnology on herbal drugs: A Review", *J. Adv. Pharm. Technol. Res.,* vol. 3, no. 3, pp. 142-146, 2012.
[http://dx.doi.org/10.4103/2231-4040.101006] [PMID: 23057000]

[4]   D. Shree, C.N. Patra, and B.M. Sahoo, "Novel Herbal Nanocarriers for Treatment of Dermatological Disorders", *Pharm. Nanotechnol.,* vol. 10, no. 4, pp. 246-256, 2022.
[http://dx.doi.org/10.2174/2211738510666220622123019] [PMID: 35733305]

[5]   J. Shi, J.H. Weng, and T.J. Mitchison, "Immunomodulatory drug discovery from herbal medicines: Insights from organ-specific activity and xenobiotic defenses", *eLife,* vol. 10, p. e73673, 2021.
[http://dx.doi.org/10.7554/eLife.73673] [PMID: 34779403]

[6]   D.M. Smith, J.K. Simon, and J.R. Baker Jr, "Applications of nanotechnology for immunology", *Nat. Rev. Immunol.,* vol. 13, no. 8, pp. 592-605, 2013.
[http://dx.doi.org/10.1038/nri3488] [PMID: 23883969]

[7]   J. H. Kim, and H. Kim, "Combination treatment with herbal medicines and Western medicines in atopic dermatitis: Benefits and considerations", *Chinese Journal of Interactive Medicine,* vol. 22, pp. 323-327, 2016.
[http://dx.doi.org/10.1007/s11655-016-2099-0]

[8]   C.T. Che, Z.J. Wang, M.S.S. Chow, and C.W.K. Lam, "Herb-Herb Combination for Therapeutic Enhancement and Advancement: Theory, Practice and Future Perspectives", *Molecules,* vol. 18, no. 5, pp. 5125-5141, 2013.
[http://dx.doi.org/10.3390/molecules18055125] [PMID: 23644978]

[9]   S. Gurunathan, M.H. Kang, M. Qasim, and J.H. Kim, "Nanoparticle-Mediated Combination Therapy: Two-in-One Approach for Cancer", *International Journal of Molecular Sciences,* vol. 19, no. 10, p. 3264, 2018.
[http://dx.doi.org/10.3390/ijms19103264] [PMID: 30347840]

[10]   T. Lyu, S. Wu, R.J.G. Mortimer, and G. Pan, "Nanobubble Technology in Environmental Engineering: Revolutionization Potential and Challenges", *Environmental Science Technology,* vol. 53, no. 13, pp. 7175-7176, 2019.
[http://dx.doi.org/10.1021/acs.est.9b02821] [PMID: 31180652]

[11]   I.L. Wang, C.Y. Hsiao, Y.H. Li, F.B. Meng, C.C. Huang, and Y.M. Chen, "Nanobubbles Water Curcumin Extract Reduces Injury Risks on Drop Jumps in Women: A Pilot Study", *Evid. Based Complement. Alternat. Med.,* vol. 2019, no. April, pp. 1-9, 2019.
[http://dx.doi.org/10.1155/2019/8647587] [PMID: 31057656]

[12]   J. Jin, L. Yang, F. Chen, and N. Gu, "Drug delivery system based on nanobubbles", *Interdisciplinary Materials,* vol. 1, no. 4, pp. 471-494, 2022.
[http://dx.doi.org/10.1002/idm2.12050]

[13]   Y. Wang, "Preparation Method and Application of Nanobubbles: A Review", *Coatings,* vol. 13, no. 9, p. 1510, 2023.
[http://dx.doi.org/10.3390/coatings13091510]

[14]   G. Senthilkumar, and J. A. Kumar, "Nanobubbles: a promising efficient tool for therapeutic delivery of antibacterial agents for the Staphylococcus aureus infections", *Applied Nanoscience,* vol. 13, pp. 6177-6190, 2023.
[http://dx.doi.org/10.1007/s13204-023-02854-x]

[15]   M. Jannesari, O. Akhavan, H.R.M. Hosseini, and B. Bakshi, "Graphene/CuO2 Nanoshuttles with Controllable Release of Oxygen Nanobubbles Promoting Interruption of Bacterial Respiration", *Biological and Medical Applications of Materials and Interfaces,* vol. 12, no. 32, pp. 35813-35825, 2020.
[http://dx.doi.org/10.1021/acsami.0c05732] [PMID: 32664715]

[16]   M. Arruebo, "Drug delivery from structured porous inorganic materials", *Wiley Interdiscip Rev Nanomed Nanobiotechnol,* vol. 4, no. 1, pp. 16-30, 2012.
[http://dx.doi.org/10.1002/wnan.132] [PMID: 21374827]

[17]   J. Guo, B. D. Mattos, B. L. Tardy, V. M. Moody, G. Xiao, H. Ejima, J. Cui, K. Liang, and J. J.

Richardson, "Porous Inorganic and Hybrid Systems for Drug Delivery: Future Promise in Combatting Drug Resistance and Translation to Botanical Applications", *Current Medicinal Chemistry,* vol. 26, no. 33, pp. 6107-6131, 2019.
[http://dx.doi.org/10.2174/0929867325666180706111909]

[18]   M. L. Rocca, A. Rinaldi, G. Bruni, V. Fruili, L. Maggi, and M. Bini, "New Emerging Inorganic–Organic Systems for Drug-Delivery: Hydroxyapatite@Furosemide Hybrids", *Journal of Inorganic and Organometallic Polymers and Materials,* vol. 32, pp. 2249-2259, 2022.
[http://dx.doi.org/10.21203/rs.3.rs-1305809/v1]

[19]   I. Gessner, and I. Neundrof, "Nanoparticles Modified with Cell-Penetrating Peptides: Conjugation Mechanisms, Physicochemical Properties, and Application in Cancer Diagnosis and Therapy", *International Journal of Molecular Science,* vol. 21, no. 7, p. 2536, 2020.
[http://dx.doi.org/10.3390/ijms21072536]

[20]   H. Ida, Y. Takahashi, A. Kumatani, H. Shiku, T. Murayama, H. Hirose, S. Futaki, and T. Matuse, "Nanoscale Visualization of Morphological Alteration of Live-Cell Membranes by the Interaction with Oligoarginine Cell-Penetrating Peptides", *Analytical Chemistry,* vol. 93, no. 13, pp. 5383-5393, 2021.
[http://dx.doi.org/10.1021/acs.analchem.0c04097] [PMID: 33769789]

[21]   S. Mazumdar, D. Chitkara, and A. Mittal, "Exploration and insights into the cellular internalization and intracellular fate of amphiphilic polymeric nanocarriers", *Acta Pharm. Sin. B,* vol. 11, no. 4, pp. 903-924, 2021.
[http://dx.doi.org/10.1016/j.apsb.2021.02.019] [PMID: 33996406]

[22]   S.M. Curley, and D. Putnam, "Biological Nanoparticles in Vaccine Development", *Front. Bioeng. Biotechnol.,* vol. 10, no. March, p. 867119, 2022.
[http://dx.doi.org/10.3389/fbioe.2022.867119] [PMID: 35402394]

[23]   L.J. Peek, C.R. Middaugh, and C. Berkland, "Nanotechnology in vaccine delivery", *Advanced Drug Delivery Review,* vol. 60, no. 8, pp. 915-928, 2022.
[http://dx.doi.org/10.1016/j.addr.2007.05.017] [PMID: 18325628]

[24]   M.R. Landry, J.M. Walker, and C. Sun, "Exploiting Phagocytic Checkpoints in Nanomedicine: Applications in Imaging and Combination Therapies", *Front Chem.,* vol. 9, no. March, p. 642530, 2021.
[http://dx.doi.org/10.3389/fchem.2021.642530] [PMID: 33748077]

[25]   M. Bartneck, H.A. Keul, G.Z. Klarwasser, and J. Groll, "Phagocytosis independent extracellular nanoparticle clearance by human immune cells", *Nano Letters,* vol. 10, no. 1, pp. 59-63, 2009.
[http://dx.doi.org/10.1021/nl902830x] [PMID: 19994869]

[26]   X. Zheng, P. Zhang, Z. Fu, S. Meng, L. Dai, and H. Yang, "Applications of nanomaterials in tissue engineering", *RSC Advances,* vol. 11, no. 31, pp. 19041-19058, 2021.
[http://dx.doi.org/10.1039/D1RA01849C] [PMID: 35478636]

[27]   S.M. Asil, and J. Ahlawat, "Application of Nanotechnology in Stem-Cell-Based Therapy of Neurodegenerative Diseases", *Applied Sciences,* vol. 10, no. 14, p. 4852, 2020.
[http://dx.doi.org/10.3390/app10144852]

[28]   L. Farzin, S. Sheibani, M. E. Moassesi, and M. Shamsipur, "An overview of nanoscale radionuclides and radiolabeled nanomaterials commonly used for nuclear molecular imaging and therapeutic functions", *J Biomed Mater Res A,* vol. 107, no. 1, pp. 251-285, 2019.
[http://dx.doi.org/10.1002/jbm.a.36550]

[29]   M. S. O. Pijera, H. Viltres, J. Kozempel, M. Sakmar, M. Vlk, D. I. Ozdemir, M. Ekinci, S. Srinivasan, A. R. Rajabzadeh, E. Ricci Junior, L. M. R. Alencar, M. A. Qahtani, and R. S. Oliveira, "Radiolabeled nanomaterials for biomedical applications: radiopharmacy in the era of nanotechnology", *EJNMMI radiopharm. chem.,* vol. 7, no. 8, 2022.
[http://dx.doi.org/10.1186/s41181-022-00161-4]

SUBJECT INDEX

A

Abiotic stress 112
Absorption 12, 155, 156, 157, 158
  photon 158
Accelerating voltage 153
Acids 54, 55, 173, 174, 190, 203, 204, 208, 211
  ascorbic 54, 190
  caffeic 208
  carboxylic 55, 211
  cinnamic 208
  nucleic 173, 174
Acidic tumor microenvironmen 181
Acoustic emission sensors 25
Activity 177, 191, 194, 195, 207, 250, 251, 253, 254, 256, 258, 259
  anti-angiogenic 177
  anti-arrhythmic 254
  anti-diabetic 191, 195, 207
  anti-inflammatory 194
  downregulating cyclooxygenase enzyme 254
  hormesis 258
  immune defense 250
  immune surveillance 251
  neuroprotective 253
Adsorption plasmon 176
Ag nanoparticles 189, 208, 210, 212, 216
Aggressive corrosion circumstances 81
Agriculture industries 113
Air 10, 33, 62, 80, 83, 85, 86, 90, 120, 147, 150, 159
  compressed 85
  corrosion 80
  injection 62
  pollution 120
Alanine aminotransferase 260
Albumin, bovine serum 190
Algorithms 228, 244
  deep learning 228
Allopathic medication process 258

Alloys 11, 16, 38, 80, 81, 82, 83, 87, 162, 163
  aluminum-copper 82
  metallic 81
Alpha-amylase inhibition test 195
Aluminum 81, 82
  alloys 82
  doping 81
  in aluminum alloys 82
  in magnesium alloys 82
  in stainless steel 81
Alzheimer's disease 254
Annealing 151, 152, 153, 157
  process 153
  temperature 151, 152, 157
Anti 170, 177, 254, 255
  -angiogenic agents 177
  -bacterial agents 170
  -cancerous agents 170
  -obesity actions 254
  -parasitic phytochemical effects 255
Anti-cancer 196, 197, 198, 199
  activity 196, 197, 199
  drug 198
Antioxidant 54, 55, 112, 129, 174, 190, 207, 208, 209, 254
  absorbed 174
  enzymatic 112
Antiseptic agents 187
Apoptosis 198, 251
Architectures 225, 227, 238
  custom Neural Network 227
  neural network 225, 238
Aromatic amino acid pathway 253
Artificial intelligence 67, 229
Atherosclerosis 254
Atomic force microscopy (AFM) 57, 203, 214
Automobile Industries 90
Automotive engine management Systems 25

www.ingramcontent.com/pod-product-compliance
Lightning Source LLC
Chambersburg PA
CBHW050815220326
41598CB00006B/222